Mycoplasma Protocols

METHODS IN MOLECULAR BIOLOGY™

John M. Walker, SERIES EDITOR

105. **Phospholipid Signaling Protocols,** edited by *Ian M. Bird, 1998*
104. **Mycoplasma Protocols,** edited by *Roger J. Miles and Robin A. J. Nicholas, 1998*
103. ***Pichia* Protocols,** edited by *David R. Higgins and James M. Cregg, 1998*
102. **Bioluminescence Methods and Protocols,** edited by *Robert A. LaRossa, 1998*
101. **Mycobacteria Protocols,** edited by *Tanya Parish and Neil G. Stoker, 1998*
100. **Nitric Oxide Protocols**, edited by *Michael A. Titheradge, 1998*
99. **Human Cytokines and Cytokine Receptors**, edited by *Reno Debets, 1998*
98. **Forensic DNA Profiling Protocols**, edited by *Patrick J. Lincoln and James M. Thomson, 1998*
97. **Molecular Embryology:** *Methods and Protocols,* edited by *Paul T. Sharpe, 1998*
96. **Adhesion Proteins Protocols**, edited by *Elisabetta Dejana, 1998*
95. **DNA Topology and DNA Topoisomerases:** *II. Enzymology and Topoisomerase Targetted Drugs,* edited by *Mary-Ann Bjornsti, 1998*
94. **DNA Topology and DNA Topoisomerases:** *I. DNA Topology and Enzyme Purification,* edited by *Mary-Ann Bjornsti, 1998*
93. **Protein Phosphatase Protocols**, edited by *John W. Ludlow, 1997*
92. **PCR in Bioanalysis**, edited by *Stephen Meltzer, 1997*
91. **Flow Cytometry Protocols**, edited by *Mark J. Jaroszeski, 1998*
90. **Drug–DNA Interactions:** *Methods, Case Studies, and Protocols,* edited by *Keith R. Fox, 1997*
89. **Retinoid Protocols**, edited by *Christopher Redfern, 1997*
88. **Protein Targeting Protocols**, edited by *Roger A. Clegg, 1997*
87. **Combinatorial Peptide Library Protocols**, edited by *Shmuel Cabilly, 1998*
86. **RNA Isolation and Characterization Protocols**, edited by *Ralph Rapley, 1997*
85. **Differential Display Methods and Protocols**, edited by *Peng Liang and Arthur B. Pardee, 1997*
84. **Transmembrane Signaling Protocols**, edited by *Dafna Bar-Sagi, 1998*
83. **Receptor Signal Transduction Protocols,** edited by *R. A. J. Challiss, 1997*
82. **Arabadopsis Protocols,** edited by *José M Martinez-Zapater and Julio Salinas, 1997*
81. **Plant Virology Protocols,** edited by *Gary D. Foster, 1997*
80. **Immunochemical Protocols (2nd. ed.),** edited by *John Pound, 1997*
79. **Polyamine Protocols,** edited by *David M. L. Morgan,* 1997
78. **Antibacterial Peptide Protocols,** edited by *William M. Shafer,* 1997
77. **Protein Synthesis:** *Methods and Protocols,* edited by *Robin Martin, 1997*
76. **Glycoanalysis Protocols,** edited by *Elizabeth F. Hounsel, 1997*
75. **Basic Cell Culture Protocols,** edited by *Jeffrey W. Pollard and John M. Walker, 1997*
74. **Ribozyme Protocols,** edited by *Philip C. Turner, 1997*
73. **Neuropeptide Protocols,** edited by *G. Brent Irvine and Carvell H. Williams, 1997*
72. **Neurotransmitter Methods,** edited by *Richard C. Rayne, 1997*
71. **PRINS and *In Situ* PCR Protocols,** edited by *John R. Gosden, 1997*
70. **Sequence Data Analysis Guidebook,** edited by *Simon R. Swindell, 1997*
69. **cDNA Library Protocols,** edited by *Ian G. Cowell and Caroline A. Austin, 1997*
68. **Gene Isolation and Mapping Protocols,** edited by *Jacqueline Boultwood, 1997*
67. **PCR Cloning Protocols:** *From Molecular Cloning to Genetic Engineering,* edited by *Bruce A. White, 1996*
66. **Epitope Mapping Protocols,** edited by *Glenn E. Morris, 1996*
65. **PCR Sequencing Protocols,** edited by *Ralph Rapley, 1996*
64. **Protein Sequencing Protocols,** edited by *Bryan J. Smith, 1996*
63. **Recombinant Proteins:** *Detection and Isolation Protocols,* edited by *Rocky S. Tuan, 1996*
62. **Recombinant Gene Expression Protocols,** edited by *Rocky S. Tuan, 1996*
61. **Protein and Peptide Analysis by Mass Spectrometry,** edited by *John R. Chapman, 1996*
60. **Protein NMR Protocols,** edited by *David G. Reid, 1996*
59. **Protein Purification Protocols,** edited by *Shawn Doonan, 1996*
58. **Basic DNA and RNA Protocols,** edited by *Adrian J. Harwood, 1996*
57. **In Vitro Mutagenesis Protocols,** edited by *Michael K. Trower, 1996*
56. **Crystallographic Methods and Protocols,** edited by *Christopher Jones, Barbara Mulloy, and Mark Sanderson, 1996*
55. **Plant Cell Electroporation and Electrofusion Protocols,** edited by *Jac A. Nickoloff, 1995*
54. **YAC Protocols,** edited by *David Markie, 1995*
53. **Yeast Protocols:** *Methods in Cell and Molecular Biology,* edited by *Ivor H. Evans, 1996*
52. **Capillary Electrophoresis:** *Principles, Instrumentation, and Applications,* edited by *Kevin D. Altria, 1996*
51. **Antibody Engineering Protocols,** edited by *Sudhir Paul, 1995*
50. **Species Diagnostics Protocols:** *PCR and Other Nucleic Acid Methods,* edited by *Justin P. Clapp, 1996*
49. **Plant Gene Transfer and Expression Protocols,** edited by *Heddwyn Jones, 1995*
48. **Animal Cell Electroporation and Electrofusion Protocols,** edited by *Jac A. Nickoloff, 1995*
47. **Electroporation Protocols for Microorganisms,** edited by *Jac A. Nickoloff, 1995*

METHODS IN MOLECULAR BIOLOGY™

Mycoplasma Protocols

Edited by

Roger Miles

King's College, London, UK

and

Robin Nicholas

Central Veterinary Laboratory, Surrey, UK

Humana Press Totowa, New Jersey

999 Riverview Drive, Suite 208
Totowa, New Jersey 07512

This publication is printed on acid-free paper. ∞
ANSI Z39.48-1984 (American Standards Institute) Permanence of Paper for Printed Library Materials.

Cover illustration: Fig. 4 from Chapter 31, "Immunological and Biochemical Characterization of Membrane Proteins," by Michael F. Duffy, Amir H. Noormohammadi, Nina Baseggio, Glen F. Browning, and Philip F. Markham.

Cover design by Patricia F. Cleary.

For additional copies, pricing for bulk purchases, and/or information about other Humana titles, contact Humana at the above address or at any of the following numbers: Tel: 973-256-1699; Fax: 973-256-8341; E-mail: humana@humanapr.com, or visit our Website at www.humanapress.com

Printed in the United States of America. 10 9 8 7 6 5 4 3 2 1

Library of Congress Cataloging in Publication Data

Main entry under title:

Methods in molecular biology™.

Mycoplasma protocols / edited by Roger Miles and Robin Nicholas
p. cm. — Methods in molecular biology ; v. 104)
Includes index.
ISBN 0-89603-525-5 (alk. paper)
1. Mycoplasma—Laboratory manuals. 2. Mycoplasma diseases—Laboratory manuals.
I. Miles, Roger (Roger J.) II. Nicholas, Robin. III. Series: Methods in molecular biology (Totowa, New Jersey) ; 104
QR352.M882 1998
579.3'28—dc21

98-4288
CIP

Preface

Over the last decade, interest in mycoplasmas has been greatly stimulated by the spread of contagious bovine pleuropneumonia (CBPP), caused by *Mycoplasma mycoides* subsp. *mycoides* SC, in Europe and Africa and the discovery of a possible association between mycoplasmas and AIDS. During this period there has also been a recognition by control authorities and international organizations, including the Office International des Epizooties, that mycoplasmas are major causes of economic loss in livestock and continue to be a problem in humans as a result of a range of chronic respiratory and urogenital diseases. However, it is likely that the true significance of mycoplasmas is still underestimated because of the difficulties in working with these extremely fastidious microorganisms. Advances in such laboratory technology as DNA amplification has enabled the detection of unrecoverable mycoplasmas in tissue samples and this will provide us with a better understanding of their role in disease. Important developments have also been seen in more conventional areas such as serological tests and media formulation enabling improved recovery and identification.

Mycoplasma Protocols provides in a single volume up-to-date and easy-to-follow methods for the detection, isolation, identification, and characterization of mycoplasmas, with a major emphasis on those of medical and veterinary significance. It also includes biochemical, genetic, and molecular techniques that will form the basis of understanding pathogenicity and, particularly, adhesion to host cells. The detection of mycoplasmas in cell cultures is also covered. The book is written as a practical guide for scientific and technical staff in medical and veterinary diagnostic and research laboratories. The contributors are all active laboratory-based research workers, many of whom have an understanding of diseases in the community or in the field, and are thus familiar with the problems faced by medical and veterinary practitioners. We are grateful to them for their enthusiasm, professional approach, and punctuality in providing their contributions.

We wish to recognize the International Organization for Mycoplasmology (IOM) which has greatly increased awareness of mycoplasmas through its biennial meetings and has provided opportunities to establish new collaborations. In the animal health field the initiation of the European Commission COST Action 826 on Ruminant Mycoplasmoses has also greatly stimulated cooperation, information and reagent exchange, and coordination of research activities. In relation to our own work, we would particularly like to acknowledge the EC and MAFF, UK for their support of EC FAIR project CT95-0711 on CBPP and for their continuing interest in mycoplasma research.

Roger Miles
Robin Nicholas

Contents

Contributors

CALISTA AGBANYIM • *Division of Life Sciences, King's College, London, UK*

SAMANTHA BAKER • *Bacteriology Department, Central Veterinary Laboratory, Surrey, UK*

HYWEL J. BALL • *Veterinary Science Division, Department of Agriculture, Stormont, Belfast, Northern Ireland*

NINA BASEGGIO • *Veterinary Preclinical Centre, University of Melbourne, Parkville, Victoria, Australia*

JOHN B. BASHIRUDDIN • *Istituto Zooprofilattico Sperimentale dell'Abruzzo e del Molise "G Corporale," Teramo, Italy*

FRANCES A. BECKER • *Microbiological Quality Control, Genetech Inc., San Francisco, CA*

SVEND BIRKELUND • *Department of Medical Microbiology and Immunology, University of Aarhus, Denmark*

JANET M. BRADBURY • *Department of Veterinary Pathology, University of Liverpool, Liverpool, UK*

GLEN F. BROWNING • *Veterinary Preclinical Centre, University of Melbourne, Parkville, Victoria, Australia*

GUNNA CHRISTIANSEN • *Department of Medical Microbiology and Immunology, University of Aarhus, Denmark*

MICHAEL F. DUFFY • *Veterinary Preclinical Centre, University of Melbourne, Parkville, Victoria, Australia*

KEVIN DYBVIG • *Department of Comparative Medicine, University of Alabama at Birmingham, Birmingham, AL*

DAVID FINLAY • *Veterinary Sciences Division, Department of Agriculture, Stormont, Belfast, Northern Ireland*

JOACHIM FREY • *Institut für Veterinärbakteriologie, Universität Bern, Bern, Switzerland*

PATRICIA FURR • *Pinner, Middlesex, UK*

Malin U. K. Heldtander • *Department of Bacteriology, National Veterinary Institute (SVA), Uppsala, Sweden*

Helmut Hotzel • *BGVV, Federal Institute, College of Veterinary Medicine, Jena, Germany*

Tanja K. Jarhede • *Department of Biochemistry, Umeå University, Umeå, Sweden*

Karl-Erik Johansson • *Department of Bacteriology, National Veterinary Institute (SVA), Uppsala, Sweden*

Paul A. Kapke • *Veterinary Medical Research Institute, College of Veterinary Medicine, Iowa State University, Ames, IA*

Philip F. Markham • *Veterinary Preclinical Centre, University of Melbourne, Parkville, Victoria, Australia*

Gerald K. Masover • *Microbiological Quality Control, Genetech Inc., San Francisco, CA*

Roger Miles • *Division of Life Sciences, King's College, London, UK*

F. Chris Minion • *Veterinary Medical Research Institute, College of Veterinary Medicine, Iowa State University, Ames, IA*

Robin A. J. Nicholas • *Bacteriology Department, Central Veterinary Laboratory, Surrey, UK*

Amir H. Noormohammadi • *Veterinary Preclinical Centre, University of Melbourne, Parkville, Victoria, Australia*

Bertil Pettersson • *Department of Bacteriology, National Veterinary Institute (SVA), Uppsala, Sweden*

J. Dennis Pollack • *Department of Medical Microbiology and Immunology, The Ohio State University, Columbus, OH*

François Poumarat • *CNEVA-Laboratorie de Pathologie Bovine, Lyon, France*

José B. Poveda • *Facultad de Veterinaria, Universidad de las Palmas de Gran Canaria, Las Palmas, Spain*

Georges A. Rawadi • *Laboratoire des Mycoplasmes, Institut Pasteur, Paris, France*

Konrad Sachse • *BGVV, Federal Institute, Jena, Germany*

Eugenio Scanziani • *Istituto di Anatomia Patologica Veterinaria e Patologica Aviare, Milano, Italy*

PAUL TAYLOR • *Virology Department, Royal Brompton National Heart and Lung Hospital, London, UK*

LEROY L. VOELKER • *Department of Comparative Medicine, University of Alabama at Birmingham, Birmingham, AL*

ÅKE WIESLANDER • *Department of Biochemistry, Umeå University, Umeå, Sweden*

DAVID WINDSOR • *Mycoplasma Experience, Reigate, Surrey, UK*

HELENA WINDSOR • *Mycoplasma Experience, Reigate, Surrey, UK*

Mycoplasma Protocols

1

Introduction

Roger J. Miles and Robin A. J. Nicholas

The term "mycoplasmas" is used to describe members of the genus *Mycoplasma* and, more generally, of the class *Mollicutes*. The study of mollicutes is frequently referred to as "mycoplasmology," and those who work with mollicutes as "mycoplasmologists." This has presumably arisen because work with mollicutes is generally carried out in specialist laboratories and by personnel whose scientific interests are predominantly or exclusively associated with these organisms. There are many regional organizations as well as a long-standing international organization for mycoplasmology. Such distinction is not normally afforded to the study of specific bacterial groups. However, the common bond that draws together those working on diverse aspects of mollicutes and their biology is the technical difficulty of work with the organisms.

Mollicutes are characterized by the absence of a cell wall and their small genome size and structural simplicity. Since they lack a cell wall, they are osmotically fragile and pleomorphic. The small genome size (<600 kbp in certain *Mycoplasma* species) places a restriction on the number of proteins that can be coded for and, as a consequence, mollicutes possess limited metabolic activities and are dependent on a vast array of nutrients from their environment. The resultant nutritional fastidiousness is a major barrier to work with mollicutes.

Mollicutes are widely distributed as pathogens or commensal organisms of a wide range of plant and animal hosts, including insects. Taxonomically, they are considered sufficiently distinct from cell-walled bacteria to be placed in a separate division, *Tenericutes*, which has four orders and eight recognized genera *(1)*. The major impetus for work with mollicutes has been their association with diseases of humans and economically important diseases of other animals and plants. However, because they lack a cell wall and their membrane compo-

From: *Methods in Molecular Biology, Vol. 104: Mycoplasma Protocols*
Edited by: R. J. Miles and R. A. J. Nicholas © Humana Press Inc., Totowa, NJ

sition may be modified by alterations in the growth medium, they are valuable organisms in which to study cell membrane structure and function. Importantly, the genomes of *Mycoplasma genitalium* and *Mycoplasma pneumoniae* have been fully sequenced ***(2,3)***, and sequencing of the *Ureaplasma urealyticum* genome is almost complete ***(4)***. The small genome size, particularly in *M. genitalium* (580 kbp), suggests that these organisms ought to became major targets of attention in attempts to obtain a complete understanding between genome sequence and cellular structure and function. However, the extent to which evolution toward a reduction in genome size has been accompanied by greater sophistication in genome organization and the multiplicity of enzyme function (*see* Chapter 10) remains to be seen.

This volume is primarily concerned with those mollicutes of medical and veterinary significance (*see* Chapters 2 and 3). These are frequently associated with protracted respiratory, arthritic, and urogenital diseases and belong to the genera *Acholeplasma*, *Mycoplasma*, and *Ureaplasma*, the most important of which is *Mycoplasma* with more than 100 species. The methods described are concerned with the detection, isolation, identification, and characterization of pathogenic mollicutes and the genetic and molecular techniques that will form the basis of understanding pathogenicity and might be applied to the development of vaccines. The detection of mollicutes in tissue-cell cultures, which are essential in many areas of medicine in relation to viral diagnosis, antibody and vaccine production, and research, is also considered (*see* Chapters 23 and 24). General methods previously described for use with cell-walled bacteria, and that can be adapted for studies of mollicutes with little or no modification, are not included.

A major aspect of disease diagnosis is the isolation of target organisms. The particularly fastidious nature of mollicutes makes this a difficult task for many species, and isolation and culture media are considered in Chapters 4–7. The crucial importance of medium quality control in diagnostic and regulatory laboratories is stressed in Chapter 8. All of the media described are undefined. Defined media have been described for a small number of *Acholeplasma* and *Mycoplasma* strains ***(5)***; however, these are not of practical value in isolation. The achievement of defined media capable of growing clinical isolates would be an important step forward in enabling the systematic study of nutritional requirements and development of improved media. However, the problem appears enormous, since it is not only necessary to provide all of the essential nutrients in available forms, but also to balance the concentrations of unknown combinations of nutrients that share common uptake mechanisms ***(6)***. An important point in relation to medium design and selection is that we may not at the present time be fully aware of the true role of mollicutes in disease because of slow-growing or currently unculturable types. Characteristics of the relatively newly isolated human species, *Mycoplasma spermatophilum*, are the difficulty of its isolation on primary culture and its slow growth rate (2–3 wk for colony development) ***(7)***.

The identification of mollicutes, following initial isolation, may be achieved by a variety of methods. Immunological methods (*see* Chapters 12–15) are the most widely used in routine laboratories, although there are many examples of serological crossreactions, for example, among the "*Mycoplasma mycoides*" cluster ***(8)***. There has been recent concern over the validity of immunological detection methods based on the use of monoclonal antibodies (MAbs), since it is now evident that many antigenic surface proteins are variably expressed. However, since all methods are (presumably) evaluated against a wide range of test strains, the use of antibody preparations against such proteins is readily avoided. Both MAbs and polyclonal antibodies can also be used in the direct detection of mollicutes, by staining techniques, in fixed tissues (*see* Chapter 16).

Biochemical characteristics (Chapters 9–11) are useful in restricting the range of possible target species for confirmation by immunological methods. They may also have a significant role to play in the subdivision of existing taxonomic groups, especially if it is possible to associate biochemical features with pathogenicity. A possible example is the association of glycerol-oxidizing ability in *M. mycoides* subspecies *mycoides* (small colony) SC with African and Australian isolates, but not with the apparently less-virulent European isolates ***(9)*** (*see* Chapter 11). Analysis of the distribution of insertion sequences in genomes may also distinguish subspecific groups (*see* Chapter 22).

Genetic methods (Chapters 17–22) for the identification and characterization of mollicutes include polymerase chain reaction (PCR) techniques, rRNA sequence analysis, RAPD fingerprinting, and DNA–DNA hybridization. rRNA sequence analysis is particularly useful in the identification of species from unusual hosts and in establishing new species. DNA–DNA hybridization has also been used to establish taxonomic groups within *Mollicutes*, for example, the subspecies of *Mycoplasma capricolum* ***(10)***. PCR has been widely applied in identification, and probes are available for a diverse range of mollicute species, including the major species associated with the contamination of tissue-cell cultures. However, we have included only a single representative PCR methodology (for *M. mycoides* subspecies *mycoides* SC, *see* Chapter 19). An important aspect of PCR techniques is that they may enable the detection of target mollicutes in clinical isolates, without the need to culture the organisms.

A major aim of medical research is to understand the molecular basis of pathogenesis and so develop strategies for combating disease. This understanding will require the use of a range of techniques to identify and manipulate the expression of target genes. Techniques for transformation and transposon mutagenesis are described in Chapters 25 and 26. Also, although transduction systems have not yet been reported, extrachromosomal elements have been demonstrated in certain species, and procedures for their identification are thus given (*see* Chapter 27). The possibility that in certain cases, pathogenicity might be enhanced by the presence of extrachromosomal elements, as in some cell-walled bacteria, should not be discounted. An area currently attracting considerable research interest is the identification of antigenic surface proteins, since this may lead to the development of vaccines based on their structure. The separation and characterization of membrane proteins (*see* Chapters 30 and 31), particularly those involved in adhesion to host cells (*see* Chapters 32 and 33), is a key step in the development of such vaccines. However, their commercial production is likely to require cloning of the relevant gene(s) and protein expression in rapidly growing mollicutes or cell-walled bacteria, such as *Acholeplasma laidlawii* and *Escherichia coli* (*see* Chapters 28 and 29). A major difficulty in obtaining expression in these organisms is that in *Mycoplasma*, *Ureaplasma*, and *Spiroplasma*, the UGA codon is used as a tryptophan codon rather than a stop codon.

This brief review of the contents has attempted only to put the techniques to be described into context, since the aim of the volume is simply to provide a detailed series of protocols of current relevance to human and veterinary medicine. A collection of general reviews on the pathogenicity and molecular biology of mollicutes was last published in 1992 ***(11)***.

References

1. International Committee on Systematic Bacteriology Subcommittee on the Taxonomy of *Mollicutes* (1995) Revised minimum standards for the description of new species of the class *Mollicutes* (Division *Tenericutes*). *Int. J. Syst. Bacteriol.* **45**, 605-612.
2. Fraser, C. M., Gocayne, J. D., White, O., Adams, M. D., Clayton, R. A. Fleischmann, R. D., et al. (1995) The minimal gene complement of *Mycoplasma genitalium. Science* **270**, 397–403.
3. Himmelreich, R., Hilbert, H., Plagens, H., Pirkl, E., Li, B. C., and Herrmann, R. (1996) Complete sequence analysis of the genome of the bacterium *Mycoplasma pneumoniae. Nucleic Acids Res.* **24**, 4420–4449.
4. Glass, J. I., Glass, J. S., Lefkowitz, E. J., Chen, E. Y., and Cassell, G. H. (1996) The ureaplasma genome project: microfactory DNA sequencing of a microbial genome. *IOM Lett.* **4**, 12.
5. Rodwell, A. W. (1983) Defined and partly defined media, in *Methods in Mycoplasmology*, vol. 1 (Razin, S. and Tully, J. G., eds.), Academic, New York, pp. 163–172.
6. Miles, R. J. (1992) Cell nutrition and growth, in *Mycoplasmas: Molecular Biology and Pathogenesis* (Maniloff, J., McElhaney, R. N., Finch, L. R., and Baseman, J. B., eds.), American Society for Microbiology, Washington, DC, pp. 23–69.

7. Hill, A. C. (1991) *Mycoplasma spermatophilum*, a new species isolated from human spermatozoa and cervix. *Int. J. Syst. Bacteriol.* **41**, 229–233.
8. Cottew, G. S., Breard, A., Da Massa, A. J., Ernø, H., Leach, R. H., Lefevre, P. C., et al. (1987) Taxonomy of the *Mycoplasma mycoides* cluster. *Isr. J. Med. Sci.* **23**, 632–635.
9. Houshaymi, B. M., Miles, R. J., and Nicholas, R. A. J. (1997) Oxidation of glycerol differentiates African from European isolates of *Mycoplasma mycoides* subspecies *mycoides* SC (small colony). *Vet. Record* **140**, 182,183.
10. Bonnet, F., Saillard, C., Bové, J. M., Leach, R. H., Rose, D. L., Cottew, G. S., et al. (1993). DNA relatedness between field isolates of *Mycoplasma* F38 group, the agent of contagious caprine pleuropneumonia, and strains of *Mycoplasma capricolum*. *Int. J. Syst. Bacteriol.* **43**, 597–602.
11. Maniloff, J., McElhaney, R. N., Finch, L. R., and Baseman, J. B. (1992) *Mycoplasmas: Molecular Biology and Pathogenesis*. American Society for Microbiology, Washington DC.

2

Medical Significance of Mycoplasmas

Paul Taylor

1. Introduction

Mycoplasmas were first isolated from humans in 1937 *(1)*, although it was not until 1951 *(2)* that selective media were used to identify mycoplasmas in the oropharynx. In 1962, an etiological link was established between what we now know to be *Mycoplasma pneumoniae* and cases of primary atypical pneumonia *(3)*. Other mycoplasmas have been isolated from predominantly respiratory and genito-urinary sites. These are outlined in **Table 1**.

A number of other mycoplasmas have been isolated infrequently from humans. These include: *Mycoplasma primatum,* which has been isolated from the genito-urinary tract; *Mycoplasma spermatophilum,* which has been isolated from human sperm and cervical specimens; *Mycoplasma pirum,* which has been isolated from peripheral blood lymphocytes from patients with AIDS; *Mycoplasma penetrans,* which has been isolated from the urine of patients with AIDS; and *Mycoplasma fermentans.* This latter mycoplasma was originally isolated from the genital tract of <1% of asymptomatic adults, but it has also been isolated from other sites after immunosuppression of the host and detected by the polymerase chain reaction (PCR) in synovial fluid from patients with rheumatoid arthritis *(4)*.

2. *Mycoplasma pneumoniae*

2.1. History

The term primary atypical pneumonia (PAP) was first applied to penicillin- and sulfonamide-resistant chest infections in 1938 *(5)* in order to distinguish the gradual nature of the clinical course from that of a "typical" pneumonia caused by, for example, *Streptococcus pneumoniae*. In subsequent years, Eaton and coworkers demonstrated the infectious nature of these PAP agents in labo-

From: *Methods in Molecular Biology, Vol. 104: Mycoplasma Protocols*
Edited by: R. J. Miles and R. A. J. Nicholas © Humana Press Inc., Totowa, NJ

Table 1
Mycoplasmas Isolated from Humans

Species	Location O, oropharynx; R, respiratory; GU, genito-urinary	Occurrence + Rare, ++ Common, +++ Widespread
M. pneumoniae	R	+
M. genitalium	GU	+
	R	+
U. urealyticum	GU	++
M. hominis	GU	++
	O	+
M. buccale	O	+
M. faucium	O	+
M. lipophilum	O	+
M. orale	O	+++
M. salivarium	O	+++
M. fermentans	Joints	+

ratory animals ***(6)***. This classical work has been reviewed and placed in context ***(7)***. Eaton's agent, as it became known, was compared with bovine pleuropneumonia by Marmion and Goodburn ***(8)*** and was finally cultured on inanimate media ***(3)***. The designation *M. pneumoniae* was applied to this organism by an international group of mycoplasma scientists in 1963 ***(9)***.

2.2. Clinical Presentation

Infection with *M. pneumoniae*, particularly in young children, may be subclinical or very mild, such as a self-limiting upper respiratory infection. Progression to more severe symptoms, such as pneumonia, may occur in childhood but is more common in young adults. The initial presenting symptoms of *M. pneumoniae* infection after a 2–3 wk incubation period may include some or all of the following: cough (frequently dry and nonproductive), fever, malaise, chills, sore throat, and headache. Subsequent development of symptoms may result in pneumonia in about 3–10% of cases or tracheobronchitis in the majority (70%) of cases, although pharyngitis and myringitis may also occur. Asymptomatic infections account for about 20% of cases.

Most respiratory symptoms of *M. pneumoniae* resolve within 4–6 wk. In some cases, however, secondary bacterial infections may occur, especially with *Haemophilus influenzae* ***(10)***. If such secondary infections remain unrecognized, this may lead to subsequent bronchiectasis or other pulmonary abnormalities.

There are a number of other less frequent complications of *M. pneumoniae* infection that occur outside the lung. Up to 7% of patients have complications involving the central nervous system (CNS), including meningoencephalitis, Guillain-Barré syndrome, hemiplegia, and acute psychosis. These CNS infections may have a 10% mortality rate with about one third of patients being left with permanent or persistent neurological deficit. Approximately 3% of cases also have sequelae affecting the skin, including: Stevens-Johnson syndrome, erythema multiforme minor, and erythema nodosum. Sequelae that affect the ear include bullous myringitis, otitis media, and otitis externa. Other nonspecific side effects of infection include hemolytic anemia, conjunctivitis, rheumatic fever, and arthritis.

2.3. Epidemiology

Although *M. pneumoniae* is a continual source of respiratory infections throughout the year, there is a predominance of outbreaks in winter months. In the UK, there is a 4-yr periodicity of epidemics. Overall *M. pneumoniae* accounts for about 15% of community acquired pneumonias, whereas in closed populations, such as military training camps, up to 50% of cases may be owing to this pathogen. In family outbreaks, the infection rate in children may be as high as 81% ***(11)***.

It has been demonstrated ***(12)*** that there are two groups of *M. pneumoniae* that differ in the amino acid composition of the P1 adhesin, which is necessary for adherence to epithelial cells of the respiratory tract. These subtypes may be responsible for different virulence potentials.

M. pneumoniae has been isolated from genito-urinary specimens from female patients attending a gynecological clinic ***(13)*** and from the urethra of one of three male sexual partners. This suggests the most likely mode of transmission is through orogenital contact.

2.4. Therapy

Tetracyclines and erythromycin both reduce disease severity of *M. pneumoniae* infection when started soon after the onset of disease. The newer macrolides, such as clarithromycin and azithromycin, have some major advantages for *M. pneumoniae* therapy including increased tolerability and lower daily dosage. Despite chemotherapy, the organism can still be recovered from the respiratory tract for up to 3 mo ***(14)***, which demonstrates the bacteriostatic nature of therapy.

3. *Mycoplasma genitalium*

3.1. History

M. genitalium was first isolated in 1981 ***(15)*** from the urethras of 3 of 13 men with nongonococcal urethritis (NGU). Its association with the respiratory tract was demonstrated by a retrospective survey of samples from which *M. pneumoniae* had already been isolated ***(16)***. There are a number of common properties between *M. genitalium* and *M. pneumoniae,* including similar biochemical characteristics, the presence of a tip attachment structure, and serological crossreactions, which makes the interpretation of diagnostic serological tests for *M. pneumoniae* very difficult.

3.2. Clinical Presentation

Since isolation from clinical samples is very difficult, the ultrasensitive detection of genomic sequences by PCR has been used to discover *M. genitalium* in the lower genital tract of 20% of women attending a genito-urinary medicine (GUM) clinic ***(17)***. PCR also detected *M. genitalium* DNA in male GUM patients with nonchlamydial urethritis (17/99, 17%) significantly more frequently than in men with chlamydial urethritis ***(18)***.

Culture or DNA probes have detected *M. genitalium* in up to 30% of men with NGU. PCR has also been used to detect *M. genitalium* DNA in eight urethral samples and four brochoalveolar lavage samples of which three also contained *M. pneumoniae* DNA ***(19)***. *M. genitalium* has been detected by PCR in synovial fluids from two patients, one with Reiter's and the other with arthritis ***(20)***.

3.3. Therapy

The antibiotic sensitivity of *M. genitalium* has been shown to be very similar to that of *M. pneumoniae* ***(21)*** with inhibition shown by tetracyclines, macrolides, and quinolones, especially sparfloxacin.

4. *Ureaplasma urealyticum*

4.1. History

The term T (tiny) strain mycoplasmas was initially applied as a result of the very small colonies that appeared on supplemented agar containing urea as substrate. They were first described by Shepard ***(22)*** from men with primary and recurrent NGU. This mycoplasma was ultimately classified as a new genus and species *U. urealyticum* ***(23)***.

4.2. Clinical Presentation

See **Table 2**.

4.3. Epidemiology

U. urealyticum has been detected in the cervix or vagina of between 40 and 80% of sexually mature, asymptomatic women and at a slightly lower rate in the urethra of sexually active men ***(28)***.

Table 2
Clinical Presentation of *U. urealyticum*

Disease	Role of *U. urealyticum*	Reference
Nonchlamydial NGU	+++	*(24)*
Prostatitis	+	
Chronic lung disease of the premature newborn (<1000g birthweight)	++	*(25)*
Arthritis in the immunosuppressed	+	*(26)*
Chorioamnionitis	+	*(27)*

+++ Strong evidence, + weak evidence.

4.4. Therapy

Most isolates are susceptible to tetracycline (approx 10% of strains *[29]* are resistant) and erythromycin, but not to clindamycin. Erythromycin is the antibiotic of choice for treating neonatal ureaplasmal infections. Because of the possibility of reinfection or development of resistance, further cultures should be obtained after completion of the initial antibiotic chemotherapy.

5. *Mycoplasma hominis*

5.1. History

The first reference to isolation of human mycoplasmas in 1937 ***(1)*** was from a Bartholin's abscess. These were probably *M. hominis.* Originally, two types of *M. hominis* were proposed, but it became clear that strain 2 was serologically indistinguishable from *Mycoplasma arthritidis,* a mycoplasma commonly isolated from rats.

5.2. Clinical Presentation

See **Table 3**. *M. hominis* has been isolated from routine blood cultures, suggesting that a bacteremia with this mycoplasma may occur after major surgery. It has been suggested ***(33)*** that empirical therapy for *M. hominis* should be considered in postcardiovascular surgery patients with mediastinitis or a sternal wound infection which are routine culture and Gram-negative. Septic arthritis is a rare presentation of *M. hominis,* which generally occurs in the immunosuppressed host ***(34)***.

5.3. Epidemiology

Colonization of the oropharynx and genital tract occurs during or shortly after birth, but after 1 or 2 yr, it is no longer isolated. Reacquisition into the lower genital tract occurs with sexual contact and increases in rate with the

Table 3
Clinical Presentation of *M. hominis*

Disease	Role of *M. hominis*	Reference
Pelvic inflammatory disease	+	***30***
Postabortal fever	+++	***31***
Postpartum fever	+++	***31***
Pyelonephritis	++	***32***

+++ Strong evidence. + Weak evidence.

number of sexual partners ***(35)***. Isolation rates from control populations of women are between 40 and 50%, whereas those attending GUM clinics exceed 90%. In men, rates of isolation can vary from about 5% in control populations to 40% in those attending GUMs.

5.4. Therapy

Most isolates are susceptible to tetracycline and clindamycin but not erythromycin. Increased resistance to tetracyclines has been demonstrated ***(36)***; consequently, clindamycin should be used in such cases.

6. Commensal Mycoplasmas

6.1. History

Mycoplasma buccale, Mycoplasma faucium, Mycoplasma lipophilum, Mycoplasma orale, and *Mycoplasma salivarium* are part of the normal flora of the human oropharynx and are generally regarded as commensal organisms. *M. orale* has been detected from between 30 and 60% of throat swabs from adults, whereas *M. salivarium* has been detected from between 60 and 80%. *M. buccale*, *M. faucium,* and *M. lipophilum* are observed in <5% of cases.

6.2. Clinical Presentation

The role of *M. orale* and *M. salivarium* in disease is limited to a few reported infections, such as septic arthritis in immunocompromised hosts. In a survey of mycoplasmas isolated from brochoalveolar lavage ***(37)*** *M. salivarium* was the most common mycoplasma isolate and was detected more frequently from HIV-positive cases (18%) than from HIV-negative (8%).

References

1. Diens, L. and Edsall, G. (1937) Observations on the L-organism of Klienberger. *Proc. Soc. Exp. Biol. Med.* **36,** 740–744.
2. Smith, P. F. and Morton, H. E. (1951) Isolation of pleuropneumonia-like organisms from the throat of humans. *Science* **113,** 623,624.

3. Chanock, R. M., Hayflick , L., and Barile, M. F. (1962) Growth on artificial medium of an agent associated with atypical pneumonia and its identification as a PPLO. *Proc. Nat. Acad. Sci.* **48,** 41–49.
4. Schaeverbeke, T., Gilroy, C. B., Bebear C., Dehais, J., and Taylor- Robinson, D. (1996) Detection of *Mycoplasma fermentans* by a PCR assay in synovial fluids from patients with rheumatoid arthritis and other rheumatic disorders. *IOM Letters* **4,** 363.
5. Reimann, H. A., (1938) An acute infection of the respiratory tract with atypical pneumonia. *JAMA* **111,** 2377–2384.
6. Eaton, M. D., Meiklejohn, G., and van Herick, W., (1944) Studies on the aetiology of primary atypical pneumonia. A filterable agent transmissible to cotton rats, hamsters and chick embryos. *J. Exp. Med.* **79,** 649–668.
7. Marmion, B. P. (1990) Eaton agent - science and scientific acceptance: a historical commentary. *Rev. Infect. Dis.* **12,** 338–353.
8. Marmion, B. P., and Goodburn G. M., (1961) Effect of an inorganic gold salt on Eaton's primary atypical pneumonia agent and other observations. *Nature* **189,** 247–248.
9. Chanock, R. M., Dienes, L., Eaton, M. D., Edward, D. G., Freundt, E. A., Hayflick, l., Hers, J. F. P., Jensen, K. E., Liu C., Marmion B. P., Morton H. E., Mufson M. A., Smith P. F., Somerson, N. L., and Taylor-Robinson D. (1963) *M. pneumoniae:* proposed nomenclature for atypical pneumonia organism (Eaton agent). *Science* **140,** 662.
10. Staugas, R. and Martin, A. J. (1985) Secondary bacterial infections in children with proved *M. pneumoniae. Thorax* **40,** 546–548.
11. Hanukoglu, A., Hebroni, S. and Fried, D. (1986) Pulmonary involvement in *M. pneumoniae* infection in families. *Infection* **14,** 1–6.
12. Su, C. J., Chavoya, A., Dallo, S. F., and Baseman, J. B. (1990) Sequence divergency of the cytadhesin P1 gene on *Mycoplasma pneumoniae. Infect. Immun.* **58,** 2669–2674.
13. Goulet, M., Dular, R., Tully, J. G., Billowes, G., and Kasatiya, S. (1995) Isolation of *M. pneumoniae* from the human urogenital tract. *J. Clin. Micro.* **33,** 2823–2825.
14. Niitu, Y., Hasegawa, S., and Kubota, H. (1974) In vitro development of resistance to erythromycin, other macrolide antibiotics and lincomycin in *M. pneumoniae. Antimicrob. Agents Chemother.* **5,** 513–518.
15. Tully, J. G., Taylor-Robinson, D., Cole, R. M., and Rose, D. L. (1981) A newly discovered mycoplasma in the human urogenital tract. *Lancet* **1,**1228–1291.
16. Baseman, J. B., Dallo, S. F., Tully, J. G., and Rose, D. L., (1988) Isolation and characterisation of *Mycoplasma genitalium* strains from the respiratory tract. *J. Clin. Microbiol.* **26,** 2266–2269.
17. Palmer, H. M., Gilroy, C. B., Claydon, E. J., and Taylor-Robinson, D. (1991) Detection of *Mycoplasma genitalium* in the genitourinary tract of women by the polymerase chain reaction. *Int. J. STD AIDS* **2,** 261–263.
18. Jensen, J. S., Orsum, R., Dohn, B., Uldum, S., Wormm A. M., and Lind, K. (1993) *Mycoplasma genitalium:* a cause of male urethritis? *Genitourin. Med.* **69,** 265–269.

19. Berbeyrac, B., Bernet-Poggi, C., Febrer, F., Renaudin, H., Dupon, M., and Bebear, C. (1993) Detection of *M. pneumoniae* and *M. genitalium* in clinical samples by polymerase chain reaction. *Clin. Infect. Dis.* **17 (Suppl. 1),** S83–89.
20. Taylor-Robinson, D., Gilroy, C. B., Horowitz, S., and Horowitz, J. (1994) *Mycoplasma genitalium* in the joints of two patients with arthritis. *Eur. J. Clin. Micro. Inf. Dis.* **13,** 1066–1069.
21. Renaudin, H., Tully, J. G., and Bebear, C. (1992) In vitro susceptibilitis of *Mycoplasma genitalium* to antibiotics. *Antimicrob. Agents Chemother.* **36,** 870–872.
22. Shepard, M. C., (1954) The recovery of pleuropneumonia-like organisms from Negro men with and without non-gonococcal urethritis. *Am. J. Syph. Gonorrhoea Vener. Dis.* **38,** 113–124.
23. Shepard, M. C., Lunceford, C. D., Ford, D. K., Purcell, R. H., Taylor-Robinson D., Razin, S., and Black, F. T. (1974) *Ureaplasma urealyticum* gen. nov., sp. Nov: proposed nomenclature for the human T (T-strain) mycoplasma. *Int. J. Syst. Bacteriol.* **24,** 160–171.
24. Taylor-Robinson, D. and McCormack, W. M. (1980) The genital mycoplasmas. *N. Engl. J. Med.* **302,** 1003–1008.
25. Cassell, G. H., Waites, K. B., Watson, H. I., Crouse, D. T., and Harasawa, R. (1993) *Ureaplasma urealyticum* intrauterine infection: role in prematurity and disease in newborns. *Clin. Micro. Rev.* **6,** 69–87.
26. Taylor-Robinson, D., Furr, P., and Webster, A. D. B. (1986) Ureaplasma urealyticum in the immunocompromised host. *Paediatr. Infect. Dis.* **5 (Suppl.),** 236–238.
27. Gray, D. J., Robinson, H. B., Malone, J., and Thompson, R. B. (1991) Adverse outcome in pregnancy following amniotic fluid isolation of *Ureaplasma urealyticum. Prenat. Diagn.* **12,** 111–117.
28. Cassell, G. H., Waites, K. B., Watson,·H. L., Crouse, D. T., and Harawasa R. (1993) *Ureaplasma urealyticum* intrauerine infection: role in prematurity and disease in new-borns. *Clin. Microbiol Rev.* **6,** 69–87.
29. Simson, J. B., Hale, J., Bowie, W. R., and Holmes, K. K. (1981) Tetracycline resistant *Ureaplasma urealyticum:* a cause of persistent nongonococcal urethritis. *Ann. Intern. Med.* **94,** 192–194.
30. Mardh, P.-A. and Westrom, L. (1970) Tubal and cervical cultures in acute salpingitis with special reference to *Mycoplasma hominis* and T-strain mycoplasmas. *Br. J. Vener. Dis.* **46,** 179–186.
31. Taylor-Robinson, D., and Munday, P. E. (1988) Mycoplasmal infections of the female genital tract and its complications, in *Genital Tract Infection in Women* (Hare, M. J., ed.) Churchill Livingstone, Edinburgh, pp. 228–247.
32. Thomsen, A. C., and Lindskov, H. O. (1979) Diagnosis of *Mycoplasma hominis* pyelonephritis by demonstration of antibodies in urine. *J. Clin. Microbiol.* **9,** 681–687.
33. Sielaff, T. D., Everett, J. E., Shumway, S. J., Wahoff, D. C., Bolman, R. M. and Dunn, D. L. (1996) *Mycoplasma hominis* infections occurring in cardiovascular surgical patients. *Ann. Thorac. Surg.* **61,** 99–103.

34. Luttrell, L. M., Kanj, S. S., Corey, G. R., Lins, R. E., Spinner, R. J., Mallon, W. J., and Sexton, D. J. (1994) *Mycoplasma hominis* septic arthritis: two case reports and review. *Clin. Inf. Dis* **19,** 1067–1070.
35. McCormack, W. M., Lee, Y-H., and Zinner, S. H. (1973) Sexual activity and vaginal colonisation with genital mycoplasmas. *Ann. Intern. Med.* **78,** 696–698.
36. Cummings, M. C. and McCormack, W. M. (1990) Increase in resistance of *Mycoplasma hominis* to tetracyclines. *Antimicrob. Agents Chemother.* **34,** 2297–2299.
37. Teel, L. D., Fineli, M. R., and Johnson, S. C. (1994) Isolation of mycoplasma species from bronchoalveolar lavages of patients positive and negative for human immunodeficiency virus. *J. Clin. Micro.* **32,** 1387–1389.

3

The Veterinary Significance of Mycoplasmas

Robin Nicholas

1. Contagious Bovine Pleuropneumonia

The reappearance of contagious bovine pleuropneumonia (CBPP) in Italy in 1990 after an absence of 100 years and the continued spread of CBPP in many parts of Africa are powerful reminders that this age-old disease is still a long way from eradication. Control of this disease was almost achieved during the 1970s when it was eradicated from Australia, restricted to a small area in eastern Africa, and was giving rise to only sporadic outbreaks in Europe *(1)*. However, as a result of civil wars and the subsequent breakdown of veterinary services, CBPP began to spread and is today the most important cattle disease in Africa, causing greater losses than rinderpest according to the Office d'International Epizooties (OIE). As a result, CBPP is designated as a list A disease. *Mycoplasma mycoides* subsp. *mycoides* small colony (SC) variant, which is the cause of CBPP, was the first mycoplasma to be isolated nearly a century ago. It differs from most other mycoplasmas in that it can be a primary cause of disease just about fulfilling Koch's postulates. The mechanism of its pathogenicity remains uncertain, but recent work confirms early findings that hydrogen peroxide production is an important factor *(2)*. Evidence has also been provided to explain why European strains were far less virulent than African strains. It showed that African, but not European, strains hydrolyzed glycerol and produced hydrogen peroxide. *M. m. mycoides* SC is a member of the "mycoides cluster," which contains six important mycoplasmas of large and small ruminants (**Table 1**). The six share biochemical, immunological, and genetic characteristics that can lead to difficulties for diagnosis. A major boost to research into CBPP was the development of sensitive and specific primers for a PCR to detect to the causative mycoplasma *(3)*. This heralded a major research initiative and led to the setting up of several European and

From: *Methods in Molecular Biology, Vol. 104: Mycoplasma Protocols*
Edited by: R. J. Miles and R. A. J. Nicholas © Humana Press Inc., Totowa, NJ

Table 1
Members of the *Mycoides* cluster

Species	Host	Disease
M. mycoides subsp. *mycoides* SC	Cattle (goat, sheep)	CBPP
M. mycoides subsp. *mycoides* LC	Goat (sheep, cattle)	CA[a], pneumonia
M. mycoides subsp. *capri*	Goat	Respiratory
M. capricolum subsp. *capripneumoniae*	Goat	CCPP
M. capricolum subsp. *capricolum*	Goat (sheep)	CA. pneumonia
Mycoplasma bovine group 7	Cattle	Arthritis, mastitis

[a]Contagious agalactia.

international groups dedicated to improving the diagnosis and control of this disease. However, because of the costs of control and a largely ineffective vaccine, CBPP is likely to continue to plague Africa for many years to come.

2. Contagious Caprine Pleuropneumonia

Contagious caprine pleuropneumonia (CCPP), caused by *Mycoplasma capricolum* subsp. *capripneumoniae* (formerly F38), has been known for many years and results in major losses in goat herds in at least 30 countries in Africa and Asia. Research into the control of CCPP has been hampered by confusion regarding the exact cause of the disease. Two other "mycoides cluster" members, *M. m. mycoides* large colony (LC) and *M. m. capri,* were for some time implicated in the etiology of the disease because they can cause a pleuropneumonia in small ruminants that resembles CCPP. This confusion was compounded by the difficulty of isolating and growing the highly fastidious F38 in vitro. Once a suitable media was developed by MacOwan ***(4)*** for F38, its role as the primary cause of CCPP was confirmed. True CCPP is characterized by its ready contagious nature to susceptible goats and histopathologically by an interstitial intralobular edema of the lung, rather than thickening of the interlobular septa, that is seen with *M. m. mycoides* LC and *M. m. capri.* Molecular biological tools are now available to distinguish these mycoplasma ***(5).*** However because of their cost, they are unlikely to be used routinely in those countries where CCPP is a problem.

3. Contagious Agalactia

M. m. mycoides LC is also one of the causative mycoplasmas of contagious agalactia, a serious disease syndrome of small ruminants that is characterized by mastitis, arthritis, keratoconjunctivitis, and occasionally abortion. However, *Mycoplasma agalactiae* is the main cause of the disease in sheep and goats, but *M. m. mycoides* LC, *Mycoplasma capricolum* subsp. *capricolum,* and *Mycoplasma putrefaciens* also contribute significantly to losses, particularly in goats

Table 2
Recent Outbreaks of Contagious Agalactia

Species isolated	Date	Place	% Mortality (morbidity)
M. agalactiae	1994	Gran Canaria	25 (50)
M. agalactiae/ *M. m. mycoides* LC	1994	USA	15 (?)
M. putrefaciens	1992	Spain	0 (<1)
M. agalactiae	1990	Spain	8 (36)
M. m. mycoides LC/ *M. m. capri*	1990	Gran Canaria	8 (20)
M. putrefaciens	1987	USA	30 (90)

(**Table 2**). The disease has been estimated to cause annual losses of as much as $30 million in European countries around the Mediterranean, mainly as a result of milk production losses. The disease has been known for nearly 200 years in Europe and was known as "mal di sito" ("disease of the place") in Italy, because it was possible for animals to become infected after grazing on contaminated pasture. The main causative organism, *M. agalactiae,* was first isolated and cultured in 1925 and was the second mycoplasma to be discovered, 27 years after the mycoplasma of CBPP. Just like CBPP, contagious agalactia has been exported to most continents of the world via clinically normal carrier animals. The disease continued to spread, entering Sardinia in 1980, Brazil in 1986, and the Canary Islands in 1992 ***(6)***. Contagious agalactia is predominantly a disease of milking sheep and goats. It often appears in a herd in the spring soon after lactation begins and probably represents the activation of latent infection. The young ruminants become infected directly at suckling, whereas the adults are contaminated via the milker's hands, milking machines, or by bedding that often provides a rich source of mycoplasmas. Like most mycoplasma infections, antibiotic control is generally ineffective, tending to promote the carrier state. In Europe, both live and inactivated vaccines have been used with mixed success. Some have provided protection from clinical disease and have been useful in endemic areas, but the problems of encouraging the carrier state still apply. Generally, the duration of immunity is short.

4. Bovine Respiratory Mycoplasmosis

Mycoplasma bovis is the most frequently occurring pathogenic bovine mycoplasma in Europe and America and is the cause of several diseases that include mastitis, genital and ocular disorders, and most importantly, pneumonia and arthritis in calves and young cattle. However, unlike the mycoplasmas causing CBPP and CCPP, *M. bovis* is almost always associated with other pathogens, such as *Haemophilus somnus, Pasteurella multicoda,* or *Pasteurella haemolytica. M. bovis* was first isolated in 1961 from severe bovine mastitis in the US and initially called *M. agalactiae* var. (later subsp.) *bovis* because of both clinical and biochemical similarities to *M. agalactiae,* the cause of contagious agalactia in small ruminants. The first isolation in Britain was made from a severe case of calf pneumonia in the south of England. Northern Ireland escaped infection until 1990, when cases of calf pneumonia were reported from which *M. bovis* was recovered; these were attributed to the relaxation of border controls within the European Community (EC) and the subsequent increase in calf importation ***(7)***. Severe arthritis, also linked to importation into the province, was reported in cattle from which *M. bovis* was isolated from joints and milk. In 1994, *M. bovis* was isolated for the first time from a group of cattle imported into the Republic of Ireland from France, most of which showed severe acute respiratory disease ***(8)***. In addition to *M. bovis,* other mycoplasmas associated with pneumonia include *Mycoplasma dispar, Mycoplasma bovirhinis,* and occasionally *Mycoplasma bovigenitalium. M. dispar* is a proven pathogen, but is rarely isolated because it requires a specialized medium for growth.

5. Mycoplasmas of Swine

Enzootic pneumonia, caused by *Mycoplasma hyopneumoniae,* occurs in all pig-rearing countries. The disease is rarely fatal, but morbidity rates, manifested by poor growth and feed conversion, are high. Annual losses in the US have been estimated to be as high as $330 million and rising ***(9)***. *M. hyopneumoniae* is capable of causing disease in its own right but is most severe in coinfections with *P. multicoda* or *Actinobacillus pleuropneumoniae.* This mycoplasma is particularly fastidious even by mycoplasma standards, but advances in diagnosis, in particular with PCR, have meant detection is now easier and certainly less subjective. Although other mycoplasmas, such as *Mycoplasma hyorhinis* and *Mycoplasma flocculare,* have been implicated in porcine disease, their main importance lies in their ability to crossreact with *M. hyopneumoniae,* thereby confusing diagnosis.

6. Avian Mycoplasmas

Much is known of the mycoplasmas of chickens and turkeys, including *Mycoplasma gallisepticum* and *Mycoplasma meleagridis,* which can cause serious respiratory disease, *Mycoplasma synoviae,* which can cause lameness and retarded growth, and *Mycoplasma iowae,* which can result in reduce hatchability. *Mycoplasma anseris* and an uncharacterized mycoplasma, *M.* sp. 1220 have been isolated from sick geese ***(10)***. A further 16 have been isolated from a wide range of domesticated species, but have not been directly linked to disease. In contrast, little is known about the mycoplasma flora of other avian species, including wild birds. Recently, three new species of mycoplasmas, *Mycoplasma buteonis, Mycoplasma falconis,* and *Mycoplasma gypis,* have been described in birds of prey in Spain suffering from respiratory disease ***(11)***. A new species in the US, *Mycoplasma corogypsi,* was isolated from the footpad abscess of a black vulture. In the UK, we have recovered an as yet unidentified mycoplasma from peregrine falcons suffering from eye lesions.

7. Mycoplasmas of Sea Mammals

In 1979 and 1980, over 400 harbor seals *(Phoca vitulina)* died along the coast of New England ***(12)***. The primary cause of death was considered to be an influenza A virus, but *Mycoplasma* species were also isolated from the respiratory tract of some affected animals. These isolates, which hydrolyzed arginine, did not react with antisera prepared against any known mycoplasmas and were classified as a new species: *Mycoplasma phocidae.* It was not known what contribution *M. phocidae* made to the overall pathogenesis, but it was believed that environmental factors and high population density may also have exacerbated the disease. Attempts to infect gray seals *(Halichoerus grypus)* and harp seals *(Phoca groenlandica)* with *M. phocidae* and/or the influenza virus failed to cause respiratory disease, suggesting the causes were multifactorial. In 1988, an epizootic of pneumonia killed 18,000 harbor seals in the North Sea. In addition to several viruses, two new species of arginine metabolizing mycoplasmas, *Mycoplasma phocarhinis* and *Mycoplasma phocacerebrale,* were isolated ***(13)***. Polyarthritis, a common condition caused by mycoplasmas, abortions, and skin lesions, were also seen in these sea mammals. Clearly further work is necessary to establish a role for mycoplasmas in disease in sea mammals.

8. New and Emerging Mycoplasmas of Unusual Species

Over the last 20 years, a dramatic decline was seen in the number of desert tortoises *(Gopherus agassizii)* in the US chiefly as a result of an upper respiratory tract disease. Infected tortoises showed clinical signs of rhinitis with clear to purulent nasal discharge, palpebral edema, dehydration, and cachexia in the late stages and a high mortality rate. In 1991, a new mycoplasma, *Mycoplasma agassizi,* was isolated from sick tortoises and shown to be the principal cause ***(14)***. A new mycoplasma, biochemically but not serologically similar to *M. capricolum* subsp. *capricolum,* was isolated from dead and sick crocodiles *(Crocodylus niloticus)* suffering swollen limb joints, progressive lameness, and paresis on a crocodile ranch in Zimbabwe ***(15)***. The disease responded partially to antibiotic therapy and was reproduced with an isolate from the affected farm. During a survey of arthritic elephants *(Elephas maximus)* in circuses and zoos, mycoplasmas were isolated from the genital organs of about 60% females ***(16)***. One isolate, *Mycoplasma elephantis,* was distinct from previously characterized mycoplasmas. Work remains to be done to determine its role in the pathogenicity.

References

1. Nicholas, R. A. J. and Bashiruddin, J. B. (1995) *Mycoplasma mycoides* subsp. *mycoides* SC variant: the agent of contagious bovine pleuropneumonia and member of the *Mycoplasma mycoides* cluster. *J. Comp. Pathol.* **113,** 1–27.
2. Houshaymi, B., Miles, R. J., and Nicholas, R. A. J. (1997) Oxidation of glycerol differentiates African from European isolates of *Mycoplasma mycoides* subsp. mycoides SC (small colony). *Vet. Rec.* **140,** 182,183.
3. Bashiruddin, J. B., Taylor, T. K., and Gould, A. R. (1994) A PCR-based test for the specific identification of *Mycoplasma mycoides* subsp. *mycoides* SC. *J. Vet. Diagn. Invest.* **6,** 428–434.
4. MacOwan, K. J. (1976) A mycoplasma from chronic coprine pleuropneumonia in Kenya. *Tropical Animal Health Production* **8,** 28–36.
5. Thiaucourt, F., Bölske, G., Leneguersh, D., Smith., D., and Wesonga., H. (1996) Diagnosis and control of contagious caprine pleuropneumonia. *Rev. Sci. Tech. Off. Int. Epiz.* **15,** 1415–1429.
6. Nicholas, R. A. J. (1996) Contagious agalactia: an update, in *Mycoplasmas of Ruminants: Pathogenicity, Diagnostics, Epidemiology and Molecular Genetics* (Frey, J. and Sarris, K., eds), EUR 16934 EC, Brussels pp. 60–62.
7. Reilly, G. A.C., Ball, H. J., Cassidy, T. D., and Bryson, T. D. G. (1993) First reported isolation of *Mycoplasma bovis* from pneumonic calves in Northern Ireland. *Vet. Rec.* **133,** 550,551.
8. Doherty, M. L., McElroy, M. C., Markey, B. K., Carter, M. E., and Ball, H. J. (1994) Isolation of *Mycoplasma bovis* from a calf imported into the Republic of Ireland. *Vet. Rec.* **135,** 259,260.
9. Done, S. H. (1997) Enzootic pneumonia (mycoplasmosis) revisited. *Pig J.* **38,** 40–61.
10. Stipkovitz, L. and Kempf, I. (1996) Mycoplasmoses in poultry. *Rev. Sci. Tech. Off. Int. Epiz.* **15,** 1495–1525.
11. Poveda, J. B., Giebel, J., Flossdorf, J., Meier, J., and Kirchhoff, H. (1994) Mycoplasma buteonis sp. nov., *Mycoplasma falconis* sp. nov. and *Mycoplasma gypsis* sp. nov., three species from birds of prey. *Int. J. Syst. Bacteriol.* **44,** 94–98.
12. Ruhnke, L. and Madoff, S. A. (1992) *Mycoplasma phocidae* sp nov., isolated from harbor seals. *Int. J. Syst. Bacteriol.* **41,** 39–44.
13. Giebel, J. Meier, A., Binder, J. Flossdorf, J. Poveda, J. B. Schmidt, R., and Kirchhoff, H. (1991) *Mycoplasma phocarhinis* sp. nov. and *Mycoplasma phocacerebrale* sp. nov. two new species from harbor seals. *Int. J. Syst. Bacteriol.* **41,** 39–44.

14. Jacobson, E. R., (1991) Chronic upper respiratory tract disease of free-ranging desert tortoises *(Xerobates agassizii). J. Wildlife Dis.* **27,** 296–316.
15. Mohan, K., Foggin, C. M., Muvavarirwa, P., Honywill, J., and Pawinda, A. (1995) Mycoplasma-associated polyarthritis in farmed crocodiles *(Crocodylus niloticus)* in Zimbabwe. *Ondespoort J. Vet. Res.* **62,** 45–49.
16. Kirchhoff, H., Schmidt, R., Lehmann, H., Clark, H. W., and Hill, A. C. *Mycoplasma elephantis* sp. nov., a new species from elephants. *Int. J. Syst. Bacteriol.* **46,** 437–441.

4

Recovery of Human Mycoplasmas

Paul Taylor

1. Introduction

There are a range of methods that can be used to recover mycoplasmas from the infected sites of humans. The current standard method is culture on supplemented agar or broth media. The prolonged time for culture particularly for *Mycoplasma pneumoniae,* has necessitated a reliance by diagnostic laboratories on serological techniques that are beyond the scope of this text but are summarized in **Table 1**.

Molecular biological methods have been applied to most clinically important mycoplasmas. The sensitivity of such techniques as the polymerase chain reaction (PCR) has not been determined for all clinical samples, some of which contain inhibitors. **Table 2** gives a comparison of the sensitivities of a number of methods for the direct detection of *M. pneumoniae.*

Although PCR could be regarded as the "gold standard" of direct detection methods, it may remain positive after the initial period of infection. The method must consequently be evaluated to determine the clinical significance of positive results. Therefore, culture remains the most sensitive of the direct detection methods for the diagnosis of mycoplasma infections of humans. Isolation methods are, however, prolonged and may be hampered by contamination with other bacteria (*see* **Notes 1** and **2**).

2. Materials

1. Broth: The liquid medium is a beef extract infusion broth containing peptone and NaCl. It may be obtained commercially as mycoplasma (or PPLO) broth (e.g., from Difco, Surrey, UK and ICN, Oxford, UK).
2. Soft agar: agar, e.g., Bacto-agar (Difco). Add to the aforementioned broth to give a concentration of about 0.8%. Commercial agar media are available (Difco, ICN).
3. Serum: (*see* **Note 3**).

From: *Methods in Molecular Biology, Vol. 104: Mycoplasma Protocols*
Edited by: R. J. Miles and R. A. J. Nicholas © Humana Press Inc., Totowa, NJ

Table 1
An Evaluation of Serological Tests for *M. Pneumoniae*[a]

Assay	Sensitivity	Specificity	Speed	Commercially available	Cost
Complement fixation test	++	++	18 h	Not in total	+
Cold agglutinins	++	+	4 h	N	+
Immunofluorescent antibody	+++	++	3 h	Y	++
Indirect agglutination	+++	++	4 h	Y	+
ELISA IgG	++	++	4 h	Y	++
ELISA IgM (μ capture)	+++	+++	4 h	Y	++

[a] + = Low, ++ = moderate, +++ = high.

Table 2
A Comparison of the Sensitivities of a Number of Methods for the Direct Detection of *M. pneumoniae*

Method	cfu/mL	Reference
PCR	1.5	*(1)*
Agar culture	100–300	*(2)*
Gen—probe	2000	*(3)*
Enzyme immunoassay	10,000	*(4)*
Immunoblot	100,000	*(5)*

4. Yeast extract: Suspend 250 g of fresh baker's yeast in 1 L of deionized water, and autoclave at 115°C for 10 min. Cool the suspension, and clarify by centrifugation at 380*g* for 10 min. Dispense the supernatant fluid in 20-mL volumes, autoclave again, and store at –20°C (*see* **Note 4**).
5. Substrates: Prepare glucose, arginine, or urea as 10% stock solutions, and sterilize by filtration.
6. Indicator: Prepare phenol red as a stock solution by dissolving 1 g of the powder (BDH Ltd) in 2.5 mL of normal NaOH and making up to 100 mL with deionized water. Coarse-filter the solution through standard filter paper, and then sterilize by filtration.
7. Inhibitors: Thallous acetate 10% (w/v) in water, store at 4°C (*see* **Note 5**); penicillin 100,000 IU/mL, store at –20°C; amphotericin B 0.5 mg/mL, store at –20°C. Sterilize by membrane filtration.
8. Final medium: Prepare agar plates by melting 70 mL of the base, cooling it to 45°C, and adding the following: 20 mL serum, 10 mL yeast extract, 10 mL sub-

strate, 0.2 mL phenol red, 0.2 mL thallous acetate, 0.2 mL penicillin, and 0.1 mL amphotericin B. The final pH of media should be: glucose 7.8, arginine 7.0, urea 6.5. Plates may be stored at 4°C for up to 4 wk. Prepare broth media by adding the same amounts of the supplements to 70 mL of broth base and dispensing the mixture in 5-mL amounts in 7-mL screw-capped (bijou) bottles. They may be stored at 4°C for up to 4 wk.

9. Diphasic medium, selective (*see* **Note 6**): Dispense mycoplasma base in 1.5-mL amounts in 7-mL screw-capped (bijou) bottles, and allow to set. Overlay with 2.5 mL of a broth medium containing: 30.0 mL mycoplasma broth base, 0.6 mL phenol red (1%), 0.4 mL thallous acetate (10%), 15.0 mL glucose (10%), 30.0 mL horse serum, 15.0 mL yeast extract (25%),1.5 mL methylene blue (0.1%), and 0.65 mL penicillin (100,000 IU/mL). Adjust the color of the final medium to gray or purple with NaOH. Occasionally, it may be necessary to add HCl if the color goes beyond purple. Store at 4°C for up to 4 wk *(6)*.
10. Diphasic medium, SP4 (*see* **Note 7**): The solid phase contains the following dissolved in 615 mL of deionized water: 3.5 g mycoplasma broth base (Difco), 10 g tryptone (Difco), 5.3 g peptone (Difco), 5 g glucose, and 8 g agar (Noble). Boil to dissolve and autoclave at 121°C for 15 min. Hold at 50°C, and add the following supplements: 50 mL CMRL 1066 medium 10X (Gibco Europe), 35 mL yeast extract (25%), 100 mL Yeastolate (Difco), 170 mL fetal calf serum (Flow Laboratories; inactivated), 10 mL penicillin (100,000 IU/mL), and 2 mL phenol red (1%). Place 1 mL of the SP4 agar in a 7-mL screw-capped (bijou) bottle, and allow to set. Overlay with 2 mL of complete SP4 broth (i.e., SP4 medium without the agar) (*see* **Note 8**).
11. A7B Agar (*see* **Note 9**): Add 35.5 g of basal agar (Oxoid CM401) to 1 L of distilled water, boil to dissolve, and distribute in 70-mL volumes. Autoclave at 121°C for 15 min, and bring to 50°C prior to aseptic addition of the following supplements: 0.5 mL CVA enrichment (Gibco), 10 mL yeast extract (25%), 20 mL horse serum (heat-inactivated), 1 mL urea (10%), 0.25 mL L-cysteine hydrochloride (4%), 1 mL penicillin G (100,000 U), 0.165 g putrescine dihydrochloride, 1 mL $MnSO_4{\cdot}H_2O$ (1.5%), 1 mL thallous acetate (0.1%), and add 0.5 mL of *N* HCl to adjust pH to 5.5.
12. Choice of specimens and sampling (*see* **Table 3**): Use complete mycoplasma broth as a transport medium for swabs. Squeeze the swabs into the fluid, which is used to inoculate the media. In culturing tissue samples, cut the fresh tissue, and push the cut edge along the surface of an agar plate. Do not be use homogenized tissue owing to the release of mycoplasma inhibitors.
13. Dienes stain: contains (g/100 mL in distilled water) methylene blue, 2.5; azure II, 1.25; maltose, 10; Na_2CO_3, 0.25; benzoic acid, 0.2.

3. Methods

3.1. Respiratory Samples

See **Fig. 1**.

Table 3
Choice of Specimens for Mycoplasma Isolation

Species	Substrate	Clinical samples
M. pneumoniae	Glucose	Sputum, brochoalveolar lavage,[a] throat swabs, pleural fluid, lung biopsies, cerebrospinal fluid
M. genitalium	Glucose	Urethral swabs, throat swabs, sputum
U. urealyticum	Urea	Vaginal swabs, cervical swabs, urethral swabs, urine (neonates: endotracheal secretions)
M. hominis	Arginine	Vaginal swabs, cervical swabs, urethral swabs, urine
M. buccale	Arginine	Throat swabs, sputum
M. faucium	Arginine	Throat swabs, sputum
M. lipohilium	Arginine	Throat swabs, sputum
M. orale	Arginine	Throat swabs, sputum
M. salivarium	Arginine	Throat swabs, sputum

[a] Ref. *(7)*.

3.2. Genital Samples

See **Fig. 2**.

3.3. Examination of Cultures

1. Examine broth and plate cultures daily for ureaplasmas; otherwise, do this twice weekly for up to 3 wk. Occasionally, for example, during epidemiological surveys, diphasic broths are reincubated for up to 6 wk; this increases the chances of detecting mycoplasmas.
2. When an acid (glucose) or alkaline (arginine and urea) change is noted in broths, subculture to agar incubated aerobically or anaerobically as described in Table 4.
3. Discard tubes with bacterial and fungal contamination, which produces a cloudy green or yellow discoloration of the broth within 24 h. Subculture 0.2 mL of broths showing a distinctive color change onto mycoplasma agar plates (*see* **Note 10**).
4. Although some mycoplasmas form tiny pinpoint colonies that can be seen with the naked eye, examine using a low-power microscope, which is essential to see the finer detail and to distinguish between colonies and artifacts. Examine the plates unopened at a magnification of ×80 or ×100. Scan the surface of the medium for what have been appropriately described as "fried egg" colonies, in which a dark central zone is usually surrounded by a lighter peripheral zone. A cross-section of a colony is shown in **Fig. 3** (*see* **Note 11**).
5. Because mycoplasma colonies grow into the agar, they cannot be removed with a loop, so the following method is suitable: cut a block of agar 5–10 mm square (containing as many colonies as possible) from the original plate with a sterile scalpel.

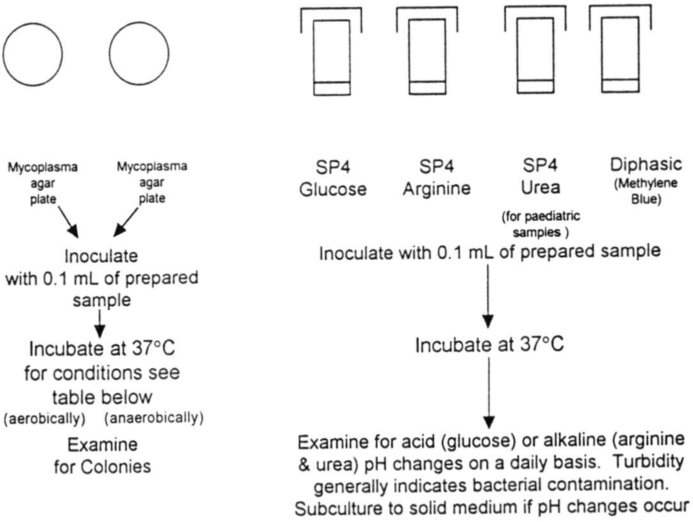

Fig. 1. Scheme for isolating mycoplasmas from respiratory samples.

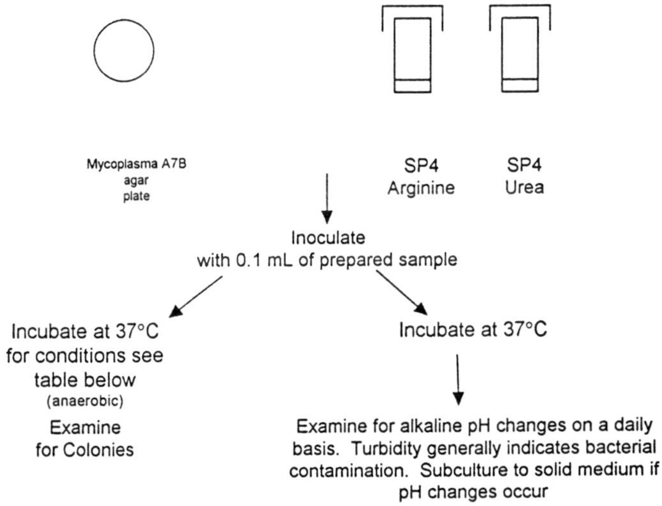

Fig. 2. Scheme for isolating mycoplasmas from genital samples.

6. Invert the block on to the surface of a fresh mycoplasma agar plate, and incubate at 37°C for 3–5 d.
7. Move the block around the agar surface, and reincubate the plate.
8. After a further 3–5 d, young colonies of mycoplasmas appear along the track of the agar block.

Table 4
Growth Characteristics of Human Mycoplasmas[a]

Species	Atmosphere for incubation	Time for appearance of colonies
M. pneumoniae	1 + 2	
M. genitalium	1 + 2	
U. urealyticum	3	1–4 days
M. hominis	1 + 2	1–4 days
M. buccale	2	
M. faucium	2	
M. lipohilium	2	
M. orale	2	
M. salivarium	2	
M. fermentans	2	

[a] 1 = Aerobic, 2 = 5%, CO_2, 95% N_2, 3 = 10–20% CO_2, 80–90% N_2.

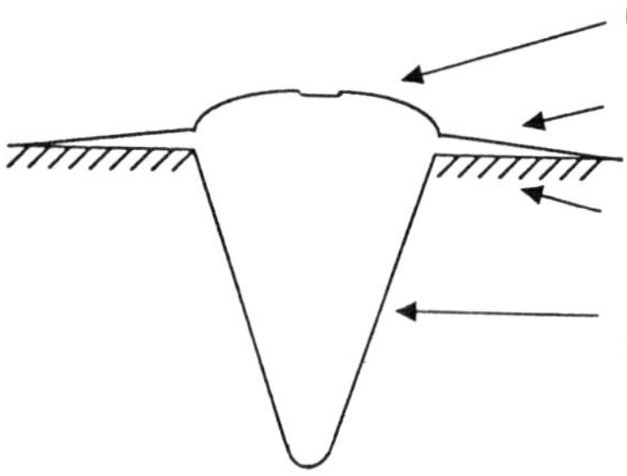

Fig. 3. Cross-section of a mycoplasma colony.

3.4. Identification

3.4.1. Staining

This method is useful for distinguishing between mycoplasma colonies and artifacts or bacterial colonies.

1. Cut a block of agar containing the colonies and about 5 mm square from the plate, and transfer, colony side up, to a microscope slide.
2. Place a drop of the stain on a coverslip, and invert over the agar block.
3. Examine the preparation with a low-power microscope (*see* **Note 12**).

3.4.2. Biochemical and Biological Reactions

Carbohydrates, amino acids, and urea are metabolized by certain mycoplasmas of medical interest (**Table 5**). This provides a convenient method of separating them into groups.

Table 5
Biochemical and Biological Properties of Mycoplasmas Isolated from Humans

	Mycoplasma pneumoniae	*Mycoplasma hominis*	*Mycoplasma orale*	*Mycoplasma salivarium*	*Mycoplasma fermentans*	*Ureaplasma urealyticum*
Growth substrates	glucose + yeast extract	arginine	arginine + yeast extract	arginine	glucose + arginine	urea + yeast extract
Colony morphology	granular*	"fried egg"	"fried egg"	"fried egg"	granular*	granular*
Growth rate	Slow	Medium	Medium	Medium	Medium	Rapid
Aerobic	+	–	–	–	–	–
Inhibition by 0.01% methylene blue	(Reduction)	+	+	+	+	+
Hemadsorption of erythrocytes (E) and HeLa cell cultures (H)	E + H	H	E	–	–	E

*no peripheral zone around colony

Seed young colonies, prepared by agar block transfer, into broths containing the substrates. Fermentation of glucose and hydrolysis of urea are indicated by the color changes of the indicators. Arginine hydrolysis is accompanied by a change to an alkaline pH (*see* **Note 13**).

3.4.3. Erythrocyte Hemadsorption

1. Pour a 10% suspension of washed erythrocytes in PBS over a plate that shows about 100 young colonies.
2. Incubate the plate at room temperature for 30 min, and remove the suspension of cells.
3. Wash the agar surface with PBS, and examine under a low-power microscope for hemadsorption of erythrocytes to the colonies.

3.4.4. Hemolysis of Erythrocytes

1. Pour an agar overlay consisting of 1 mL of 25% washed erythrocytes and 2 mL of mycoplasma agar base (at 45°C) carefully over the plate on which the colonies are growing.
2. Incubate the plate aerobically at 37°C after the agar has set.
3. Refrigerate for 30 min before examining at 24 and 48 h (*see* **Note 14**).

3.4.5. Growth Inhibition

The above biochemical and biological tests may aid in the identification of colonies of mycoplasmas, but serological methods are necessary for specific identification.

1. Prepare a 3- to 7-day old culture of the isolate by seeding it with young colonies with dense growth obtained by the block transfer method.
2. Impregnate paper discs, about 5 mm in diameter, with hyperimmune specific antisera (complement fixation titer not <1:320). Dried disks give similar results.
3. Flood a dry mycoplasma agar plate with the broth culture, and remove the excess fluid. Absorb the small amount of broth remaining for 15–30 min. Apply the disks (up to 5) around the periphery of the plate, and incubate at 37°C, under appropriate atmospheric conditions, for 5–7 d.
4. A zone of inhibition appears round the disk containing the antiserum homologous to the strain of mycoplasma.

3.4.6. Immunofluorescent Staining

This is the most reliable method for the identification of mycoplasmas. It has the advantage over the growth inhibition tests in that pure cultures (and therefore subcultures) are not required. In addition, individual species in mixed cultures may be identified. The results are available within a few hours and correlate well with those of other identification tests (*see* **Note 15**)

1. Boil rubber or neoprene bungs, 10 mm in diameter, in distilled water, and dry before use.

2. Place the bottom of a bung gently on the top of the chosen colony. Lift carefully and press onto a microscope slide. (A number of replicates may be made from the same bung by pressing harder on each successive slide.)
3. Discard the bung into disinfectant.
4. Fix the impression by passing the slide through a Bunsen flame.
5. Stain by the indirect immunofluorescence method using specific sera from humans or animals, and antispecies FITC conjugates. Both sera and FITC may require adsorption with pig liver powder and washed yeast cells if media constituents cause nonspecific fluorescence.
6. Examine under UV microscopy, which should show bright green fluorescence with homologous sera.

4. Notes

1. The culture of mycoplasmas from humans remains restricted to a small number of diagnostic laboratories. Although some ready-made commercial media are available, they have not improved the overall use of culture methods, which remain more of an art form rather than a science. Once molecular techniques for the detection of mycoplasmas have been firmly established and their significance confirmed, it is likely that they will replace most culture methods.
2. Broth and agar media for mycoplasma media should be supplemented with a range of growth factors, including horse serum, yeast extract, substrates (glucose, arginine, or urea), an indicator, and inhibitors to discourage the growth of other organisms. Components like yeast extract should be carefully monitored to ensure maximum recovery of mycoplasmas.
3. This provides lipid growth factors and cholesterol, necessary for the growth of all human mycoplasmas. Each batch of serum should be tested for its ability to support the growth of particular mycoplasmas. Although human serum has been used in the culture of mycoplasmas affecting humans, specific antibody may be present; horse and fetal calf sera are therefore preferred. The serum should be stored at –20°C or below.
4. Some species, including *M. pneumoniae* and *Mycoplasma orale,* grow poorly in the absence of yeast extract. This supplement provides a heat-stable, low-mol-wt mixture of vitamins, purines, pyrimidines, and other essential nutrients. Several commercial preparations are available, but freshly prepared extract is recommended, since its use allows greater control over preparation and quality. To avoid contamination of other media and cell cultures with this yeast, the extract should be prepared in a separate room.
5. Care should taken with thallous acetate, since it is toxic.
6. A selective diphasic medium consisting of an agar base and a broth phase was described by Kraybill and Crawford in 1965 ***(8)***. The incorporation of methylene blue in the broth phase inhibits all respiratory mycoplasma apart from *M. pneumoniae.* It has been demonstrated ***(9)*** that this diphasic medium is a more sensitive primary isolation system than agar inoculation.

7. SP4 medium was originally described for spiroplasma isolation but was shown ***(10)*** to support the growth of *M. pneumoniae* from clinical samples more rapidly and sensitively than conventional diphasic medium. Subsequently, a comparison with horse serum broth demonstrated the increased sensitivity of this medium.
8. Methylene blue inhibits the growth of *M. pneumoniae* in this medium ***(11)***.
9. This is a modification of the medium described by Shepard ***(12)***. It incorporates manganous sulfate that in the presence of ammonia, produced in the metabolism of urea by ureaplasmas, forms a brown manganous dioxide reaction, allowing colonies of ureaplasmas to be distinguished from other mycoplasmas.
10. *M. pneumoniae* usually grows in the selective diphasic broth within 8–12 days. Ureaplasmas and *Mycoplasma hominis* are more rapid growers and may form colonies within 2–4 d. Acid, produced from glucose, changes the color of the medium from blue to clear light green. The color of the agar plug changes from blue to yellow.
11. It will become apparent that the central zone corresponds to growth of the colony down into the medium, whereas the peripheral zone is surface growth. Colonies of *M. pneumoniae* growing in agar media in primary culture lack the peripheral zone, but this is formed after several agar to agar subcultures. Colony size varies from 10–55 μm. Artifacts, such as serum precipitates, fat globules, and cell nuclei (the so-called pseudocolonies), are distinguished from mycoplasma colonies by focusing on different planes. Careful attention to media preparation will avoid the appearance of at least some of these artifacts.
12. Mycoplasma colonies stain dark blue, and the stain is retained for over 24 h. Ureaplasma colonies stain a greenish blue. The stain is eliminated from bacterial colonies within 30 min.
13. Other properties that may assist in identification include growth rates, aerobic reduction of tetrazolium, and inhibition by 0.01% methylene blue.
14. Clear (β-) hemolytic plaques are produced by *M. pneumoniae.* Other respiratory mycoplasmas give incomplete (α) plaques, which are similar and take longer to develop.
15. A number of methods have been proposed for the initial preparation of the colonies. These include the staining of colonies growing on agar blocks or on agar-coated slides and colony impressions obtained by hot-water fixation. The problems associated with these techniques are background fluorescence, necessity to subculture, and the loss of specificity as a result of heat inactivation. They have been overcome by the use of rubber bungs to transfer colonies to microscope slides.

References

1. Tjhie, J. H. T., van Kuppeveld, F. J., Rosendaal, R., Melchers, W. J. G., Gordijn, R., Walbodmers, J. M. M., Meijher, C. J. L. M., and van der Brule, A. J. C. (1994) Direct PCR enables detection of *Mycoplasma pneumoniae* in patients with respiratory disease. *J. Clin. Micro.* **32,** 11–16.
2. Kenny, C. E., Kaiser, G. G., Cooney, M. K., and Foy, H. M. (1990) Diagnosis of Mycoplasma pneumoniae pneumonia: sensitivities and specificities of serology with lipid antigen and isolation of the organism on soy peptone medium for identification of infections *J. Clin. Micro.* **25,** 2087–2093.

3. Harris, R., Marmion, B. P., Varkanis, G., Kok, T., Lunn, B., and Martin, J. (1988) Laboratory diagnosis of Mycoplasma pneumoniae infection. 2. Comparison of methods for the direct detection of specific antigen or nucleic acid sequences in respiratory exudates. *Epidemiol. Infect.* **101,** 685–694.
4. Kok, T-W., Varkanis, G., Marmion, B. P., Martin, J., and Esterman, A. (1988) Laboratory diagnosis of Mycoplasma pneumoniae infection. 1 Direct detection of antigen in respiratory exudates by enzyme immunoassay. *Epidemiol. Infect.* **101,** 669–684.
5. Cimolai, N., Schryvers, A., Bryan, L. E., and Woods, D. E. (1988) Culture-amplified immunological detection of Mycoplasma pneumoniae in clinical specimens. *Diagn. Micro. Infect. Dis.* **9,** 207–212.
6. Smith, C. B., Friedwell, W. T., and Chanock, R. M. (1967) Shedding *M. pneumoniae* after tetracycline and erythromycin therapy. *N. Engl. J. Med.* **276,** 1172–1175.
7. Lehtomaki, K., Kleemola, M., Tukianen, P., Katanen, M.-L., and Latinen, L. A. (1987) Isolation of *M. pneumoniae* from bronchoalveolar lavage. *J. Infect. Dis.* **155,** 1339–1341.
8. Kraybill, W. H. and Crawford, Y. E. (1965) A selective medium and colour test for *Mycoplasma pneumoniae. Proc. Soc. Exp. Biol. Med.* 1**18,** 965–970.
9. Craven, R. B., Wenzel, R. P., Calhoun, A. M., Hendley, J. O., Hamory, B. H., and Gwaltney J. M. (1976) Comparison of the sensitivity of two methods for isolation of *M. pneumoniae. J. Clin. Micro.* **4,** 225–226.
10. Tully, J. G., Rose, D. L., Whitcome, R. F., and Wanzel, R. P. (1979) Enhanced isolation of *M. pneumoniae* from throat washings with a newly modified culture medium. *J. Infect. Dis.* **139,** 478–482.
11. Senterfit, L. B. (1984) Laboratory diagnosis of mycoplasma infections. *Isr. J. Med. Sci.* **20,** 905–907.
12. Shepard, M. C. and Combs, R. S. (1979) Enhancement of *Ureaplasma urealyticum* growth on a differential agar medium (A7B) by a polyamine, putrescene. *J. Clin. Micro.* **10,** 931–933.

5

Recovery of Mycoplasmas from Animals

Robin Nicholas and Samantha Baker

1. Introduction

Of the 30 or so mollicute species that have been isolated from small and large ruminants, only a handful have been shown to cause disease in their own right. This chapter will concentrate on methods for sampling, transporting, and isolating those pathogenic mycoplasmas, including *Mycoplasma mycoides* subsp. *mycoides* small colony (SC), *Mycoplasma capricolum* subsp. *capripneumoniae*, *Mycoplasma bovis*, *Mycoplasma agalactiae*, and *M. m. mycoides* large colony (LC), which cause economically important diseases, such as contagious bovine pleuropneumonia (CBPP), contagious caprine pleuropneumonia (CCPP), and contagious agalactia. Although DNA amplification techniques are being used with ever-increasing frequency for the detection and identification of mycoplasmas (*see* Chapters 18–20), the isolation of these organisms by conventional techniques is still required by most national and international authorities to confirm disease outbreaks.

With the exception of *M. c. capripneumoniae*, the cause of CCPP, and *M. dispar*, a cause of respiratory disease in calves, the majority of pathogenic mycoplasmas are not intrinsically difficult to grow, and most general purpose media (Eaton's, Friis modified, Chanock's, or Hayflick's) will suffice. What may complicate isolation, however, is bacterial contamination, the heavy presence of antibiotics in the clinical samples, and/or where there is overgrowth by less important but rapidly growing mollicutes, including acholeplasmas. Various strategies are available to counter these problems, such as the use of antibacterial agents like thallium acetate, in addition to the usual range of nonmycoplasmastatic antibiotics. The use of selective inhibitors, such as nisin *(1)* which is capable of suppressing acholeplasmas and cell-walled bacteria with-

From: *Methods in Molecular Medicine, Vol. 104: Mycoplasma Protocols*
Edited by: R. J. Miles and R. A. J. Nicholas © Humana Press Inc., Totowa, NJ

out affecting mycoplasma growth, promises important advances in the development of selective media.

Since the development of a media for *M. c. capripneumoniae* by MacOwan and Minette in 1976 ***(2)*** for the first time, several others ***(3,4)*** have been reported. A high serum concentration (about 25%) and the addition of pyruvate appear to be essential for the successful growth of this difficult mycoplasma. A commercially available medium (Mycoplasma Experience, Reigate, UK) is also now available.

M. dispar is a particularly fastidious and slow-growing mycoplasma, especially on solid media. It is easily overgrown by *M. bovirhinis*, a commonly occurring mycoplasma of little significance that is often present in the same samples. A selective media incorporating antisera has been reported that suppresses *M. bovirhinis* while promoting the growth of *M. dispar* ***(5)***.

2. Materials

See Chapter 4 for details of constituents, e.g., preparation of fresh yeast extract.

1. Transport medium: dissolve 2.5 g of heart infusion broth (HIB) in 70 mL of distilled water by heating at 80°C. Add to HIB 10 mL of 25% fresh yeast extract, 20 mL of horse serum, 0.2 mL of 10% thallium acetate solution, and 0.5 mL of penicillin (200,000 IU/mL). Adjust pH to 7.6 and sterilize by filtration (*see* **Note 1**).
2. General purpose growth medium (Eaton's): dissolve 21 g of Difco (Molesey, UK) PPLO broth base (without crystal violet) in 700 mL of distilled water. To PPLO broth, add 100 mL of freshly prepared yeast extract, 200 mL of unheated horse serum, 10 g of glucose, 0.5 mL of penicillin (200,000 IU/mL), 12.5 mL of 0.2% phenol red, and 0.02 g of DNA. Adjust pH to between 7.6–7.8, and sterilize by filtration. Prepare solid medium by adding 10 g of LabM agar no. 1 (Bury, UK), or agar of equivalent quality and dispense into sterile Petri dishes.
3. Growth medium for *M. c. capripneumoniae* (H25P): dissolve 17.5 g of Bacto-PPLO broth (without crystal violet) in 650 mL of glass-distilled water, and sterilize by autoclaving at 121°C for 30 min. This comprises part A. Mix 250 mL of horse serum (inactivated at 56°C for 30 min) with 100 mL of fresh yeast extract, 4 mL of 50% glucose, 4 mL of 0.5% phenol red, and 8 mL of 25% sodium pyruvate (part B). Aseptically add part B to part A, and adjust pH to 7.8 with NaOH or HCl. For primary isolation, add 4 mL of 5% thallium acetate solution and 250 mg of ampicillin to the growth medium. Prepare solid medium as for Eaton's above.
4. Growth medium for *M. dispar* (Friis medium modified): combine 500 mL modified Hank's balanced salt solution with 8.2 g Bacto-brain heart infusion (Difco), 8.7 g Bacto-PPLO broth without crystal violet (Difco), and 750 mL distilled water. Autoclave, and then add 60 mL freshly prepared yeast extract, 4.5 mL of phenol red (0.5% solution), 250 mg bacitracin, and 250 mg methicillin. Finally, add horse serum and porcine serum to a final concentration of 10% of each. Adjust to pH 7.4 with NaOH, and sterilize by filtration. Prepare solid medium as for Eaton's above.

3. Method

3.1. Sample Collection and Transport

Table 1 shows the range of samples that can be taken for the isolation of the pathogenic mycoplasmas. The normal bacteriological procedures apply to sample-taking. To ensure optimal recovery, fresh samples of milk and synovial fluid must be taken. Lung lavage techniques have been advocated for detecting invading mycoplasmas in the lower respiratory tract ***(6)*** but require veterinary assistance.

3.1.1. Nose, Eye, or Ear Swabs

1. Prewet sterile cotton swabs in transport media, and then insert deep into nasal passage or ear canal or gently swab the surface of the eye.
2. Place swab in transport media, and snap off handle (*see* **Note 2**).

3.1.2. Lung Samples

1. Locate lesions and sterilize exterior of organ by searing with hot instrument, flaming, or boiling ***(6)***.
2. Where possible, use a fresh set of sterile instruments for each tissue.
3. Aseptically remove small pieces of tissue (1–3 cm^3) from the interface between consolidated and unconsolidated area.
4. Place each piece of tissue in separate sterile screw-capped jar containing transport medium.
5. Where lesions are encapsulated, take or scrape sample from internal surface using scalpel blade, and place in transport medium.

3.1.3. Milk Samples

1. Cleanse the tip of the animal's teat.
2. Discard the initial stream of milk.
3. Fill a sterile tube with the next stream of milk.
4. Allow milk to stand.
5. If a cream layer separates, place two drops of the layer into a broth using a Pasteur pipet.
6. If there is clotting, use a portion of the clot in preference to the supernatant liquid.
7. If no layer develops, use the whole milk (*see* **Note 3**).

3.1.4. Pleural Fluid

Pleural fluid is the sample of choice for the diagnosis of CBPP and CCPP, but is only present in animals in the acute phase of the disease ***(4, 7)***. Mycoplasmas can be recovered from this sample in pure culture and in high numbers.

1. Identify animal in the acute phase of disease by clinical signs.
2. Kill animal by humane means, and position carefully for postmortem examination (avoid raising carcass vertically) (*see* **Note 4**).

Table 1
Recovery of Mycoplasmas from Ruminants

Species	Growth in vitro	Host	Sample from live animal	Sample from dead animal
M. m. mycoides SC	Good	Cattle	Nasal swab, nasal discharge, bronch-alveolar washings, pleural fluid	Lung lesions, pleural fluid, broncho-pulmonary lymph nodes
M. c. capripneumoniae	Fastidious	Goats	Nasal swabs, ear swabs[a]	Lung lesions, pleural fluid, mediastinal lymph nodes
M. agalactiae	Good	Sheep/goats	Milk, joint fluid, ocular swabs, nasal and ear swabs	Udder and associated lymph nodes, joint fluid, lung lesions
M. bovis	Good	Cattle	Nasal swabs, ocular swabs, joint fluid, milk, bronch-alveolar washings, pleural fluid	Lung lesions, joint fluid, udder
M. m. mycoides LC	Good	Goats	Milk, joint fluid, ocular swabs, nasal swabs	Udder and associated lymph nodes, joint fluid, lung lesions
M. c. capricolum	Good	Goats/sheep	Milk, joint fluid, ocular swabs, nasal swabs	Udder and associated lymph nodes, joint fluid, lung lesions
M. dispar	Fastidious	Cattle	Nasal swabs, broncho-alveolar washings, pleural fluid	Lung lesions

[a]Theoretically feasible but we are unaware of any successful isolations of *M.c. capripneumoniae* by this route

3. Open chest and remove aseptically at least 10 mL of straw-colored pleural fluid.
4. Lyophilize for long-term storage or international transport.

3.1.5. Transport

1. Send samples to laboratory as quickly as possibly, preferably the same day.
2. Keep samples cool (about 4°C).
3. If microbiological examination cannot be performed immediately, store samples and whole or parts of organs in deep-freeze for up to several months.
4. For international transport, where freezing during transport is not possible, lyophilize samples and send (*see* **Notes 5** and **6**).

3.2. Isolation of Mycoplasmas from Samples

When isolating mycoplasmas from ruminants, it is advisable to use at least two media, a general-purpose and specialist media, to maximize the chance of recovery.

3.2.1. Isolation on Solid and Liquid Media

1. Make 10-fold dilutions (10^{-1}–10^{-6}) of liquid sample (pleural fluid, nasal exudate, synovial fluid, and so forth) or tissue homogenate (*see* **Note 7**) in appropriate medium (*see* **Note 8**).
2. Deposit, spread a few drops of each sample on the solid medium, and dispense a 10% (v/v) inoculum into liquid medium.
3. In addition, make a direct impression on the solid medium with the cut surface of a lung lesion or lymph node without spreading it.
4. Streak swabs (if available) directly onto solid medium.
5. Incubate broths (optimally with gentle shaking) and plates at 37°C in a humidified atmosphere with 5% CO_2 (*see* **Note 9**).
6. Examine broths daily for signs of growth or changes of pH indicated by a color change in the media (*see* **Note 10**).
7. Examine plates after 2–3 d under 35× magnification for the typical "fried egg" appearance (*see* **Note 11**).

3.2.2. Subculturing

For most serological identification tests, it is necessary to grow the mycoplasma on solid medium. Subculture from broth should be carried out immediately when growth is apparent or pH change is seen (*see* **Note 12**). If neither occurs, subculture intervals can be extended to 7 d.

1. For broth-to-broth subculturing, place a 10% (v/v) inoculum of incubated broth into the new broth, using a pipet.
2. For broth-to-plate subculturing, place plate on level surface, and carefully add a single drop (about 25 µl) of incubated broth onto agar. Allow drop to soak in before moving plate (use 4 cultures/9-cm plate).
3. For plate-to-broth subculturing, cut out an agar block containing colonies with a sterile spatula OR pick up a well-separated single colony by taking out an agar plug with a Pasteur pipet, and drop into the fresh broth.

4. For plate-to-plate subculturing, cut out an agar block with colonies, and place face down on the new plate; carefully slide block on the surface.

3.2.3. Contamination

Contamination will be seen as excessive turbidity in the broth.

1. Filter-sterilize by passing 1 mL of the contaminated broth using a sterile syringe through a 0.45 μm filter enabling several drops to fall into a fresh broth and/or plate.
2. Repeat as necessary.

4. Notes

1. For the transport of clinical samples containing *M. m. mycoides* SC, the Office des International Epizooties recommends the incorporation of 0.3% of agar into the medium.
2. Some workers advocate discarding the swab after agitating in the transport media, because it reduces bacterial contamination. However, it also reduces the chances of isolating organisms where they are scarce.
3. Milk that appears normal (and therefore likely to contain only small numbers of mycoplasmas) can be incubated whole after the addition of ampicillin to between 1 and 10 mg/mL, and subcultured after 1–2 days. This procedure increases the levels of contamination with ampicillin-resistant organisms, but will occasionally result in isolation of a mycoplasma from milk, which proved negative by the standard procedure.
4. It is possible for a veterinary surgeon to take a few milliliters of pleural fluid from the acutely affected live animal by puncturing the thoracic cavity in its sloping part between the seventh and eighth ribs.
5. Freeze-drying lung homogenates from CCPP-affected goats before transport overseas has proven successful for the recovery of *M. c. capripneumoniae.*
6. Many countries require a special import license to be obtained in advance for any biological material, especially for tissues that contain animal pathogens, including mycoplasmas.
7. Tissue samples are best chopped with scissors and then shaken vigorously or pulverized in medium (10% w/v).
8. Dilution of the samples has a number of benefits: first, it reduces the effects of mycoplasmacidal substances, including antibiotics released by the tissues, and second, it reduces bacterial contamination and the overgrowth by less important, but more exuberant mollicutes.
9. An atmosphere of CO_2 is unnecessary for *M. m. mycoides* SC, since air is adequate.
10. Broths should be examined against a good, even light. Bacterial contamination will be seen as gross turbidity evident within 24 h. Mycoplasma growth will appear between 3 and 5 d, and is usually seen as a very fine cloudiness, usually described as "opalescence"; it may be necessary to compare with an uninoculated broth to see the growth, particularly in the case of *M. c. capripneumoniae. M. m. mycoides* grows well and usually produces "whirls" from the bottom of the tube when

shaken. If a film appears on the surface of the liquid medium, which also has an orange color the culture may be *M. bovis* if isolate is from cattle and *M. agalactiae* from small ruminants.

11. Mycoplasmas grow into the agar (*see* Chapter 4), which makes the use of a loop for subculturing unsatisfactory. The strongest growers can be seen readily with the naked eye, reaching 1 mm or even, in the case of *Acholeplasma laidlawii* for instance, 1.5 mm, but in order to see smaller colonies or detailed or larger colonies, it is necessary to use a plate microscope. All members of the *M. mycoides* cluster except *M. c. capripneumoniae* grow within 3 d, producing colonies of between 1 and 3 mm. *M. ovipneumoniae* can be suspected when isolated from small ruminant lung when the colonies are centerless and do not stick to the agar surface. In general, however, colonies may vary greatly in size in a single culture and within different cultures of the same strain, for example, in having small or large centers or a granular or smooth appearance. Occasionally, colonies of a pure culture on the same plate will differ in appearance, perhaps because of age. It follows, then, that colonial morphology is relatively valueless as a typing aid.
12. Some mycoplasmas, notably *M. m. mycoides*, will die rapidly at a pH much below 7.0, and in the case of *M. m. mycoides*, which produces acid rapidly in the medium, subculturing at more than 3-d intervals is likely to result in loss of the strain. We normally subculture three times before rejecting the material as negative. Previous subcultures are held, so that at the end, the primary broth will have been incubated for 4 wk. Plates are incubated for a week.

References

1. Abu-Amero, K. K., Halablab, M. A., and Miles, R. J. (1996) Nisin resistance distinguishes mycoplasma spp. and provides a basis for selective growth media. *Appl. Environ. Microbiol.* **62**, 3107–3111.
2. MacOwan, K. J and Minette, J. E. (1976) A mycoplasma from acute contagious caprine pleuropneumonia in Kenya. *Trop. Anim. Health Prod.* **8**, 91–95.
3. Bölske, G. (1988) Survey of mycoplasma infections in cell cultures and a comparison of detection methods. *Zentralblat. Bakteriol. Hyg. A* **269**, 331–340.
4. Thiaucourt, F., Bölskc, G., Leneguersh, B. Smith, D., and Wesonga, H. (1996) Diagnosis and control of contagious caprine pleuropneumonia. *Rev. Sci. Tech. Off. Int. Epiz.* **15**, 1415–1429.
5. Friis, N. F. (1979) Selective isolation of slow growing acidifying mycoplasmas from swine and cattle. *Acta. Vet. Scand.* **20**, 607–609.
6. Laak, E. A. ter, Wentink, G. H., and Zimmer, G. M. (1992) Increased prevalence of *Mycoplasma bovis* in the Netherlands. *Vet. Q.* **14**, 100–104.
7. Provost, A., Perreau, P., Breard, A., Goff, C. Le., Martel, J. L. and Cottew, G. S. (1987) Contagious bovine pleuropneumonia. *Rev. Sci. Tech. Off. Int. Epiz.* **6**, 625–679.

6

Recovery of Mycoplasmas from Birds

Janet M. Bradbury

1. Introduction

The methods used for recovering mycoplasmas from birds are broadly similar to those described for other animals and for humans. Mycoplasmas are important causes of disease and loss of production in intensively reared poultry, particularly in those that are under environmental stress *(1)*. *Mycoplasma gallisepticum* affects chickens and turkeys and is the most important of the pathogenic avian mycoplasmas. It is especially serious in broiler (meat-type) chickens in which it often acts synergistically with other agents, such as respiratory viruses or pathogenic strains of *Escherichia coli* to provoke chronic respiratory disease. In laying birds, it may also cause loss of egg production. *Mycoplasma synoviae* affects chickens and turkeys and, under some circumstances, can contribute toward respiratory disease, whereas in heavy birds, it may cause synovitis and arthritis. *Mycoplasma meleagridis* is specific for its host, the turkey, in which it can cause reduced hatchability and reduced viability of young birds and can also cause disease of the air sacs. A fourth avian mycoplasma pathogen is *Mycoplasma iowae,* but its economic effects appear to be largely confined to the turkey embryo, in which some strains are lethal. The chicken and turkey industries have invested heavily in the production of primary breeding stock that are free from these pathogens, and they also maintain expensive biosecurity and serological surveillance programs to guard against reinfection that will prejudice their home and export trade.

There are numerous other avian mycoplasmas, some of which may be pathogenic under certain circumstances, but their economic importance is unknown as yet. It is reasonable to speculate that ducks and geese under intensive management may have their own pathogenic mycoplasma flora. There is also interest in game birds, such as pheasant, partridge, and quail, because they are

From: *Methods in Molecular Biology, Vol. 104: Mycoplasma Protocols*
Edited by: R. J. Miles and R. A. J. Nicholas © Humana Press Inc., Totowa, NJ

susceptible to some of the poultry mycoplasmas and therefore might be involved in their transmission. Although wild birds have not been strongly implicated in the transmission of the pathogenic mycoplasmas to commercial poultry, a disease problem has arisen recently in house finches in the US owing to *M. gallisepticum* infection ***(2,3)***. The infection has caused such severe conjunctivitis that it has killed large numbers of birds and has also been found in other species of song bird. At the time of writing, it appears that the strains affecting finches are different from those infecting domestic poultry.

Not surprisingly, other birds, such as ostriches and birds of prey, appear to have their own unique mycoplasma flora. Extrapolation from the situations seen in other avians suggests that such birds, when kept in captivity under intensive and stressful conditions, might succumb to mycoplasma disease.

Table 1 shows the avian mycoplasma species reported so far, together with their main host(s) and major in vitro metabolic activity. Most diagnostic and research work focuses on the commercially important hosts (chickens and turkeys) and the pathogenic mycoplasmas mentioned above. DNA detection methods using PCR have been developed for these, and some (e.g., *M. gallisepticum* and *M. synoviae*) are in use as commercial kits. Such methods can be very helpful if rapid results are required for a valuable flock, which has given some suspect positive reactions in serological screening. However, there is still considerable demand from the industry for mycoplasma isolation. Recovery of the causal organism allows for further epidemiological studies and for antimicrobial sensitivity tests to be performed.

Provided that suitable quality-control measures are adopted, the procedures described below should enable the isolation of the four recognized pathogenic avian *Mycoplasma* species and also the numerous nonpathogenic *Mycoplasma* and *Acholeplasma* species. The medium ***(4)*** is modified from that originally described by Edward ***(5)***, and includes swine serum rather than horse serum. Studies so far indicate that this formulation is equally as good for avian mycoplasmas as the more refined SP-4 formulation ***(6)***, but it should be noted that there are occasional reports of "atypical" strains of *M. gallisepticum* or *M. synoviae,* which are extremely difficult to recover from birds, and it is also possible that there are strains that are genuinely uncultivable in present media formulations. Isolations of avian ureaplasmas have also been reported occasionally in some countries. The media used for this purpose contain urea as a substrate and are adjusted to low pH ***(7)***. The isolation and cultivation of ureaplasmas is described in Chapter 7.

2. Materials

1. Test strains for quality control of media: Recent field isolates of *M. gallisepticum*, *M. synoviae*, *M. meleagridis,* and *M. iowae* (*see* **Note 1**).

Table 1
Mycoplasmas Found in Birds[a]

Species	Hosts	Glucose/arginine metabolism[b]
Mycoplasma anatis	Duck	+/–
Mycoplasma anseris	Goose	–/+
Mycoplasma buteonis	Buzzard	+/–
Mycoplasma cloacale	Turkey, goose	–/+
Mycoplasma columbinasale	Pigeon	–/+
Mycoplasma columbinum	Pigeon	–/+
Mycoplasma columborale	Pigeon	+/–
Mycoplasma corogypsi	Vulture	+/–
Mycoplasma falconis	Falcon	–/+
Mycoplasma gallinaceum	Chicken	+/–
Mycoplasma gallinarum	Chicken, turkey	–/+
Mycoplasma gallisepticum	Chicken, turkey, pheasant, partridge, songbirds[c]	+/–
Mycoplasma gallopavonis	Turkey	+/–
Mycoplasma glycophilum	Chicken	+/–
Mycoplasma gypis	Vulture	–/+
Mycoplasma iners	Chicken, turkey	–/+
Mycoplasma iowae[d]	Turkey, chicken	+/+
Mycoplasma imitans	Goose, duck, partridge	+/–
Mycoplasma lipofaciens	Chicken, turkey	+/+
Mycoplasma meleagridis	Turkey	–/+
Mycoplasma pullorum	Chicken	+/–
Mycoplasma sturni	Starling (european)	+/–
Mycoplasma synoviae	Chicken, turkey	+/–

[a]Some *Acholeplasma* and *Ureaplasma* species are also found in birds, but their significance is unknown.
[b]Glucose-fermenting and arginine-hydrolyzing species produce acid and alkaline change, respectively, in broth medium containing a suitable pH indicator.
[c]Several other species have also been isolated sporadically from songbirds.
[d]This species also includes serovars J, K, N, Q, and R.

2. Yeast extract: 250 g active dry baker's yeast in 1 L of distilled, deionized water. Heat, with stirring, to boiling point, and centrifuge at 3500*g* to remove debris. Adjust pH to 8.0 using NaOH, pass through a sterilizing filter, and store at –20°C (*see* **Note 2**).
3. Stock solutions: Prepare 1, 10, 10, and 5% w/v aqueous solutions of nicotinamide adenine dinucleotide (NAD), glucose (pH 7.8–8.0), arginine (pH 7.0), and thallium acetate, respectively (*see* **Note 3**). Sterilize by filtration and store at –20°C.
4. Phenol red indicator: Grind 0.1 g of phenol red with 2.82 mL 0.1*M* NaOH in a pestle and mortar, and make up to 100 mL with water. Sterilize by autoclaving at 8–10 lb/in^2 for 30 min.

5. Penicillin solution: Use a sterile disposable syringe to add the required volume of buffered distilled water (pH 7.2–7.4) to a vial of penicillin powder. Allow to dissolve, and store at 4°C for a maximum of 1 wk.
6. Swine serum: Either purchase from a reputable source or prepare in the laboratory from pooled slaughterhouse blood (*see* **Note 4**). Heat at 56°C for 45 min. Store at –20°C.

3. Methods

3.1. Preparation of Media

1. Growth media: All ingredients should be of the best possible grade. Make mycoplasma broth in two parts, A and B. For 100 mL of final product: Part A Dissolve 1.47 g Difco PPLO broth powder (Difico, West Molesey, Surrey, UK) in 70 mL distilled deionized water, and autoclave at 15 lb/in^2 for 15 min; part B assemble the following sterile ingredients: 15 mL swine serum, 10 mL yeast extract, 1 mL of 1% NAD solution, 1 mL of 10% glucose solution, 1 mL of 10% arginine solution, 1 mL of 5% thallium acetate solution, 100,000 U penicillin, 2 mL of 0.1% phenol red solution. Prepare 100 mL of mycoplasma agar in the same manner, except add 1 g of purified agar to part A before autoclaving and omit phenol red from part B. Bring both parts to 56°C in a water bath for mixing and pouring into 5-cm diameter Petri dishes (*see* **Note 5**). Allow to solidify and store at 4°C. If there is excessive moisture on the agar surface, the plates can be left at room temperature overnight or incubated at 37°C for approximately half an hour.
2. To ensure consistency in the quality of different batches of medium, some quality-control procedures should be carried out (*see* **Note 6**).

3.2. Collecting the Specimens

1. Swabs (*see* **Note 7**) can be used on live birds to collect tracheal, oropharyngeal, eye, nasal, or cloacal samples (*see* **Note 8**).
2. Specimens can also be taken from dead birds, and can include exudates, body fluids, or tissues (*see* **Note 8**). Dip dead birds up to the neck in a suitable disinfectant to dampen the feathers and reduce bacterial contamination, taking care not to allow disinfectant to enter the mouth, eyes, or nares. Open the birds using aseptic technique, and examine the thoracic and abdominal cavities for lesions, such as tracheitis, congestion of the lungs, and airsacculitis. If there is a history of lameness or evidence of swelling of the tibiotarsal-tarsometarsal (hock) joint, examine the joint cavity for excess fluid and evidence of synovitis or arthritis, and remove appropriate specimens.
3. If it is feasible, inoculate the specimens on to mycoplasma agar and into mycoplasma broth (*see* **Subheading 3.3.1., step 1**) while on the farm or in the post-mortem room. Otherwise, pack the specimens suitably for cooled transport to the laboratory by the fastest means (*see* **Note 9**).

3.3. Culturing the Specimens

1. Inoculate specimens onto mycoplasma agar and into broth. For swabs, rub them over the agar surface, and then agitate in broth. The swab can be retained in the

broth for incubation to enhance mycoplasma growth ***(8)***, but this practice is also likely to enhance growth of any contaminating bacteria. It should also be noted that rayon swabs on aluminum sticks have been found inhibitory to *M. gallisepticum* ***(8)***. Some clinical specimens may themselves be inhibitory to mycoplasma growth (*see* **Note 10**), and for these, it is advisable to prepare dilutions. Cut small pieces of tissue, and rub the freshly cut surface over the agar. Pass soft tissues through a syringe (minus needle), and make an approx 1:10 (w/v) dilution in broth. Harder tissues, such as articular cartilage, can be macerated coarsely in broth. Prepare 10-fold dilutions to 10^{-3} and inoculate each on to mycoplasma agar (*see* **Note 11**) retaining all the dilutions for incubation.

2. Incubate cultures at 37°C (*see* **Note 12**). Plates should preferably be incubated in a CO_2 incubator maintaining 5% CO_2 (*see* **Note 13**). Examine broths daily for color change in the pH indicator and plates every 2–3 d for colony growth.
3. Subculture immediately from any broths showing a color change (*see* **Note 14**), or after 1 wk if there is no color change. Subculture into fresh broth (1:10 v/v) and/or onto mycoplasma agar. The aim is to produce discrete mycoplasma colonies in sufficient numbers to identify the species by a method such as immunofluorescence or immunoperoxidase staining (*see* **Note 15**).

4. Notes

1. Strains for quality control should be organisms that are not too well adapted to the laboratory medium, i.e., organisms of low in vitro passage level. It is also preferable to select the more fastidious strains, which have proven to be slow-growing or which show only sparse growth. These should be grown up in batches and frozen at the lowest available temperature. The number of colony-forming units in a thawed aliquot of each strain is then established ***(9)***.
2. The quality of the yeast extract is very important, particularly for isolating *M. iowae*. Commercial formulations have sometimes proven less successful for primary isolation than "home-made" extracts.
3. NAD is added because *M. synoviae* is reported to require this for primary isolation. The use of thallium acetate is not permitted in some countries, and ampicillin (1.0 mg/mL) can be used as a substitute.
4. Serum should be mycoplasma-free and should not be from antibiotic-treated animals. When obtaining serum from slaughterhouse blood, the blood should be collected in large clean vessels, which should be only half-filled. It is held, covered, overnight at room temperature and then the serum is removed as quickly as possible the following day, keeping samples cool at all times. Centrifugation may be necessary to remove excess red blood cells. Pass through a clarifying filter, and sterilize by filtration. Heat inactivation can be extended up to 1 h, especially if the serum is in volumes of 1 L or more. In the absence of a source of swine serum, horse serum can be substituted, but the former is preferred for isolating avian mycoplasmas.
5. One hundred milliliters of agar will make approx 11 plates. Using less agar per plate is a false economy, because the success of isolation may be compromised.

6. Quality control of new batches of media is important, since some batches of ingredients, especially the serum and yeast extract, have occasionally proven inhibitory to mycoplasma growth. Aliquots of the selected test strains of known viable count (*see* **Note 1**) are thawed, and viable counts performed using the new medium ingredients. The results are compared to those obtained with the previous acceptable batch.
7. Studies have shown that swabs that have been predipped in mycoplasma broth enhance recovery and survival of small numbers of mycoplasmas ***(10)***.
8. The nature of the specimen depends to a large extent on the purpose of the work. For routine diagnosis, tracheal swabs are probably the most commonly submitted samples, and they can also be processed for PCR. If the birds have clinical signs and are to be sacrificed, specimens should be collected from affected tissues. If, for example, the purpose of isolation is experimental study of pathogenesis, samples should then be taken from many sites and may even include brain, liver, kidneys, and so forth. Such sites would not normally be cultured for routine diagnostic purposes, although *M. gallisepticum* is occasionally recovered from the brains of birds with nervous signs.
9. The speed and temperature of transport to the laboratory can profoundly affect the success of isolation, particularly if small numbers of the organism are present ***(10)***.
10. Tissues may contain substances inhibitory to the growth of mycoplasmas ***(11)***; therefore, dilutions are recommended, especially if the birds have been treated with antibiotics. Respiratory tract exudates may contain growth-inhibiting antibodies, which could affect isolation unless dilutions are made. The same may apply to synovial fluid in *M. synoviae* infection of the joint.
11. One agar plate can be used for up to four dilutions by marking the base of the plate into sectors and inoculating each sector with 30 μL of the appropriate sample.
12. Although the body temperature of birds is approx 42°C, all the recognized avian mycoplasmas appear to grow well at 37°C.
13. If a CO_2 incubator is not available, incubate plates in an airtight container in which a candle has been lit and allowed to burn up the oxygen. High relative humidity should be maintained by including wet filter paper or the equivalent.
14. Cultures will die if they become too acid or too alkaline. *M. synoviae* is particularly sensitive to acid pH.
15. The advantages of using immunofluorescence for identification of isolates are described in Chapter 14. Pure cultures are needed if isolates are to be identified by growth or metabolism inhibition tests, and the cloning procedure not only takes extra time but may fail to single out the desired pathogen if it is in mixed culture.

References

1. Jordan, F. T. W. (1996) Avian Mycoplasmosis, in *Poultry Diseases,* 4th ed. (Jordan, F. T. W. and Pattison, M., eds.), W. B. Saunders Company, London, pp. 81–93.

2. Luttrell, M. P., Fischer, J. R., Stallknecht, D. E., and Kleven, S. H. (1996) Field investigation of *Mycoplasma gallisepticum* infections in house finches *(Carpodacus mexicanus)* from Maryland and Georgia. *Avian Dis.* **40,** 335–341.
3. Ley, D. H., Berkhoff, J. E., and McLaren, J. M. (1996) *Mycoplasma gallisepticum* isolated from house finches *(Carpodacus mexicanus)* with conjunctivitis. *Avian Dis.* **40,** 480–483.
4. Bradbury, J. M. (1977) Rapid biochemical tests for characterization of the *Mycoplasmatales. J. Clin. Microbiol.* **5,** 531–534.
5. Edward, D. G. ff. (1947) A selective medium for pleuropneumonia-like organisms. *J. Gen. Microbiol.* **1,** 238–243.
6. Tully, J. G. (1995) Culture medium formulation for primary isolation and maintenance of mollicutes, in *Molecular and Diagnostic Procedures in Mycoplasmology*, vol. 1, *Molecular Characterization.* (Razin, S. and Tully, J. G., eds.), Academic, San Diego, pp. 33–39.
7. Shepard, M. C. (1983) Culture media for ureaplasmas, in *Methods in Mycoplasmology,* vol. 1, *Mycoplasma Characterization.* (Razin, S. and Tully, J. G., eds.), Academic, New York, pp. 137–146.
8. Zain, Z. M. and Bradbury, J. M. (1995) The influence of type of swab and laboratory method on the recovery of *Mycoplasma gallisepticum* and *Mycoplasma synoviae* in broth medium. *Avian Pathol.* **24,** 707–716.
9. Rodwell, A. W. and Whitcomb, R. F. (1983) Methods for direct and indirect measurement of mycoplasma growth, in *Methods in Mycoplasmology,* vol. 1, *Mycoplasma Characterization.* (Razin, S. and Tully, J. G., eds.), Academic, New York, pp. 185–196.
10. Zaini, Z. M. and Bradbury, J. M. (1996) Optimising the conditions for isolation of *Mycoplasma gallisepticum* collected on applicator swabs. *Vet. Microbiol.* **49,** 45–57.
11. Taylor-Robinson, D. and Chen, T. A. (1983) Growth inhibitory factors in animal and plant tissues, in *Methods in Mycoplasmology,* vol. 1, *Mycoplasma Characterization* (Ruzin, S. and Tully, J. G., eds.), Academic, New York, pp. 109–114.

7

Cultivation of Ureaplasmas

Patricia M. Furr

1. Introduction

The existence of ureaplasmas, known formerly as T-strains, was first recognized in 1954 *(1)*. Subsequently, it was found that these organisms possess the enzyme urease *(2)*, a feature that made them unique among mollicutes and eventually led to their reclassification *(3)* within the new genus, *Ureaplasma*.

The fragile nature of ureaplasma cells requires that, to achieve optimal recovery from clinical specimens, particular attention has to be paid to the manner in which the specimens are collected *(4)*, transported, and stored, and to the quality of the growth medium used. The selection of the transport medium is governed by availability, cost, and most importantly, whether the specimen is to be additionally examined for microorganisms other than ureaplasmas. The latter consideration is important, since antibiotics such as streptomycin or gentamicin, frequently included in transport media for other organisms, such as viruses or chlamydiae, may prove deleterious to the viability of the ureaplasmas. Viral transport medium comprised of a nutrient broth supplemented with 10% heat-inactivated fetal calf serum, or indeed, Stuart's bacterial transport medium can be used, but neither is ideal for ureaplasmas; more suitable media, such as sucrose phosphate broth (2SP) *(5)* or preferably standard liquid medium (SLM) *(6)*, are given in **Subheading 2.**

Ureaplasmas are nutritionally fastidious and, as with other mollicutes, it may never be possible to produce a medium that gives optimum results with all species and strains. However, there are many formulae available. Those described in this chapter have been proven to support the growth of ureaplasmas from both human and animal sources. In preparing media, it is the quality of each of the individual components that is important. Where the suppliers of

From: *Methods in Molecular Biology, Vol. 104: Mycoplasma Protocols*
Edited by: R. J. Miles and R. A. J. Nicholas © Humana Press Inc., Totowa, NJ

components are indicated in **Subheading 2.**, this is done as a guide. Equivalent products from other suppliers may be equally suitable.

2. Materials

1. Specimens: vaginal, urethral, or throat swabs (*see* **Note 1**); urine (*see* **Note 2**); synovial fluid; biopsy and tissue samples (*see* **Note 3**).
2. Transport media.
   a. Sucrose phosphate broth (2SP): sucrose, 48.46 g; K_2HPO_4, 2.088 g; KH_2PO_4, 1.088 g. Dissolve each component in about 100 mL of distilled water and make up to 1 L. Adjust the pH to 7.0 with 1N NaOH or 1N HCl, if necessary. Dispense into convenient volumes and autoclave at 10 lb psi for 15 min. Store at +4°C. Prior to use, add fetal calf serum (10% v/v; Gibco BRL, Life Technologies Ltd., Paisley, UK), which has been inactivated by heating at 56°C for 30 min. Distribute into small aliquots (e.g., 1.8 mL) and store at –20°C.
   b. Standard liquid medium (SLM). To make 1 L of medium, mix the following components: heat-sterilized PPLO broth base (Difco, West Molesey, Surrey, UK) (2.1% w/v), 700 mL (*see* **Note 4**); filter-sterilized benzyl penicillin (10^5 U/mL), 10 mL (*see* **Note 5**); filter-sterilized thallous acetate (2.5 % w/v), 10 mL (*see* **Note 6**); heat-inactivated donor horse serum (Imperial Laboratories, Andover, UK), 200 mL; heat-sterilized yeast extract (25% w/v), 100 mL (*see* **Note 7**); heat-sterilized phenol red (0.1% w/v; *see* **Note 8**), 20 mL. Adjust the pH of the complete medium using sterile 1N HCl (autoclaved at 15 psi for 15 min) to either pH 6.2–6.5 for the subsequent detection of ureaplasmas only, or to pH 6.8–7.0, for the detection of both mycoplasmas and ureaplasmas. Dispense the complete medium aseptically in the required volumes and store at –20°C or below (*see* **Note 9**).
3. Growth media.
   a. Urea broth. This is a modification of the SLM transport medium. Add 10 mL of a 10% w/v solution of urea (energy substrate; *see* **Note 10**) to each 1 L of SLM. Check the pH, and if necessary, adjust with sterile 1N HCl to pH 6.2–6.5. Use the medium immediately or store at –20°C until needed. Ureaplasma growth produces a characteristic color change in the medium from yellow to magenta (*see* **Note 11**).
   b. Urea agar medium. This is again a modification of SLM but is usually prepared in smaller volumes, as it cannot be stored frozen. To make 200 mL of medium, add 2.0 mL penicillin (10^5 U/mL, filter sterilized; *see* **Note 5**), 2.0 mL thallous acetate (2.5 % w/v, filter-sterilized; *see* **Note 6**), 20 mL yeast extract (25% w/v, heat-sterilized; *see* **Note 7**), 40 mL heat inactivated horse serum, 4.0 mL phenol red (0.1% w/v, heat-sterilized; *see* **Note 8**), 2.0 mL urea solution (10% w/v, filter sterilized; *see* **Note 10**), and 10 mL HEPES (1 M, heat sterilized; *see* **Note 12**) to 140 mL Mycoplasma agar base (*see* **Note 13**), which has been autoclaved and allowed to cool to 56°C. Mix the medium and dispense as required (using 5 cm plastic Petri dishes 6 mL per plate is required).

c. U4 broth *(7)*. To make 100 mL of medium, aseptically mix 4.0 mL Hanks' balanced salts solution (×10) (Gibco BRL), 20 mL Hartley digest broth (sterilized by autoclaving at 15 psi for 15 min; Oxoid, Basingstoke, UK), 2 mL phenol red (0.1% w/v, heat-sterilized; *see* **Note 8**), 48 mL sterile deionized water, unheated fetal calf serum (Gibco BRL) 15 mL, 0.5 mL ampicillin (200 mg/mL; *see* **Note 14**), 0.5 mL thallous acetate (5% w/v; *see* **Note 6**), 0.25 mL urea solution (20% w/v; *see* **Notes 10, 11**), 10 mL yeast extract (prepared by the procedure used at Compton; *see* **Note 15**), 1.0 mL magnesium sulfate solution (0.025%; *see* **Note 16**). Adjust the pH to 6.0–6.2 with sterile 1N HCl. Ureaplasma growth is denoted by a change in color of the broth from yellow to pink.

d. U4 agar. The majority of the components of this agar medium are as used in U4 broth. Mix 4.0 mL Hanks' balanced salts solution (×10), 20 mL Hartley digest broth, and 1.2 g HEPES with 44 mL deionized water. Add 0.8 g purified agar (Oxoid) or agarose and autoclave at 10 lb psi for 10 min. Allow to cool to 56°C and then add: fetal calf serum (20 mL), yeast extract (10 mL; prepared by the procedure used at Compton), ampicillin (0.5 mL), thallous acetate (0.5 mL), putrescine dihydrochloride (1.0 mL, 1% w/v, Sigma; *see* **Note 17**), magnesium sulfate solution (1.0 mL), and L-cysteine hydrochloride solution (1.0 mL, 0.9% w/v; *see* **Note 18**). Adjust the pH to 6.0–6.2 with sterile 1N HCl, and pour the agar into dishes.

e. Bromothymol blue broth (*8*). To make 100 mL of broth, mix 2.1 g PPLO broth base (without crystal violet; Difco), 0.1 g yeast extract (Difco), 1.0 mL bromothymol blue (0.4% w/v) and 90 mL deionized water. Autoclave at 15 lb psi for 15 min, cool to 50°C and supplement with 10 mL normal horse serum (Imperial Laboratories), 0.25 mL urea solution (0.1% w/v; *see* **Note 9**), 0.1 mL GHL (Glycine Histidine Lysine) tripeptide (Sigma, Poole, Dorset, 20 μg/mL solution; *see* **Note 19**), and 1.0 mL ampicillin solution (100 mg/mL; *see* **Note 14**). Adjust the pH to 6.0 with sterile 1N HCl. Ureaplasma growth in this broth is detected by a color change from yellow to green.

f. Bromothymol blue agar. Mix 2.1 g PPLO broth base (without crystal violet; Difco), 0.1 g yeast extract (Difco), 1.0 mL bromothymol blue (0.4% w/v), 1.19 g HEPES, 90 mL deionized water, and 0.75 g agar No. 1 (Oxoid). Autoclave at 15 lb psi for 15 min, cool to 50°C, and add the supplements listed for bromothymol blue broth. Nystatin (final concentration 50 U/mL) may also be added (*see* **Note 20**). Adjust the pH to 6.0 with sterile 1N HCl before dispensing.

3. Methods

3.1. Transportation and Storage of Specimens

Ideally, keep specimens cool on wet ice, transport immediately to the laboratory, and culture. If an overnight delay occurs, store at +4°C. Where samples are to be sent by post, freeze and pack in 'cardice' at –70°C. Alternatively, for small tissue samples or the contents of swabs, dispatch unfrozen in a suitable transport medium. Sucrose phosphate broth (2SP) or preferably standard liquid medium (SLM) are both suitable. The contents of swabs are expressed into 1.8 mL of medium (*see* **Note 1**). Tissues are dispatched in 1–10 mL medium according to the size of the sample.

3.2. Isolation of Ureaplasmas

1. Inoculate specimens (*see* **Notes 1-3**) into 1.8 mL enrichment broth (three suitable media are given in Section 2). Use 0.2 mL of liquid sample (urine, synovial fluid, transport medium containing fluid from expressed swabs, and so on) or 0.2 g of homogenized tissue. (*See* **Note 21**).
2. Using the same medium, prepare a series of 10-fold dilutions (as far as 10^{-3}) of the inoculated broth (*see* **Note 22**).
3. Incubate all inoculated broths at 37°C.
4. Observe for an appropriate color change, which will depend upon the medium selected and is caused by ammonia production from the hydrolysis of urea. Since ureaplasmas prefer microaerophilic conditions, the color change is seen initially at the bottom of the culture. Examine cultures regularly (several times a day) since ureaplasmas lose viability at high pH/NH_3 concentration. Most ureaplasma cultures will give a color change within 24 h.
5. To confirm the presence of ureaplasma and isolate strains, inoculate onto agar medium (*see* **Subheading 3.3.**). Biochemical tests for the confirmation of isolates as *Acholeplasma*, *Mycoplasma* or *Ureaplasma* species are given in Chapter 9.

3.3. Inoculation and Incubation of Agar Plates

1. Dry agar plates at 37°C prior to inoculation.
2. Spread the inoculum (100 µL for 5 cm dia plates) over the surface of the agar and allow to dry.
3. Incubated in an aspirator or McIntosh and Fildes jar containing an atmosphere of 5% CO_2 in 95% N_2 at 37°C for 72 h.
4. Examine microscopically for tiny colonies. These are frequently not the typical "fried egg" colonies associated with mycoplasma cultures. In media with HEPES buffer, the rate of rise in pH of the agar around colonies is slowed; hence, bigger colonies are formed. In media with magnesium sulphate, ureaplasmas (but not other mollicutes) appear as small brown colonies. It is also possible to distinguish ureaplasma from mycoplasma colonies using a manganous chloride plus urea reagent (*see* **Note 23**).

3.4. Preparation of Stock Cultures and Estimation of Viable Cell Numbers

1. To preserve cultures for experimental work, use broth cultures in which the medium has only just changed color. Dispense into small aliquots (0.5–1.0 mL) and store immediately at –70°C or in liquid nitrogen. The aim is to produce a number of inocula that will give reproducible growth when subsequently used. It

is always preferable to prepare and store large batches of a single inoculum, as this will reduce the work need to standardize different batches with respect to cell concentration and metabolic activity.

2. To recover frozen cultures, thaw rapidly by shaking gently in a water bath at 37°C.
3. To estimate the numbers of viable ureaplasmas in stored inocula or clinical or other samples, prepare a series of 10-fold dilutions (normally to a dilution of 10^{-8}) in urea broth; it is convenient to do this by adding 0.2 mL quantities of cell suspensions to 1.8 mL of medium contained in small glass vials of 2.5 mL capacity (*see* **Note 21**). Incubate the cultures, and determine the highest dilution at which the characteristic medium color change occurs. This dilution is deemed to contain one color-changing unit (ccu).

4. Notes

1. The contents of the swab should be expressed into 1.8 mL transport medium and the swab discarded. It is not advisable to break off the swab and leave it in the broth, especially if only small numbers of organisms are present, as resins in the wood or cotton may be toxic.
2. Urine specimens may be centrifuged and concentrated (×10), although this is not usually necessary.
3. Biopsy and tissue samples may be shredded gently with scalpel blades. If the piece of tissue is large, a tissue grinder may be used, but care should be taken to keep the material cool as enzymes released during maceration may be deleterious to the ureaplasmas.
4. PPLO broth base is prepared as a 2.1% w/v solution in deionized water, dispensed in the required amounts, autoclaved at 15 lb psi for 15 min and stored at +4°C.
5. Penicillin has no effect on ureaplasmas, as these organisms do not possess a cell wall. SLM is usually prepared with 1000 U penicillin/mL but can be used at twice this concentration depending on the nature of the specimen. It may be prepared in advance, filter-sterilized (0.22 µm membrane filter), and stored frozen at –20°C or below.
6. Thallous acetate is prepared as a 2.5 w/v solution in deionized water, filter-sterilized (0.22 µm membrane filter), and stored frozen at –20°C.
7. Yeast extract is available commercially, but many laboratories prefer to prepare their own. A suitable method is to mix 125 g of dried yeast with deionized water to form a paste and steam for 90 min. After cooling, the solution is centrifuged at 12,000*g* for 45 min at +4°C. The supernatant fluid is removed and may be frozen at –20°C at this stage until required. Before use, the pH is adjusted to 8.0 with 1N NaOH. The extract is then filtered through a 0.8 µm membrane, dispensed in aliquots and sterilized by autoclaving at 15 lb psi for 15 min. The extract may be stored frozen at –20°C.
8. The phenol red solution (0.1 % w/v) is prepared by dissolving 1 g in 28.2 mL 0.1 N NaOH and diluting to 1 L with deionized water. The final solution is autoclaved at 15 psi for 15 min.

9. SLM may be prepared and dispensed in advance but if possible should be frozen (–20°C) until required to preserve the efficacy of the components. Small quantities may be kept at +4°C for short periods.
10. Urea is prepared as a 10% w/v solution in deionized water, filter-sterilized (0.22 μm membrane filter), dispensed and stored frozen at –20°C.
11. The high rate of urea hydrolysis by ureaplasmas leads to a rapid rise in pH, hence, change in color of the broth.
12. HEPES is prepared as a 1M solution (238.3 g per L deionized water). This is dispensed in 5 mL aliquots, autoclaved at 15 lb psi/15 min and stored at –20°C. When HEPES is incorporated in the medium, at a final concentration of 50 m*M*, little or no color change will be observed.
13. Mycoplasma agar base is prepared by adding 2 g of agar to 140 mL of Mycoplasma broth base (Oxoid) or PPLO broth (without crystal violet, Difco). Allow to soak for 30 min before autoclaving at 15 lb psi for 20 min. After allowing to cool and adding other medium components, it is stored at +4°C until needed.
14. Ampicillin may replace penicillin in ureaplasma growth medium. Ampicillin has a broader antibacterial spectrum and is more appropriate for specimens known to contain large numbers of cell-walled bacteria. Ampicillin solutions are sterilized by membrane filtration (0.22 μm filter) and stored frozen at –20°C or below.
15. To prepare yeast extract, using the Compton procedure, 25 g dried yeast is suspended in 100 mL of deionized water in a large flask. The suspension is then autoclaved at 10 lb psi for 5 min, allowed to cool, and centrifuged at 12,000*g* for 1 h. The supernatant is autoclaved again for 10 min at 10 lb psi and stored frozen at –20°C until required.
16. Magnesium sulphate is prepared as a 0.025% w/v solution in deionized water, dispensed and autoclaved at 10 lb psi for 10 min. It may be stored at +4°C.
17. Putrescine dihydrochloride is prepared as a 1% w/v solution in deionized water, sterilized by membrane filtration (0.22 μm filter) and stored at -20°C.
18. L-cysteine hydrochloride is prepared as a 0.9% w/v solution in deionized water, autoclaved at 10 lb psi for 10 min and stored at +4°C.
19. GHL tripeptide is prepared as a 20 μg/mL solution in deionized water and sterilized using a 0.22 μm filter.
20. Nystatin (stored in the dark) is prepared as a 50,000 U/mL solution in deionized water and sterilized using a 0.22 μm filter. It is incorporated in the broth to reduce the risk of fungal contamination.
21. It is important to have only a small air space above cultures and tightly fitting caps; otherwise CO_2 may diffuse from the medium causing a nonspecific alkaline color change. Glass vials are preferred, as most plastic containers allow gaseous diffusion. Nondisposable glass vials should be acid-washed.
22. Dilutions of the initially inoculated broth are made to reduce the concentration of potential inhibitors (e.g., residual antibiotics, antibody and inhibitors present in swabs) in clinical specimens. Dilution also reduces the presence of contaminating bacteria that are not always controlled by the penicillin and thallous acetate components of the medium.

23. The reagent contains 1% w/v urea and 0.8% w/v manganous chloride. It is dispensed in small aliquots and stored at –20°C. A few drops of the solution are added to the surface of a block of agar containing suspected ureaplasma colonies. Ureaplasma, but not other mollicute, colonies turn rapidly brown.

References

1. Shepard, M. C. (1954) The recovery of pleuropneumonia-like organisms from negro men with and without nongonococcal urethritis. *Amer. J. Syph.* **38**, 113–124.
2. Purcell, R. H., Taylor-Robinson, D., Wong, D. C., and Chanock, R. M. (1966) Color test for the measurement of antibody to T-strain mycoplasmas. *J. Bacteriol.* **92**, 6–12.
3. Shepard, M. C., Lunceford, C. D., Ford, D. K., Purcell, R. H., Taylor-Robinson, D., Razin, S., and Black, F. T. (1974) *Ureaplasma urealyticum* gen. nov. sp. nov.: Proposed nomenclature for the human T (T-strain) mycoplasma. *Int. J. System. Bacteriol* **24**, 160–171.
4. Furr, P. M. and Taylor-Robinson, D. (1981) The inhibitory effect of various antiseptics, analgesics and lubricants on mycoplasmas. *J. Antimicrob. Chemother.* **8**, 115–119.
5. Smith, T. F., Weed, L. A., Petterson, G. R., and Segura, J. W. (1977) Recovery of chlamydia and genital mycoplasma transported in sucrose phosphate buffer and urease color test medium. *Health Lab. Sci.* **14**, 30–34.
6. Edward, D. G. (1947) A selective medium for pleuropneumonia-like organisms. *J. Gen. Microbiol.* **1**, 238–247.
7. Howard, C. J., Gourlay, R. N., and Collins, J. (1978) Serological studies with bovine ureaplasmas (T-mycoplasmas). *Int. J. Syst. Bacteriol.* **28**, 473–477.
8. Robertson, J. A. (1978) Bromothymol blue broth: Improved medium for the detection of *Ureaplasma urealyticum* (T-strain mycoplasma). *J. Clin. Microbiol.* **7**, 127–132.

8

Quality-Control Testing of Mycoplasma Medium

David Windsor and Helena Windsor

1. Introduction

Mycoplasmas lack many of the biosynthetic pathways of higher organisms; consequently, their growth and survival depend on an external supply of a wide variety of nutrients. Because they are very fastidious in their nutritional requirements, a complex growth medium is required for their culture. Although the exact requirements of most mycoplasma species are still incompletely defined, most can be grown in the laboratory in both liquid and solid agar media formulated to comprise as a minimum: a broth base, yeast extract, serum source, and solutions of pure chemicals. Details of components added to any particular medium will vary, according to the species of mycoplasma it is intended to culture (*see* Chapters 4–6). It is clear that such a large number of components must inevitably introduce a degree of variability to the growth-promoting properties of the mycoplasma media between formulation lots. Consequently, it is of paramount importance that the introduction of any new component into a formulation and the preparation of media lots should always be accompanied by a quality validation with mycoplasma strains that the particular formulation is intended to support. This quality-control (QC) testing should cover all components, including preparations of pure chemicals (e.g., glucose and argininc) bccausc although intrinsically these chemicals are unlikely to vary between lots supplied by the same manufacturer in the same way that broth base, serum, and yeast extract might, it is still necessary to exclude the possibility that errors may have occurred during their preparation (*see* **Note 1**).

Correctly validated mycoplasma medium is extremely important in various practical situations. Mycoplasmas are the causative agents of various human and animal diseases, and their isolation and identification can be a necessary part of disease management. In the laboratory environment, several species of

From: *Methods in Molecular Biology, Vol. 104: Mycoplasma Protocols*
Edited by: R. J. Miles and R. A. J. Nicholas © Humana Press Inc., Totowa, NJ

mycoplasma are often found as common contaminants of cell cultures, and correct identification may assist in the treatment of the infected cell line. Mycoplasmas can also occasionally contaminate vaccines and other cell-culture-derived biologicals. In all these circumstances, the presence or absence of mycoplasmas can be verified by placing samples from animal tissue, cell cultures, vaccines, or the biologicals under investigation into culture medium designed to support mycoplasma growth. Testing for the absence of mycoplasma is an essential part of the QC procedures of laboratories involved in the testing of vaccines and other cell-derived biologicals, and in laboratories concerned with the production and maintenance of cell lines. It is important to note that licensing authorities stipulate cultural testing of human and animal vaccines and biologicals for the absence of mycoplasma, as when properly applied, this is the most sensitive method for detecting mycoplasma contamination. Representative strains of mycoplasma species that a particular medium is formulated to support are selected and used for QC testing.

For QC testing, new components are combined, substituted into the existing formulations, and compared with the lots in current use. To test each new component separately is unnecessary since component failures are rare. However, testing of individual new components is required when the initial QC test on the medium containing combined new components proves unsuitable for any test organism (i.e., it does not adequately support the growth of the intended species of mycoplasma).

Once a newly formulated medium containing the replacement components has been passed by QC testing, lots of this formulation can be produced and used in the laboratory. A further QC test should be conducted on this medium after the last day of its usage to confirm that the particular mycoplasma species whose growth it was designed to support would have been detected if present. This is an important stage in the QC testing of mycoplasma medium, because it confirms that the laboratory's medium assembly processes and procedures are all adequate: the various components have been correctly added to the formulation; the media containers are of an adequate quality; there has been no excessive use of heat during preparation of agar media; and finally, the storage conditions have not adversely affected the stability of the medium.

These two stages in the QC testing of the mycoplasma medium (i.e., the assessment of new components and then subsequent retrospective validation of the lots after the last day of their use) can be accomplished by similar methods. As part of this QC testing, a firmly established standard operating procedure (SOP) for the reproducible dilution of test organisms is an absolute requirement for the correct interpretation of test results.

In conclusion, whether a laboratory is using mycoplasma medium to test for the absence of mycoplasma in cell cultures, vaccines, or other biologicals, or to confirm the absence or presence of mycoplasma in pathological samples, prior application of QC testing to all media produced by that laboratory will provide confidence that any particular mycoplasma species would be detected, if present.

2. Materials

1. Strains: since the media to be tested are intended for isolation procedures, it is important that the test organisms have not become media-adapted. Low-passage strains, representing those mycoplasma species that the particular formulation is intended to support, should therefore be used wherever possible (*see* **Note 2**). Mycoplasmas causing particular disease states in humans or animal species are limited in number, so, for example, with respiratory disease in a specific host, QC validation can be restricted to two or three test strains. With cell cultures, however, a wide range of mycoplasmas can cause infection, and liquid and solid mycoplasma media require screening for their ability to support the growth of several mycoplasma species before acceptance. For suggested QC organisms for cell culture, *see* **Note 3**.
2. Broth base: prepare a broth base in routine laboratory use for use as a diluent.
3. Petri dishes 60 × 15 mm (Falcon 1016 [Marathon, London, UK]).
4. 1-mL pipets (Falcon 7506).
5. 10-mL pipets (Corning Costar 4100 [High Wycombe, Buckinhamshire, UK).
6. Glycerol (Merck, R & L Slaughter, Upminster, Essex, UK) Analar product no. 10118.
7. Microtiter plates (Dynex Technologies M24, Billinghurst, Sussex, UK).
8. Finnpipette Stepper with 5.0-mL sterile tips.
9. Pressure film (Falcon 3074).
10. Bijoux bottles (Greiner 189170).
11. Universal container (Greiner 201170, Stonehouse, Gloustershire, UK).
12. Sterile pastette 0.02-mL dropper (Technical Services Consultants 50-dropper, TS/126H).
13. 5% Carbon dioxide/95% nitrogen dioxide (British Oxygen Company: 120150 [London, UK]).
14. Plate microscope.

3. Methods

All methods are conducted under aseptic conditions.

3.1. Preparation of Test Strains

1. Isolates of mycoplasma strains can be obtained from within the laboratory or from other laboratories that have isolate collections. Alternatively, American Type Culture Collection (ATCC) or National Collection (NC) cultures may be used initially, with these being replaced with low-passage strains as isolates are encountered. Isolations should be typed using specific antisera.
2. Subcultures should be kept to a minimum. However, it is necessary to identify the strain positively and to ensure the purity of the culture. Initial subcultures to establish isolation, serological identification by growth inhibition, and possible

cloning will all increase the subculture level. In practice, a subculture level of ten can be regarded as an acceptable limit (*see* **Note 4**).

3. Once the requirements of strain selection have been satisfied, 10 mL of an actively growing broth culture should be prepared for freezing by the addition of 1 mL of glycerol:water (50:50).
4. Transfer 0.1-mL aliquots (*see* **Note 5**) of this prepared broth culture into individual wells of a flexible microtiter tray from which the individual wells can be cut.
5. After filling the wells, the surface of the tray should be blotted with a tissue and then covered with pressure film, which should be firmly applied (*see* **Note 6**).
6. Label the plate with the species name, strain, and date.
7. Freeze at –70°C, trim off the microtiter tray skirt, and store as a card in a suitable rigid container. Handled carefully with gloves and –70°C frozen gel packs, the count will remain stable for several years.

3.2. Dilution of Test Strains

The procedure for dilution of test strains prior to inoculating solid or liquid medium should be defined with an SOP to ensure reproducibility of the number of microorganisms in the inoculum. This is extremely important, because the acceptability of media lots for use in isolation work depends on their ability to support a certain number of colony-forming units (CFU). Suspensions of microorganisms used to control media lots must therefore be diluted to give reproducible counts of the microorganisms within the range of 50–200 CFU/inoculated volume. A further 10-fold dilution of each test strain is prepared for inoculation into liquid medium only. Inoculation with this low dilution acts as a stringent test for liquid medium, and ensures that it is sensitive to low numbers of CFU.

1. Distribute 9.0-mL amounts of broth base into a sufficient number of sterile plastic universals and label.
2. For the particular microorganism to be tested, cut a single well from a microtiter plate of frozen test strain (*see* **Subheading 3.1.**).
3. Transfer the frozen aliquot to a culture room.
4. Remove plastic sealing film with forceps, and allow the aliquot to thaw. Transfer the contents of the well into the initial 9.0 mL of broth base using a sterile pastette, and then rinse out the well with the broth base.
5. Replace the cap on the plastic universal containing the initial dilution of test strain, and mix by inversion 10 times. This is taken as a 10^{-2} dilution.
6. For subsequent dilutions, use a fresh 1-mL pipet, take fluid from the universal into the pipet, and expel three times before adjusting the meniscus to the –0.1 mL mark, with the tip of the pipe remaining in contact with the inner wall of the universal.
7. Into the next 9.0 mL of broth base, expel the liquid to the 0.9 mL mark and discard the pipe (i.e., a total of 1 mL will be delivered).
8. Close the cap on the universal, and mix as before by inversion. This will give a 10^{-3} dilution.

9. Repeat **steps 6–8**, until the required dilutions are achieved. By following a standard known set of dilution steps for each test strain, inocula containing 50–200 CFU/0.02 mL and 5–20 CFU/0.02 mL can be achieved (*see* **Note 7**).
10. Use the test strains immediately after dilution to avoid any loss of viability.

3.3. Selection of Numbers of Test Sets

QC validation of a new component may involve testing in both solid and liquid formulations: solid medium is filled into 60 × 15 mm Petri dishes (~8 mL/plate); liquid medium is used as 2-mL test volumes contained in bijoux bottles. The actual number of test sets required will depend on the number of species of mycoplasma that a particular medium is designed to detect, with 1 test set required/species (*see* **Note 8**).

1. When applying QC testing to new components, substitute in the existing formulation, and compare with the lots already in use. For solid media, QC testing with one strain will involve two agar plates (one of existing formulation and one of the new formulation). For liquid media, QC testing with one strain involves three bijoux of test formulation and three bijoux of the existing formulation. Inoculate both agar plates with 0.02 mL containing 50–200 CFU of the test strain. Inoculate one bijou each of the test media and the existing formulation with 0.02 mL containing 50–200 CFU of the test strain, and inoculate one bijou each of the test media and the existing formulation with 0.02 mL containing 5 to 20 CFU of the test strain. The remaining bijou of each medium serves as a control against which color changes can be assessed.
2. For QC tests applied to media lots after the last day of use, the structure of the test sets is different, since no control media are required. Inoculate agar plates with 50–200 CFU, giving a count of at least 50% of the expected count (*see* **Subheading 3.6.**). Inoculate two bijoux of the liquid medium (one with 50–200 CFU and one with 5–20 CFU), with a third bijou containing uninoculated medium, which acts as a standard against which color changes can be assessed.

3.4. Inoculation of Media

1. Label the required number of agar plates and liquid medium bijoux with the species name, date, and dilution (*see* **Subheading 3.3.**).
2. Take up the suspension of test strain as prepared in **Subheading 3.2.**, using a pastette 0.02-mL dropper, and inoculate the required medium with 1 drop/culture.

   Solid medium: Inoculate each agar plate with 0.02 mL containing 50–200 CFU of the test strain.

   Liquid medium: inoculate 0.02 mL volumes containing 50–200 CFU, and 5–20 CFU of the test strain into each of the two bijoux containing 2 mL volumes of liquid medium (*see* **Note 9**).
3. Solid medium: allow to dry with the Petri dish lids tilted, and cover with a Perspex sheet to protect from dust. Liquid medium: close the liquid culture caps, thereby making the bijoux airtight.
4. Incubate the cultures.

3.5. Incubation and Inspection of Media

1. Place the plates containing the cultures in the solid agar medium under the required incubation atmosphere, generally 5% carbon dioxide/95% nitrogen.
2. Incubate both liquid and solid cultures at 36 ± 1°C.
3. Phenol red is commonly included in mycoplasma liquid media as a pH indicator, and cultures are inspected daily for color changes compared to the uninoculated controls. A change in color to yellow (acid) indicates glucose fermentation, and a change to deep red (alkaline) indicates arginine hydrolysis. Changes in color will typically occur during an incubation period of 14 d. Record the day of change on the culture container.
4. Solid cultures are inspected after the relevant incubation time for the organism to be detected, and the colonies counted when growth is complete. The time taken for complete growth to occur will vary depending on the species, e.g.:

Acholeplasma laidlawii, *Mycoplasma arginini*, *Mycoplasma hyorhinis* cultivar α, and *Mycoplasma synoviae*	5 d
Mycoplasma orale, *Mycoplasma hyopneumoniae*, *Mycoplasma pneumoniae*	7–10 d

3.6. Assessment of QC Tests and Acceptance of Media

3.6.1. Solid Medium

1. After growth of colonies is complete on the solid medium, scratch the back of the agar plate with sufficiently close parallel lines to facilitate counting at a magnification of ×25.
2. Scan the whole plate using the lines to ensure complete coverage of the inoculated area.
3. Include the colonies touching the scratched lines in a consistent manner.
4. When carrying out QC testing on new medium components, counts on the test medium should be similar to parallel counts on the existing medium. In practice, variation of growth-promoting properties of medium components is unavoidable, and a new constituent(s) may be accepted if the count on the new formulation is over 75% of that on the existing medium.
5. When carrying out QC testing on medium lots where there are no parallel platings (*see* **Subheading 3.3.**), a count of 50% of the cumulative average for a given mycoplasma suspension is regarded as acceptable.
6. Where medium failures occur, the test should be repeated to confirm the failure prior to investigation.

3.6.2. Liquid Medium

1. Medium being QC tested for acceptability of new components can be accepted into use when the color changes occur within 48 h of those occurring in cultures of existing medium inoculated in parallel.
2. Medium lots being QC tested where parallel cultures are not inoculated (*see* **Subheading 3.3.**) are accepted as satisfactory, if color changes occur within the time

period of sample incubation in that medium. However, significantly delayed color changes should be investigated.

4. Notes

1. When large volumes of media are to be prepared, samples of broth, yeast extract, and serum could be tested for acceptability prior to purchasing large lots of these medium components from the manufacturer, using the QC procedures described in this chapter. This approach would not be necessary before purchasing pure chemicals from established manufacturers.
2. *M. hyorhinis* cultivar α is a cell-culture-adapted strain of *M. hyorhinis*, which can rapidly become media-adapted on suitable media. With this organism, ATCC 29052 may be used, but passaged in Vero cells and not on mycoplasma media.
3. Suggested QC organisms for cell culture testing are: *M. orale*, *M. arginini*, *M. fermentans*, *M. hyorhinis* cultivar α, and *A. laidlawii*. All these are Group I pathogens, according to the Advisory Committee on Dangerous Pathogens' "Categorisation of Pathogens According to Hazard and Categories of Containment " (1984) HMSO.
4. Depending on the animal species of origin, site of isolation material, speed of growth of the mycoplasma strain, and the general colonial appearance, selection of a suitable test strain may not require colonial cloning (e.g., *M. hyopneumoniae* from porcine lung samples), provided serological identity is clearly established. From sites, such as chicken tracheas, cloning three times of a particular colony type is advisable before serological identification.
5. Use the Finnpipette Stepper to measure 0.1-mL aliquots.
6. The back of a horn spatula is suitable for ensuring the pressure film is firmly applied to the microtiter trays.
7. When using a new frozen test strain, it will be necessary to ascertain the value of the initial count in order to establish the number of dilution steps required to arrive at a final dilution with a count of 50–200 CFU/0.02 mL. This is achieved as follows: the frozen 0.1-mL aliquot is diluted three times, first into 9.0 mL of broth base, followed by two subsequent dilutions each of 0.1 mL into 9.9 mL; these three steps represent dilutions of 10^{-2}, 10^{-4}, and 10^{-6}, respectively. The three dilutions are then plated, and from the resulting count, the number of dilution steps required can be calculated. Fine adjustments to the count are made using single-log (1 + 9) or half-log (1 + 2.3) dilution steps.
8. It may be necessary to use more than one type of medium, if the requirements of the species to be isolated are incompatible, e.g., mycoplasmas and ureaplasmas.
9. When liquid media alone are being tested, the inoculum of 50–200 CFU/0.02 mL is always inoculated onto agar to confirm that the colony count is acceptable.

9

Biochemical Characteristics in Mycoplasma Identification

José B. Poveda

1. Introduction

Mycoplasmas are the smallest known free-living organisms, and adapted to a special mode of life as commensal organisms or opportunistic pathogens. Because of their small genome size, they have limited biosynthetic capacity, which means they lack many biochemical pathways found in the Eubacteria. They are highly adapted to their host, which provides most of its nutritional requirements for growth. For this reason, there are only a few biochemical properties that can be investigated in the diagnostic laboratory. Consequently, identification of mycoplasmas is greatly reliant on serological tests based on the recognition of structural membrane proteins by specific antiserum. However, preliminary biochemical characterization can reduce the battery of sera required for final serological identification.

The Subcommittee on the Taxonomy of *Mollicutes* in 1995 recommended standard techniques by which *Mollicutes* may be identified and classified ***(1)***. As a first step in the identification procedure, the sterol requirement of the *Mollicutes* can be determined using a rapid and simple indirect method: sensitivity to digitonin ***(2)***. The test differentiates the sterol-requiring *Mollicutes* (*Genus Mycoplasma, Ureaplasma, Entomoplasma, Spiroplasma,* and *Anaeroplasma*) from the nonsterol-requiring *Mollicutes* (*Genus Mesoplasma, Acholeplasma,* and *Asteroleplasma*). Sodium polyethanol sulfonate may also be used in addition to or instead of digitonin.

The biochemical procedures in mycoplasma identification have been standardized by Aluotto et al. ***(3)*** and constitute the basis for the protocols described here. They involve the hydrolysis of urea, fermentation of glucose, hydrolysis of arginine, phosphatase activity, film and spots production, tetrazolium

From: *Methods in Molecular Biology, Vol. 104: Mycoplasma Protocols*
Edited by: R. J. Miles and R. A. J. Nicholas © Humana Press Inc., Totowa, NJ

reduction, liquefaction of coagulated serum, and hydrolysis of casein. The genus *Ureaplasma* can be recognized by a specific characteristic: urease activity. Urea hydrolysis increases the alkalinity of the medium owing to ammonia production, which is detected by the pH indicator. It is important to assess the urease activity of colonies on agar plates ***(4)***, because complex broth media may contain other substances that can be metabolized and so increase the pH. The fermenting species of the genus *Mycoplasma* catabolize glucose and other carbohydrates yielding acid products, which result in a decrease in pH. The nonfermenting mycoplasmas, and a small group of fermenting species hydrolyze arginine producing ammonia with an increase in pH. These pH changes can be detected by a pH indicator.

Phosphatase activity is seen in a third of mycoplasmas. The test is based on the liberation of phenolphthalein on solid medium, and its reaction with sodium hydroxide produces a red color. Film and spots production on solid and liquid media indicates a lipolytic activity in some, but not all, strains of a particular mycoplasma, and is influenced by the medium composition. The reduction of tetrazolium to an insoluble brick red formazan is a capacity of numerous species of genus *Mycoplasma.* Proteolytic activity can be tested by the ability of Mollicutes to digest coagulated serum and casein. These properties are found in a very small number of species, such as *Mycoplasma bovirhinis, M. capricolum* subsp. *capricolum, M. mycoides* subsp. *mycoides,* and *M. mycoides* subsp. *capri.*

Finally, the interaction between Mollicutes and red blood cells, such as hemagglutination, hemadsorption, and hemolysis ***(5)***, are properties that are useful in identification and characterization. There is some evidence that such activities may be related to pathogenicity.

2. Materials

1. Pure cultures of the strains under investigation are obtained by triple-cloning procedure (*see* Chapters 4–6).
2. Test control organisms: *M. bovirhinis* PG43^T, *M. arthritidis* PG6^T, *Acholeplasma laidlawii* PG8^T, *Ureaplasma urealyticum* D960^T, *M. gallinarum* PG16^T, *M. m. mycoides* Y-goat, and *M. pneumoniae* FH^T.
3. Heart infusion broth: Dissolve 25 g of dehydrated HIB (Difco, West Molesey, UK) in 1000 mL of double-distilled water. Adjust the pH to 7.6, and autoclave. Store at 4°C.
4. Heart infusion agar: Dissolve 40 g of dehydrated HIA (Difco) in 1000 mL of double distilled water. Adjust the pH to 7.6 . Divide among four 0.5.L flasks (2 × 300 mL and 2 × 148 mL) and autoclave. Store at 4°C.
5. Horse serum: 750 mL. Sterilize by filtration and inactivate at 56°C for 30 min. Store at –20°C.
6. Yeast extract stock solution 10% (w/v): Dissolve 20 g of dehydrated yeast extract (Oxoid, London, UK) in 200 mL of double-distilled water. Adjust pH to 7.6,

sterilize by filtration, and dispense 10-mL vol into appropriate screw-cap tubes. Store at –20°C.

7. DNA solution 0.2% (w/v): Dissolve 0.2 g of deoxyribonucleic acid (Sigma, Poole, UK) in 100 mL of double-distilled water. Adjust the pH to 7.6, sterilize by filtration, and dispense aseptically 1-mL vol into sterile small aliquots. Store at –20°C.
8. Liquid standard medium (20% of serum): 296 mL of sterile HIB, 4 mL of sterile DNA solution, 80 mL of sterile horse serum, and 20 mL of sterile yeast extract solution. Aseptically dispense 2 mL into appropriate tubes. Store at 4°C.
9. Liquid standard medium (10% of serum): 336 mL of sterile HIB, 4 mL of sterile DNA solution, 40 mL of sterile horse serum, and 20 mL of sterile yeast extract solution. Aseptically dispense 2 mL into appropriate tubes. Store at 4°C.
10. Solid standard medium: 296 mL of sterile HIA, 4 mL of sterile DNA solution, 80 mL of sterile horse serum, and 20 mL of sterile yeast extract solution. Aseptically distribute 10 mL into small Petri dishes (60 mm in diameter).
11. Digitonin solution: Add 75 mg of digitonin to 5 mL (95% ethanol) heating at 56°C for 30 min.
12. Sterile filter disks (6 mm in diameter) impregnated with 25 µL of the ethanolic solution of digitonin: Dry disks at 37°C for 6 h, and store at 4°C.
13. Urea test medium: 146 mL of HIB, 2 mL of DNA solution, 20 mL of horse serum (inactivated), 10 mL of yeast extract solution, 20 mL of urea solution 10% (w/v), and 2 mL of phenol red solution 0.5% (w/v). Adjust the pH to 7.0, and sterilize by filtration. Aseptically dispense 2 mL into appropriate tubes. Store at 4°C.
14. Test medium without urea: 166 mL of HIB, 2 mL of DNA solution, 20 mL of horse serum (inactivated), 10 mL of yeast extract solution, and 2 mL of phenol red solution 0.5% (w/v). Adjust the pH to 7.0 and sterilize by filtration. Aseptically dispense 2 mL into appropriate tubes. Store at 4°C.
15. Paraffin (liquid): Dispense 5 mL into screw-cap tubes and autoclave.
16. Rapid reagent for urease detection: Dissolve 1 g of urea and 0.8 g of $MnCl_2 \cdot 4H_2O$ in 100 mL of distilled water. Sterilize by filtration. Store at 4°C.
17. Set tubes of standard range of pH: 84 mL of HIB, 10 mL of horse serum, 5 mL of yeast extract solution , and 1 mL of phenol red solution 0.5% (w/v). Dispense 5-mL vol into appropriate tubes. Adjust the pH of the first tube to 5.6, and increase the following ones by 0.2 pH units each time up to 8.4. Sterilize by filtration, and distribute 2 mL into sterile tubes. Store at 4°C.
18. Glucose test medium: 146 mL of HIB, 2 mL of DNA solution, 20 mL of horse serum (inactivated), 10 mL of yeast extract solution, 20 mL of glucose solution 10% (w/v), and 2 mL of phenol red solution 0.5% (w/v). Adjust the pH to 7.6, and sterilize by filtration. Aseptically dispense 2 mL into appropriate tubes. Store at 4°C.
19. Test medium without glucose: 166 mL of HIB, 2 mL of DNA solution, 20 mL of horse serum (inactivated), 10 mL of yeast extract solution, and 2 mL of phenol red solution 0.5% (w/v). Adjust the pH to 7.6, and sterilize by filtration. Aseptically dispense 2 mL into appropriate tubes. Store at 4°C.
20. Arginine test medium: 146 mL of HIB, 2 mL of DNA solution, 20 mL of horse serum (inactivated), 10 mL of yeast extract solution, 20 mL of arginine solution

10% (w/v), and 2 mL of phenol red solution 0.5% (w/v). Adjust the pH to 7.0, and sterilize by filtration. Aseptically dispense 2 mL into appropriate tubes. Store at 4°C.

21. Test medium without arginine: 166 mL of HIB, 2 mL of DNA solution, 20 mL of horse serum (inactivated), 10 mL of yeast extract solution, and 2 mL of phenol red solution 0.5% (w/v). Adjust the pH to 7.0, and sterilize by filtration. Aseptically dispense 2 mL into appropriate tubes. Store at 4°C.
22. Stock solution of phenolphthalein diphosphate: Add 0.5 mL of the sodium salt of phenolphthalein diphosphate to 50 mL of distilled water. Adjust the pH to 7.6, and sterilize by filtration. Store at 4°C.
23. Phosphatase test medium: 146 mL of sterile HIA, 2 mL of sterile DNA solution, 40 mL of sterile horse serum (inactivated at 56°C 60 min), 10 mL of sterile yeast extract solution (inactivated at 56°C 60 min), and 2 mL of sterile phenolphthalein diphosphate solution 1% (w/v). Aseptically distribute 10 mL into small Petri dishes (60 mm in diameter).
24. Stock solution of tetrazolium: Add 1 g of 2,3,5-triphenyltetrazolium chloride to 50 mL of distilled water. Adjust the pH to 7.6, and sterilize by filtration. Store at 4°C.
25. Tetrazolium reduction test medium : 146 mL of sterile HIA, 2 mL of sterile DNA solution, 40 mL of sterile horse serum (inactivated), 10 mL of sterile yeast extract solution, and 2 mL of sterile tetrazolium solution 2% (w/v). Aseptically distribute 10 mL into small Petri dishes (60 mm in diameter).
26. Serum digestion test medium: 150 mL of sterile horse serum, 40 mL of sterile HIB, and 10 mL of sterile yeast extract solution. Dispense 5-mL vol into appropriate screw-cap tubes, and heat in a slanted position at 85°C for 90 min. Store at 4°C.
27. Casein digestion test medium: Dissolve 8 g of skimmed milk (Difco) in 80 mL of distilled water. Adjust the pH to 7.6, and autoclave (at 115°C for 10 min) Prepare agar base dissolving 2 g of agar no. 1 (Oxoid) in 120 mL of distilled water. Adjust the pH to 7.6, and autoclave. Add the milk solution to the melted agar, and dispense the mixture in 10 mL volumes into appropriate tubes. Store at 4°C.
28. PBS stock solution (10X): 80 g NaCl, 2 g KH_2PO_4, 29 g $NaHPO_4 \cdot 12H_2O$, and 2 g KCl made up to 1 L with distilled water. Store the solution at 4°C.
29. PBS working solution: Add 100 mL of PBS (10X) to 900 mL of distilled water. Adjust pH to 7.2, and autoclave.
30. Red blood cells: Guinea pig erythrocytes washed three times in PBS and resuspended to 10% concentration. Store at 4°C.

3. Methods

3.1. Cell Adaptation to Growth Media

The results of the biochemical tests of the strains under investigation depend on a good growth on test media (yielding 10^7–10^8 CFU/mL). Because of this the organisms must be adapted to grow rapidly in both standard liquid media (with 10–20% horse serum) and in standard solid medium (*see* **Note 1**).

1. Prepare the mycoplasmas under investigation and the test control organisms by mixing 1 mL of each growth culture with 1 mL of standard broth medium (20% serum), and incubate at an appropriate temperature 24–96 h.
2. Transfer the cultures every 24–72 h in standard broth medium with 20% serum (200 µL into 2 mL) until they are completely adapted to this medium. The same procedure must be carried out in standard broth medium with 10% of serum (200 µL into 2 mL) until the cultures are totally adapted to this new medium.
3. Inoculate agar plates with 200 µL from the broth cultures, and incubate at an appropriate temperature and atmospheric condition. Examine the plates daily with a stereomicroscope. If there is a good concentration of colonies with a characteristic appearance, the adaptation process has been completed.

3.2. Sensitivity to Digitonin

Test control organisms: Positive: *M. bovirhinis* PG43^T, negative: *A. laidlawii* PG8^T.

1. Before use, place the plates of standard medium at 37°C for 1 h to dry.
2. Prepare dilutions of cultures of the test and control organisms in standard broth medium (20% of serum) to obtain approx 10^5 CFU/mL. Three serial 10-fold dilutions are sufficient for strains with a medium capacity of growth. However, slow-growth strains usually need dilutions of 10^{-2}, and the most vigorous strains need 10^{-4} dilutions.
3. Add 200 µL of the diluted culture, so it completely covers the agar plate surface, and then remove the excess fluid. Use sterile forceps to place a digitonin disk in the center of the inoculated area.
4. Incubate the plates under optimal growth conditions for each organism.
5. Examine the plates at 24-h intervals with a stereomicroscope, and measure the inhibition zones around the digitonin disk. Compare the results with the control test organisms. The sterol-requiring mollicutes are inhibited by 1.5% digitonin, showing inhibition zones of 5–20 mm. However, some nonsterol-requiring mollicutes may also show a nonspecific small clear zone of inhibition to digitonin (0.5–3 mm) (*see* **Note 2**).

3.3. Hydrolysis of Urea

Test control organisms: positive: *U. urealyticum* D960^T, negative: *M. arthritidis* PG6^T.

1. Prepare fresh cultures of the test and control organisms in standard broth medium (10% serum).
2. Inoculate duplicate tubes of urea test medium with 200 µL of these fresh cultures and also tubes of test medium without urea (control of inoculated medium without substrate).
3. Add 200 µL of sterile standard broth medium to two tubes of urea test medium (control of uninoculated medium with substrate).
4. Cover the surface of one set of tubes with 0.5 mL of sterile liquid paraffin, and incubate at an appropriate temperature.

5. Read the results daily for 2 wk comparing with the inoculated and noninoculated control tubes. The pH will be either compared with the standard pH set of tubes or measured on the last day. A rise of 0.5 pH unit or more means a positive test (*see* **Note 3**).

3.4. Urease Activity of Colonies

Test control organisms: positive: *U. urealyticum* D960^T, negative: *M. arthritidis* PG6^T.

1. Prepare fresh cultures of the test and control organisms in standard broth medium (20% serum).
2. Dilute the cultures using serial 10-fold dilution in standard broth medium (20% serum) up to 10^{-4}.
3. Add 200 µL of 10^{-4} dilution of the cultures to plates of standard medium, and incubate for 72 h at 37°C in a moist atmosphere containing 5% CO_2 in a jar, or in a CO_2 incubator.
4. Cut small pieces (10 mm) of agar block with colonies from the plates, and place on slides. Observe the colonies with the stereomicroscope, and then add one drop of the rapid reagent for urease detection. The urease-positive colonies change immediately to brownish color owing to the liberation of ammonia and the precipitation of manganese dioxide.

3.5. Fermentation of Glucose

Test control organisms: positive: *M. bovirhinis* PG43^T, negative: *M. arthritidis* PG6^T.

1. Prepare fresh cultures of the test and control organisms in standard broth medium (10% serum).
2. Inoculate duplicate tubes of glucose test medium with 200 µL of these fresh cultures and also tubes of test medium without glucose (control of inoculated medium without substrate).
3. Add 200 µL of sterile standard broth medium into two tubes of glucose test medium (control of uninoculated medium with substrate).
4. Cover the surface of one set of tubes with 0.5 mL of sterile liquid paraffin, and incubate at an appropriate temperature.
5. Read the results daily for 2 wk comparing with the inoculated and noninoculated control tubes. The pH will be compared with the standard pH set of tubes or measured on the last day. A decrease of 0.5 pH units or more means a positive test (*see* **Note 4**).

3.6. Hydrolysis of Arginine

Test control organisms: positive: *M. arthritidis* PG6^T, negative: *M. bovirhinis* PG43^T.

1. Prepare fresh cultures of the test and control organisms in standard broth medium (10% serum).

2. Inoculate duplicate tubes of arginine test medium with 200 µL of these fresh cultures and also tubes of test medium without arginine (control of inoculated medium without substrate).
3. Add 200 µL of sterile standard broth medium into two tubes of arginine test medium (control of uninoculated medium with substrate).
4. Cover the surface of one set of tubes with 0.5 mL of sterile liquid paraffin, and incubate at an appropriate temperature.
5. Read the results daily for 2 wk, comparing with the inoculated and noninoculated control tubes. The pH will be compared with the standard pH set of tubes or measured on the last day. A rise of 0.5 pH units or more means a positive test (*see* **Note 5**).

3.7. Phosphatase Activity

Test control organisms: positive: *M. arthritidis* PG6^T, negative: *M. bovirhinis* PG43^T.

1. Prepare fresh cultures of the test and control organisms in standard broth medium (20% serum).
2. Before use, place the plates of phosphatase test medium at 37°C for 30 min to dry the surface moisture.
3. Inoculate duplicate plates of phosphatase test medium with a drop of these fresh cultures, and incubate for 7 and 14 d, respectively, under optimal growth conditions for each organism. Also incubate two noninoculated plates as a medium control.
4. After 7 d, flood the surface of the first set of plates with 5 *N* NaOH. The appearance of a red or pink color indicates a positive result. Repeat the procedure at the 14th d of incubation (*see* **Note 6**).

3.8. Film and Spot Production

Test control organisms: positive: *M. gallinarum* PG16^T, negative: *M. arthritidis* PG6^T.

1. Prepare fresh cultures of the test and control organisms in standard broth medium (20% serum).
2. Before use, place the plates of standard medium at 37°C for 30 min to dry the surface moisture.
3. Inoculate the plates of standard medium with 200 µL of these fresh cultures, and incubate at an appropriate temperature in a moist atmosphere for 2 wk. Also incubate two noninoculated plates as a medium control.
4. Examine the plates every 48 h with a stereomicroscope for the appearance of spots around the colonies. On the last day, flood the plates with distilled water, and observe the separation of the film from the agar's surface in the positive test (*see* **Note 7**).

3.9. Tetrazolium Reduction

Test control organisms: positive: *M. bovirhinis* PG43^T, negative: *M. arthritidis* PG6^T.

1. Prepare fresh cultures of the test and control organisms on standard solid medium.
2. Before use, place the tetrazolium plates at 37°C for 30 min to dry the surface moisture.
3. Use a sterile scalpel to cut small blocks of agar with dense colony growth, and inoculate duplicate tetrazolium plates, inverting the agar blocks and sliding on agar surface.
4. Incubate one set of plates aerobically and the other anaerobically for 2 wk at an appropriate temperature in a moist atmosphere. Also incubate two noninoculated plates as a medium control.
5. Examine the plates at 72-h intervals with a stereomicroscope to detect a pink or red color in the positive test (*see* **Note 8**).

3.10. Liquefaction of Coagulated Serum

Test control organisms: positive: *M. mycoides* subsp. *mycoides* Y-goat, negative: *M. arthritidis* $PG6^T$.

1. Prepare fresh cultures of the test and control organisms on standard solid medium.
2. Prior to use, invert the serum digest tubes, place at 37°C for 1 h, and aseptically remove the excess fluid.
3. Use a sterile scalpel to cut small blocks of agar with dense colony growth and inoculate the serum digest tubes, inverting the agar blocks and sliding over the inclined surface. Incubate at an appropriate temperature for 1 mo. Also incubate two noninoculated tubes as a medium control.
4. Examine frequently for liquefaction at the highest portion of the inclined tubes and the collected liquid on the base of the tubes.

3.11. Casein Digestion

Test control organisms: positive: *M. mycoides* subsp. *mycoides* Y-goat, negative: *M. arthritidis* $PG6^T$.

1. Prepare fresh cultures of the test and control organisms on plates of standard medium. Inoculate each plate at three different sites with running drops of stock cultures. After absorption of drops, incubate at an appropriate temperature in a moist atmosphere for 24–72 h.
2. Liquefy the casein digestion test medium in a boiling water bath, and aseptically add some drops over the plates containing three bands of compact growth, covering the surface. After solidification, incubate at an appropriate temperature for 2 wk. Also incubate two noninoculated plates with casein digestion medium over the surface as a medium control.
3. Observe the plates daily for the appearance of clear bands around the lines of growth.

3.12. Hemadsorption

Test control organisms: positive: *M. pneumoniae* FH^T, negative: *M. bovirhinis* $PG43^T$.

1. Prepare cultures of the test and control organisms on plates of standard medium with a minimum of 40–60 well-separated colonies.
2. Before use, harvest the red blood cells by centrifugation. Remove the supernatant, and resuspend the cells at a 10% concentration in PBS.
3. Prepare a 0.5% suspension of red blood cells in PBS, and add 1 mL onto the plates with colonies. Incubate the plates for 30 min at 37°C.
4. Remove the excess red blood cell suspension, and wash carefully with PBS inclining the plates.
5. Examine the plates with a stereomicroscope to detect adsorption of red blood cells to single colonies (*see* **Notes 9** and **10**).

4. Notes

1. There are some difficulties when the biochemical tests are performed with complex media because other substances can be catabolized, altering the pH. Consequently, fastidious strains isolated on media supplemented with specific nutritional requirements must be tested on modified test media. For example, if the requirements are b-NAD and cystein hydrochloride (in the FM4 medium for *Mycoplasma synoviae*) both substances will have to be incorporated into the test medium formulations.
2. The digitonin test is an indirect method for preliminary assessment of the sterol requirement of mollicutes. Furthermore, experience has revealed the importance of performing this technique using agar plates with 20% serum to prevent weak inhibition seen with some nonsterol-requiring mollicutes. Nevertheless, organisms with inhibition zones <3 mm must be investigated with a more specific method comparing the growth responses in medium containing 15–20% fetal bovine serum or in serum-free media with or without 0.04% Tween 80 ***(6)***.
3. The urea hydrolysis test can give false-positive reactions with some arginine-positive organisms that can metabolize small concentrations of this amino acid contained in the medium. False positives will be detected, because the same change of pH will be seen in the inoculated tubes of test medium without urea. However, these weak reactions are not equivalent to the change produced by the control positive organism *(U. urealyticum)* where the pH is higher (nearly to 8.2). For confirmation, the urease activity of colonies should be examined.
4. With the glucose fermentation test, it should be noted that some nonfermenting organisms, such as *Mycoplasma agalactiae* may show a slight decrease in pH. It is produced by metabolism of substrates other than glucose. In this case, a similar change of pH indicator will be detected in inoculated tubes of test medium without glucose. Both weak reactions are not comparable to the change of color produced by the control positive organism *(M. bovirhinis),* where the pH level decreases to almost 5.5.
5. The investigation of arginine hydrolysis can be difficult with organisms that also ferment glucose including certain strains of *M. capricolum* subsp. *capricolum.*
6. It is important to inactivate the endogenous phosphatase activity in horse serum and yeast extract solution at 56°C for 1 h for use in the phosphatase test medium

to inactivate the endogenous phosphatase activity in both substances. False-positive reactions caused by the spontaneous degradation of sodium phenolphthalein diphosphate can arise, although this can be detected in the the control plates.

7. The productien of film and spots is associated with the lipolytic activity of mycoplasmas. Some species, such as *Mycoplasma agalactiae* and *Mycoplasma bovis,* show this characteristic, whereas others such as *Mycoplasma hyopneumoniae* and *Mycoplasma flocculare* show a weak reaction. Usually the test is carried out on agar plates with 20% of horse serum. In those cases in which swine serum is essential in the media, it is necessary to include 10% of egg yolk emulsion.
8. The tetrazolium reduction test can show some degree of variability, especially under aerobic conditions. This is more evident, for example, with field strains of *M. gallinarum* of which about one-fifth are unable to reduce this substance under aerobic conditions.
9. There is some intraspecies variation in the hemadsorption capacity of mycoplasmas, which can be partly overcome by keeping the isolated strains at a low passage level.
10. The hemadsorption capacity of mycoplasmas must be investigated with different types of erythrocytes (sheep, chicken, human, bovine, porcine, and equine). For example, some species adsorb chicken, but not guinea pig or other red blood cells, whereas others species adsorb guinea pig, but not chicken erythrocytes.

References

1. Subcommittee on the Taxonomy of *Mollicutes* (1995) Revised minimum standards for description of new species of the class *Mollicutes* (Division *Tenericutes*). *Int. J. Syst. Bacteriol.* **45,** 605–612.
2. Freundt, E. A., Andrews, B. E., Erno, H., Kunze, M., and Black, F. T. (1973) The sensitivity of *Mycoplasmatales* to sodium-polyanetholsulfonate and digitonin. *Zentralbl. Bakteriol. Parasitenkd. Infektioskr. Hyg. Abt.* 1 Orig. Reihe A **225,** 104–112.
3. Aluotto, B. B., Wittler, R. G., Williams, C. O., and Faber, J. E. (1970) Standardized bacteriologic techniques for the characterization of Mycoplasma species. *Int. J. Syst. Bacteriol.* **20,** 35–58.
4. Shepard, M. C. and Howard, D. R. (1970) Identification of "T" mycoplasmas in primary agar cultures by means of a direct test for urease. *Ann. NY Acad. Sci.* **174,** 809–819.
5. Gardella, R. S. and Del Giudice, R. A. (1983) Hemagglutination, hemadsorption, and hemolysis. *Methods in Mycoplasmology* **1,** 379–384.
6. Rose, D. L., Tully, J. G., Bové, J. M., and Whitcomb, R. F. (1993) A test for measuring growth responses of mollicutes to serum and polyoxyethylene sorbitan. *Int. J. Syst. Bacteriol.* **43,** 527–532.

10

Enzyme Analysis

The Rationale and Use of Enzyme Assays in Assigning Function to Gene Nucleotide Sequences and the Procedures for the Assay of Three Enzymatic Functions Conserved in Mollicutes

J. Dennis Pollack

1. Introduction

An issue of some biological currency pertains to the identification of the functional potential of cells as deduced from the nucleotide sequence of their genes. In effect, nucleotide sequences of genes assigned in some confidence to expressed proteins and function in one cell are taken, when found in the genomes of other cells, to be indicators of the presence of the same function. However, such comparisons have never been found to be perfectly identical. The level of dissimilarity often casts some doubt on the identification or functional assignment of genes. The assignment of enzymatic function by comparison of gene nucleotide sequences alone was recognized as questionable and, hence, only putative even by the first workers to sequence a genome completely *(1,2)*. One of the problems in assigning gene sequences was characterized by these workers as being the result of low similarities between known sequences from other species and those being analyzed. The matches might be so low that they were not detectable as similar.

The range of nucleotide similarity between two sequences from different cells that are putatively identified as having the same function often varies widely. Lack of similarity may be related to portions of their nucleotide sequences not essential for enzyme activity, that is, those nucleotide sequences lying outside of the shorter essential sequences coding for specific enzymatic function (e.g., the enzyme catalytic center). Using programs or strategies that

From: *Methods in Molecular Biology, Vol. 104: Mycoplasma Protocols*
Edited by: R. J. Miles and R. A. J. Nicholas © Humana Press Inc., Totowa, NJ

seek to recognize enzyme catalytic centers or motifs with knowledge of codon usage may result in higher similarity values.

The level of doubt in assigning functional identity after DNA sequencing may be exacerbated by other circumstances, some of which are unusual. For example, in *Acholeplasma laidlawii,* a pyrophosphate-dependent nucleoside kinase has been reported ***(3,4)***. This activity is unknown in any other cell; therefore, there is some possibility that the gene nucleotide sequence for this function may be sufficiently unlike any sequence already reported for any other nucleoside kinase, and data bank comparisons may be of no avail. Although the sequence for this unique activity would be known, its identity might go unrecognized and be without assignment; when identified, it could be characterized as a novel gene ***(2)***.

A somewhat reverse situation may occur. *Mycoplasma genitalium* is reported to have a gene coding for 6-phosphogluconate dehydrogenase (6PGD) ***(2)***. However, the same strain has no detectable 6PGD activity ***(5)***. Of course, the enzyme assays may not be sensitive enough to detect activity, they may be inappropriate, or the cells were not expressing the activity at the time of harvest for a variety of reasons. We suggest that it is also possible that the gene assignment was incorrect or, more likely, that it was correct, but the enzyme is not expressed, perhaps because of its presence in an operon with faults in the operator or repressor genes. There is even no surety that the gene was ever expressed, but even if previously expressible and functional, the step or path may now be inoperative for a variety of reasons. Further, the presence of a gene even with some identifiable expression product does not guarantee function because of the possibility of posttranslational alteration that modifies or negates such activity.

The results of single-enzyme assays alone may be metabolically misleading. The identified activity may have no metabolic role. We should have some knowledge of the metabolically linking or related activities in order to establish the cellular role of an enzyme. A single gene, even if the expressed product is functional, may be a remnant of a pathway whose other coding elements have been lost or become inoperative, rendering its own functional potential useless. It has no apparent substrate to act on, and if it did, the recognized product of its action has no obvious place to go. Isocitrate dehydrogenase (ICD) activity reported in *Anaeroplasma intermedium* may be an example of such an orphan enzyme ***(6,7)***. ICD activity was not found in any other *Mollicutes* ***(6)***. In the TCA-deficient mollicutes, the role of ICD activity is presently unknown. However, speculatively, ICD could play some role in maintaining NAD/NADH levels or be involved in trans-aminations involving α-ketoglutamate.

Even when the gene sequences coding for the entire pathway, and its metabolic linkages are identified, it is necessary to determine not only if the genes are expressed and functional, but also the stoichiometry and yield of the sequential reactions. More data are needed to establish the pathway's role in the cell's metabolism. The entire sequence might be an example of a pathway no longer in use or essential to the cell. An orphan pathway, whose coding elements are present or even expressed, but in levels or activities too low to be metabolically useful. A question arises: Are we interested in what the organism might have done or might do, or what it does?

Of greater certainty is the observation that some *Mycoplasma* spp. may express functional multienzyme proteins. Malate dehydrogenase (MDH) enzyme activity has been reported in all *Mollicutes,* including *M. genitalium* G37 and *Mycoplasma pneumoniae* M129 ***(6)***, but no nucleotide sequence in these same two mycoplasmas whose genomes have been completely sequenced has been assigned to this activity ***(2,8)***. We have reported our analyses indicating that the lactic dehydrogenase (LDH) gene has a unique sequence of three nucleotides that results in the expression of a single protein with both MDH and LDH activities ***(9)***. This unique sequence is reported in both *M. genitalium* G37 and *M. pneumoniae* M129 ***(2,8)***. Analysis of only gene nucleotide sequence without enzymatic assay may never have revealed the presence of the MDH/LDH gene as an example of one gene coding for the expression of one protein with two functions. Therefore, gene sequence is an uncertain indicator of identity, and is no indicator of gene expression or function. Obviously, if the presence of sequences that code for more than one activity are not or cannot be detected and identified by genetic analyses, enzymatic assays as a companion approach may be able to do so.

Enzymatic analyses and gene nucleotide sequencing are mutually confirmatory analytical approaches. Enzymatic analyses confirm the expression and function of genes putatively identified by nucleotide analysis and can indicate the presence of genes and metabolic pathways not recognized. The other, gene nucleotide sequencing, which is much better appreciated, also has the potential of revealing the presence of enzyme activities and metabolic pathways not imagined. For example, both the presence of a gene sequence and functional activity for ATP-dependent 6-phosphofructokinase (6PFK) were reported for *M. genitalium* G37 ***(2,5)*** and *M. pneumoniae* M129 ***(6,8)***. Gene nucleotide analyses also indicated that there was a 1PFK gene in both. This was an enzyme or function whose presence had not been imagined by us ***(2,8)***. Subsequent analyses of these two *Mycoplasma* species, and five others now indicate that they all have both 6- and 1PFK activity ***(10)***. The consequence of this observation is not trivial, because the ability to process 1-phosphofructose by 1PFK infers that there is a functional phosphoenolpyruvate:sugar (PTS) transferase system for fructose transport that was not thus far recognized. Therefore, the opinion that *M. genitalium* may be limited to the use of glucose as an energy source may not be correct, unless the EIIBC component for the uptake of fructose identified in the *M. genitalium* genome is not functional ***(2)***.

Our functional enzymatic analyses of different *Mollicutes* has revealed the presence of about 130 cytoplasmic activities, of the approx 200 theorized to be found in the cytoplasm. All are not found in any one mollicute. Three will be discussed in detail in this chapter; these are: phosphoglycerate kinase (PGK), pyruvate kinase (PK), and MDH. The reason for selecting these enzymes is that they are highly functionally conserved, that is, found in all *Mollicutes*. Their ubiquitous presence in the *Mollicutes* suggests they may be useful in the phylogenetic analysis of the class. We postulate that the two kinases are of particular importance, since most if not all of the ATP generated by any of the cytochrome-less *Mollicutes* is generated by substrate phosphorylation only at these two sites. The MDH is equally conserved as well as its companion LDH, because they are borne in the same protein, as noted above. Therefore, MDH/LDH is also of special import, because it is an example of a multienzyme protein—a product of one gene that has two functions. Procedures for the determination of four other mollicute enzyme activities were previously reported ***(11)***.

Not described in this chapter are two other general techniques for the assay of enzymes that markedly increase analytical sensitivity: fluorometric analyses and "cycling" techniques. Both of these procedures have considerable general applicability when sensitivity of detection is an issue, whether because of limited enzyme sample or low activity for any reason. In general, both techniques involve pyrimidine nucleotides in the reaction assay. Dehydrogenases are excellent candidates for these procedures because by appropriate manipulations, the levels of NAD, NADH, NADP, and NADPH are quantitatively determined with great sensitivity. Using cycling techniques, sensitivity approaches 10^{-17} mol. Description and examples of fluorometric and cycling techniques are described in great detail by Passonneau and Lowry ***(12)***.

2. Materials

2.1. Growth of Cells

With few exceptions, we now grow all mollicutes in a modification of Edward's medium, designated SNEP, containing (per liter) ***(13)***: 20 g of Mycoplasma Broth base (BBL, Cokeysville, MD), 2 g of Bacto-yeast (Difco Laboratories, Detroit, MI), 11 g of *N*-2-hydroxyethyl-piperazine-*N'*-2-ethanesulfonic acid (HEPES) (Research Organics, Cleveland, OH), 10 g of Bacto-peptone (Difco), and 0.1 g of herring sperm DNA (Sigma, St. Louis, MO); the final pH is 7.6. After autoclaving, add (per liter) 14 mL of a separately autoclaved 50% (w/v) glucose solution and 6 mL of a separately autoclaved 41% (w/v) K_2HPO_4 solution. Horse serum (heated) (Hyclone, Logan, UT or Oxoid) is added to a concentration of 2–10% (v/v) depending on the strain. One hundred micrograms of penicillin per mL medium are also added (*see* **Note 1**.)

2.2. Collection and Washing of Cells and Preparation of Cell Free Extracts

1. κ Buffer 1X at 4°C: 1 m*M* HEPES, pH 7.5, 0.155 *M* NaCl, 2 m*M* β-mercaptoethanol.
2. κ Buffer 1X at 4°C, but with 4 m*M* β-mercaptoethanol, and 100 μ*M* phenylmethylsulfonyl fluoride, for dialysis only (10% v/v glycerol, optional; *see* **Note 2**).
3. Water bath at 37°C.
4. Distilled water at 37°C.
5. High-speed centrifuge at 4°C and precooled rotor.
6. Ultracentrifuge at 4°C and precooled rotor.
7. 0.22-μm filter units.
8. Sephadex G-200, small bed volume (0.5–1 mL) in a small column, equilibrated with κ buffer.

2.3. Assay of PGK (EC 2.7.2.3)

1. Reagent A (premix): 300 m*M* triethanolamine-HCl, pH 7.5, 50 m*M* $MgCl_2$, 2 m*M* EDTA, 150 m*M* $(NH_4)_2SO_4$, 500 m*M* NaCl.
2. Reagent B: 10 m*M* ATP in water.
3. Reagent C: glyceraldehyde-3-phosphate dehydrogenase (from Baker's yeast, <0.01% PGK activity) (Sigma G2647) diluted with water to 0.5 U/25 μL.
4. Reagent D: 0.15 m*M* NADH.
5. Reagent E: 10 m*M* 3-phosphoglyceric acid in water.
6. Reagent F: 3-phosphoglycerate kinase (from Baker's yeast) (positive control) (Sigma P7634) diluted with water to 1 U/25 μL.
7. Mollicute cytoplasmic fraction.
8. Recording spectrophotometer at 25°C, quartz cuvets preferred.

2.4. Assay of PK (EC 2.7.1.40)

1. Reagent A (premix): 500 m*M* HEPES, pH 7.5, 75 m*M* $MgCl_2$, 750 m*M* KCl, 30 m*M* ADP.
2. Reagent B: lactic dehydrogenase (type II, rabbit muscle, <0.01% MDH activity) (Sigma L2500) diluted with water to 10 U/50 μL.
3. Reagent C: 0.15 m*M* NADH.
4. Reagent D: 100 m*M* phosphoenolpyruvate.
5. Reagent E: pyruvate kinase (positive control) (type II, rabbit muscle) (Sigma P1506) diluted with water to 2 U/25 μL.
6. Mollicute cytoplasmic fraction.
7. Recording spectrophotometer at 25°C, quartz cuvets preferred.

2.5. Assay of MDH (EC 1.1.1.37)

1. Reagent A (κ buffer): 1 m*M* HEPES, pH 7.5, 0.155 M NaCl, 2 m*M* β-mercaptoethanol.
2. Reagent B: 0.15 m*M* NADH.
3. Reagent C: 10 m*M* oxalacetic acid (dissolved in 100 m*M* HEPES, pH 8.0, prepared immediately before use and kept on wet ice).
4. Reagent D: malic dehydrogenase (positive control) (bovine heart, very low LDH activity) (Sigma M9004) diluted with water to 1 U/25 μL.
5. Mollicute cytoplasmic fraction.
6. Recording spectrophotometer at 25°C, quartz cuvets preferred.

3. Methods

3.1. Growth of Cells

Four liters of culture (*see* **Note 3**) are harvested to produce the final cell-free or cytoplasmic fraction. In order to obtain 4 L of culture, begin with a series of two starter cultures: the original inoculum is an SNEP adapted actively growing culture that is inoculated into 40 mL of SNEP that is transferred the next day into 400 mL. The following day, the 400 mL go into 4 L. Most cultures are incubated for 24–28 h at 30 or 36°C, depending upon the strain. Some *Mollicutes* require longer incubation. For optimal cell disruption by the preferred technique, osmotic lysis (described in **Subheading 3.2.**), cells should be in the mid- to late logarithmic phase. All media are temperature equilibrated before inoculation (*see* **Note 4**).

3.2. Collection and Washing of Cells and Preparation of Cell-Free Extracts

An osmotic lysis cell-disruption technique is described (*see* **Note 5** for alternative methods).

1. A frozen washed cell pellet from 4 L of culture is thawed at 37°C for 5–10 min, and then suspended by rapid and repeated pipeting, using a 10-mL glass pipet in 10–20 mL 37°C 1/20 or 1/50 water diluted 1X κ buffer. (*See* **Note 6**.)
2. The suspension is diluted to about 50–75 mL with the same warm diluted κ buffer and incubated in a 37°C water bath for 30 min with frequent pipeting.
3. The crude lysate (**step 2**) is treated by sequential differential centrifugation to obtain a cell-free fraction or cytoplasmic and membrane fractions ***(11)***. If only a cytoplasmic fraction is desired, the crude lysate is centrifuged at approx 350,000*g* (Beckman rotor 60Ti, 55,000 rpm) for 2 h rather than at 144,000*g* as previously recommended ***(11)***. At these greater forces, the supernatant is substantially free of ATPase activity ***(10)***.
4. The resultant cold supernatant or cytoplasmic fraction is then filtered through a 0.22-μm filter. The filtered fraction is brought to approx 1X κ buffer by addition of an appropriate amount of cold 10X κ buffer. In some cases, subsequent filtra-

tion at 4°C through a column of Sephadex-G200, equilibrated with cold 1X κ buffer, removes some remaining membrane particles. However, this latter step dilutes the preparation. Fractions can be stored at –80°C for some months without apparent deterioration of the enzymes for which assays are described in this chapter, if one part of glycerol is added before freezing.

5. The cytoplasmic fraction after **step 4** (excluding glycerol addition and freezing) is generally dialyzed before use. Dialysis is always necessary if ATP/ADP is a component of the succeeding enzyme assay(s), because *Mollicutes* have significant levels of these nucleotides in their cytoplasm ***(14)***. In order to dialyze these nucleotides out of approx 10–20 mL cytoplasmic fraction, dialysis with stirring against three to four changes (about 5 h each) of 2 L of κ buffer-for-dialysis at 4°C is essential.
6. If a membrane fraction is desired, the crude lysate (**step 2**) is centrifuged at 2000*g* for 3–5 min to bring down unlysed cells, which are then discarded. This step may be repeated. The supernatant contains membranes, ribosomes, and cytoplasm, and a smaller fraction of unbroken cells.
7. This supernatant is centrifuged at approx 30,000*g* for 30 min to bring down membranes and any remaining unbroken whole cells. The precipitate from this centrifugation is called precipitate-2. The supernatant containing ribosomes and cytoplasm is called supernatant-2.
8. The supernatant-2 (**step 6**) can be processed for a cytoplasmic fraction as described in **steps 3**, **4**, and **5**.
9. Precipitate-2 (**step 7**) is processed further to obtain a membrane fraction. This precipitate is again treated with about 20–40 mL of 37°C 1/20 or 1/50 diluted κ buffer for 30 min at 37°C to lyse any remaining whole cells and centrifuged at approximately 30,000*g* for 30 min. This precipitate contains mostly membranes and is repeatedly washed (two to five times) by centrifugation in the same manner, but with 4°C 1X κ buffer. Removal of only the top three-quarters of the pelleted membranes during this washing sequence and discarding the bottom quarter of the pellet produces a much cleaner final membrane fraction.
10. Cell-free or cytoplasmic preparations are generally used immediately or after dialysis without storage. Samples are stored at –80°C after adding an equal volume of glycerol.

3.3. Spectrophotometry

All of the described assays are performed in a recording spectrophotometer (*see* **Note 7**) and depend on the quantitative estimation of pyrimidine nucleotides: NAD, NADH, NADP, and NADPH (*see* **Note 8**). At 100 μ*M*, NADH or NADPH has an absorbance of 0.627 at 340 nm in a 1-cm light path. The oxidized forms do not absorb at 340 nm. The molar extinction coefficient (e) for NADH and NADPH at 340 nm is 6270. Therefore, depending on the spectrophotometer used, the useful range of measurement of these reduced nucleotides at 340 nm in a 1-cm light path is between 10 and 300 μ*M*.

3.4. Assay of PGK (EC 2.7.2.3)

PGK is a monomeric enzyme that catalyzes the reversible reaction

$$\text{1,3-diphosphoglycerate (DPG)} + \text{MgADP} \longleftrightarrow \text{MgATP} + \text{3-phosphoglycerate} \quad (1)$$

The reaction as illustrated in the direction of ATP formation requires the unstable reactant DPG. When the assay is run in this direction, the DPG is generated by an auxilliary system using G3P dehydrogenase. Rather than using this approach, the assay method described measures the PGK activity in the opposite direction and is a modification of the assay reported by Barber et al. ***(15)***. In this procedure, the coupled reaction sequence also in the presence of added glyceraldehyde-3-phosphate dehydrogenase is as follows:

$$\text{3-Phosphoglyceric acid (PGA)} + \text{ATP} \longleftrightarrow \text{DPG} + \text{ADP} \quad (2)$$

$$\text{DPG} + \text{NADH} \longleftrightarrow \text{NAD}^{+} + \text{glyceraldehyde-3-phosphate (G3P)} \quad (3)$$

1. Add 100 μL reagent A, 100 μL reagent B, 25 μL reagent C, mollicute cytoplasmic fraction, and water to a volume of 850 μL in a cuvet. Mix thoroughly.
2. Zero the cuvet in the spectrophotometer at 340 nm.
3. Add 50 μL reagent D. Mix quickly. Record for 3–8 min.
4. Add to the cuvet 100 μL reagent E to start the reaction. Mix quickly. Record for 5–10 min or longer.
5. Subtract the rate determined before addition of reagent E (**step 3**) from the rate determined after its addition (**step 4**). Report after calculation the μmol NADH or ATP consumed/min/mg mollicute cell protein.
6. For positive control reactions, add 25 μL of reagent F instead of mollicute cytoplasmic fraction in **step 1** (*see* **Notes 9–11**).

3.5. Assay of PK (EC 2.7.1.40)

PK is one of the major regulatory enzymes of glycolysis and catalyzes the essentially irreversible reaction.

$$\text{Phosphoenolpyruvate (PEP)} + \text{ADP} \longrightarrow \text{ATP} + \text{pyruvate} \quad (4)$$

Like PGK and other members of the triose arm of glycolysis, PK activity is found in all *Mollicutes*, and as suggested in **Subheading 1.** may be one of only two major metabolic sites possessed by *Mollicutes* where ATP is generated. In *Spiroplasma citri,* PK and ATP-dependent phosphofructokinase constitute an operon as in *Bacillus stearothermophilus* ***(16,17)***.

PK activity is measured by coupling pyruvate formation to LDH activity and the oxidation of NADH at 340 nm, essentially following the procedure of Bucher and Pfleiderer ***(18)***. The coupled reaction sequence is:

$$\text{PEP} + \text{ADP} \longrightarrow \text{ATP} + \text{pyruvate} \quad (5)$$

$$\text{Pyruvate} + \text{NADH} \longleftrightarrow \text{lactate} + \text{NAD}^{+} \quad (6)$$

1. Add 100 µL reagent A, 50 µL reagent B, mollicute cytoplasmic fraction, and water to a volume of 850 µL in a cuvet. Mix thoroughly.
2. Zero the cuvet in the spectrophotometer at 340 nm.
3. Add 50 µL reagent C. Mix quickly. Record for 3–8 min.
4. Add to cuvet 100 µL reagent D to start the reaction. Mix quickly. Record for 5–10 min or longer.
5. Subtract the rate determined after addition of reagent C (**step 3**) from the rate determined after the addition of reagent D (**step 4**). Report after calculation the µmol NADH consumed or ATP synthesized/min/mg mollicute cell protein.
6. For positive control reactions, add 25 µL of reagent E instead of mollicute cytoplasmic fraction in **step 1**. (*See* **Notes 9–11**.)

3.6. Assay of MDH (EC 1.1.1.37)

MDH activity, like PK and PGK, has been found in all *Mollicutes* except *Ureaplasma* spp. This suggests that MDH may also have some value in phylogenetic analyses. In cells of higher animals, there are two major forms of MDH: mitochondrial and cytoplasmic. The mitochondrial form of MDH has at least five conformational forms. There are no reports suggesting more than one form of MDH in *Mollicutes*.

MDH is generally classified as a tricarboxylic acid (TCA) cycle enzyme. Since mollicutes are devoid of the TCA cycle, the role of MDH is less obvious. We speculate that its activity is important in regulating the levels of NAD/NADH, and modulating aspartate amino-transferase activity and the production of aspartate. MDH activity is also of particular interest, because in some mollicutes, MDH and LDH activities are both suspected of being contained on a single protein product coded for by a single gene ***(9)***.

Follow the order of addition of reagents strictly as indicated.

1. Add mollicute cytoplasmic fraction and reagent A to 900 µL in a cuvet. Mix thoroughly.
2. Zero the cuvet in the spectrophotometer at 340 nm.
3. Add 50 µL reagent B. Mix quickly. Record for 3–8 min.
4. Add 50 µL reagent C to start the reaction (*see* **Note 12**). Mix quickly. Record for 5–10 min or longer.
5. Subtract the rate determined after addition of reagent B (**step 3**) from the rate determined after the addition of reagent C (**step 4**). Report after calculation the µmol NADH consumed or malate synthesized/min/mg mollicute cell protein.
6. In order to obtain maximum rates, in addition to trials varying the amount of enzyme, it is also advisable to test the reaction with various amounts of OAA (reagent C; *see* **Note 12**). Try amounts of reagent C less than the prescribed 50 µL.
7. For positive control reaction, add 25 µL reagent D instead of mollicute cytoplasmic fraction in **step 1**. (*See* **Notes 9–11**.)

3.7. Calculation of Enzyme Reaction Rates

The determination of the rate of the enzyme reaction is paramount. However, the use of essentially crude gemisches (i.e., cytoplasmic fractions as described above) makes accurate and reliable determinations of enzymatic rates and other enzymatic parameters very problematic. Enzyme kinetic experiments are generally determined using highly purified enzymes in well-controlled conditions. In our practice, generally reproducible data are obtained by running replicate assays with varying amounts of mollicute fractions. Data are generally reported as specific activity, i.e., μmol of reduced or oxidized pyrimidine nucleotide consumed or formed/min/mg cell protein. The highest specific activity using an enzyme preparation from one batch of cells is averaged with the analagous data from two to four different batches of cells to obtain a mean and standard deviation. Valuable discussions of enzyme purification, assays, and kinetics, supplementing these necessarily brief comments are found in a number of excellent texts *(12,19–21)*.

All of the assays described in this chapter include an essential correction step. This step requires that the reaction mixture containing the mollicute cell fraction and NADH, but minus the starting compound be mixed together first and assayed. The starting reactant should never be the pyrimidine. In addition to some mechanistic reasons, this control and prescribed sequence of additions to the cuvet are required because all mollicutes, except for acholeplasmas, and perhaps ureaplasmas have a cytoplasmic NADH oxidase activity. This means that there is a competing reaction that uses NADH. Therefore, it is necessary if NAD/NADH is involved to determine first the rate of this competing reaction and subtract it from the total rate of NADH oxidation after the starting reactant is added to the cuvet.

4. Notes

1. The SNEP medium recommended is useful for the growth of most *Mollicutes* and is used by us for the growth of all species, except *M. pneumoniae, M. genitalium,* some *Spiroplasma* spp., the anaerobes, and *Ureaplasma* spp. SNEP is an improvement over Edward's-type media, probably because the Bacto-peptone neutralizes or inactivates some unknown growth-inhibitory effects of different batches of horse serum (probably free fatty acids). In our experience, free fatty acids are inhibitory to the growth of *Mollicutes* at levels approaching 6–12 μg/mL medium. The amount varies with the composition of the medium, particularly the concentration of serum and albumin. Serum deteriorates even on frozen storage. We find the free fatty acid content of horse serum increases with frozen storage, presumably by hydrolysis of triglycerides. Also, keep the phosphate solution as a separate autoclaved additive to the autoclaved basal medium or a precipitate will form.
2. In protracted experiments, 10% (v/v) glycerol may be included in the κ buffer-for-dialysis fluid to protect the enzyme being assayed. At this concentration, glyc-

erol is without effect on the assays. However, we do not use glycerol routinely, because all our glycerol samples contain contaminants that interfere with silver staining of PAGE gels of enzymes during further purification.

3. Four liters of cells are the prescribed amount of culture, because this generally yields enough final cytoplasmic extract to do sufficient replicates of all the assays described, the protein analyses, and leaves some material to store at –80°C when mixed with one part glycerol.
4. Obtaining the largest cell crop in 24 h is strongly dependent on successful temperature equilibration of the media before inoculation. Four liters of broth in a 6-L Erlenmeyer flask at room or cooler temperatures require at least 6–8 h at 36°C in a forced-air incubator to come to 36°C. Therefore, equilibrate overnight if possible.
5. Our experience in breaking mollicute cells for enzyme assays of their cytoplasmic fractions is restricted to osmotic lysis and sonication ***(11,22,23)***. There are other techniques for breakage of mollicute cells, some specifically designed for isolation of membranes. Among these are glycerol loading prior to hypotonic shock ***(24)***, and digitonin- or dicyclohexylcarbodiimide-induced lysis ***(25,26)***. More general techniques for mollicute cell breakage are: French-pressure cell, freezing and thawing, X-press, or alumina grinding ***(22)***. The choice of procedure is important, because excessive breakage of the mollicute cell membrane results in particle sizes too small to sediment at 144,000*g* for 2 h ***(11,22,23)***.

   One advantage of certain techniques (e.g., osmotic lysis) is that membrane and cytoplasmic fractions may both be obtained from the same cell pellet without any concern that additives may interfere with subsequent enzyme assays. Some techniques have specific intrinsic properties that may not be desirable, for example; alumina grinding removes ribosomes, and sonication may disrupt subunit enzymes. We have found that osmotic lysis, essentially as first described by Razin ***(27,28)*** produced satisfactory membrane and cytoplasmic fractions for enzyme studies of over 60 different species of *Mollicutes,* including species often thought to be refractory to osmotic lysis, like *M. pneumoniae* ***(29)***.

   When sonication is the chosen disruption method, it is performed with particular attention to cooling. The washed cell pellet from 4 L of culture is overlayed with about 10–20 mL 1X κ buffer devoid of β-mercaptoethanol. Sonication is performed at maximum wattage, for three to four cycles of 15–45 s duration, while the sample is held continuously in wet ice. The probe is cooled in cold buffer for 35–60 s between cycles. The sonicated preparation is centrifuged and treated to obtain a final cell-free or cytoplasmic fraction as for osmotic lysis, described above. Although convenient, sonication is not generally recommended for the production of membrane fractions. Sonication results in many smaller membrane fragments than osmotic lysis and passage through a small column of Sephadex G-200 may be additionally necessary, especially when assaying for kinases. That is because the ATPases of *Mollicutes* are localized in their membranes and may significantly interfere with assays involving ATP ***(22)***.

6. Harvested cells must be washed with cold κ buffer by forcibly pipeting wash fluid at the pellet. Do not pipet the cells at this stage, or they will break at the pipet orifice, resulting in loss of cytoplasm with the discarded wash fluids. β-mercaptoethanol must be left out of the wash fluids during sonication, because it breaks down, forming some black precipitate, presumably sulfides. We find that κ buffer is superior to 0.25*M* NaCl as a wash fluid. We consider plain 0.25*M* NaCl hypertonic; it is also devoid of any buffering capacity that is sometimes needed if the culture is acid and lacking in reducing power.
7. There is usually enough 340-nm illumination in most spectrophotometric visible light sources to perform the assays. To remove contaminating material from cuvets they should be briefly cleaned in a solution of 95% ethanol and either 6 *N* HCl or reagent HNO_3 (1:1), and then washed in distilled water. The cleaning solution can be reused. Do not use alkaline solutions to clean quartz cuvets.
8. NADH and NADPH solutions, as well as other solutions, should be kept on wet ice during use and should be discarded after about 6 h. Premixes, as identified, may be stored at –20°C for several months without apparent deterioration. Premixes and κ buffer should be kept at room temperature during experimentation, in order to compensate partially for the cooler reagents. It is best to perform the assays at a controlled temperature. We use 25, 30, or 37°C, depending on literature recommendations.
9. Test each enzyme assay using positive control enzymes in order to be sure that the reaction is performing satisfactorily. Compare enzyme rate printed on the package with the rate you obtain, preferably at the same reaction temperature. The rates should be approximately equal. Discard solutions of positive control enzymes after 4 h, unless their stability has been predetermined. If high dilution is necessary, enzyme inactivation may be reduced by suspending the enzyme in κ buffer containing 100 mg fatty acid free or fatty acid reduced albumin/100 mL.
10. In our hands, commercially produced enzymes in ammonium sulfate (usually about 2–3*M*) appear to have a longer shelf life at 4–10°C than some other preparations. We use Sigma Company enzymes exclusively. The specific enzyme types we have recommended were chosen because of their low conflicting background activities.
11. Sometimes the crude cytoplasmic extract of *Mollicutes* is inhibitory. This has not been the case in the three assays described here with any of the *Mollicutes* we have studied. However, to test for this possibility, add positive control enzyme to a cuvet containing a complete reaction mixture, including cytoplasmic fraction after it has been shown to be negative. Redetermine the reaction rate. If the original reaction mixture in the cuvet is not inhibitory, an appropriate rate should be observed.
12. Special attention must be given to the preparation and use of oxalacetic acid (OAA) solutions in the MDH assay. We assay for MDH activity at physiological pHs around neutrality in the thermodynamically favorable direction, i.e., in the direction of malate and NAD^+ production. This requires the use of OAA as a reactant. OAA in neutral solution at 25°C deteriorates to pyruvate at about 5%/h.

Therefore, the solution should be made in small quantities, kept on ice, and used as quickly as possible. However, our own experiments indicate that conversion to pyruvate is not a significant problem, but it should be understood that since all crude mollicute cytoplasmic extracts contain LDH activity, if formation of pyruvate occurs, the presence of mollicute LDH, and the NADH in the reaction mixture would result in a false-positive reaction, indicating the presence of MDH activity. This error occurs because the mollicute LDH oxidizes NADH in the presence of the contaminating pyruvate to lactate and NAD^+. This error cannot be corrected for in the described method for MDH. We can test our solutions of OAA for pyruvate conversion by omitting the cytoplasmic extract and substituting a sample of purified LDH (25 µL reagent B in Assay of PK, **Subheading 2.4.** [LDH Sigma L2500]). This LDH in combination with NADH, buffer, and water is reacted with the OAA test solution. Any oxidation of the NADH at 340 nm, dependent on the presence of LDH, should be taken to indicate the presence of pyruvate and be corrected for or another sample of OAA prepared. It must be re-emphasized that we have never encountered this necessity. Nevertheless, there are alternate MDH assays for obviating this concern ***(12,30,31)***. One procedure assays MDH in the direction of malate oxidation ***(12)***. It involves the production of OAA in the presence of glutamate, aspartate aminotransferase, and NAD^+, and follows at 340 nm the reduction of NAD^+ rather than the oxidation of NADH.

Acknowledgment

This study was supported by Public Health Service grant RO1–A133193 from the National Institutes of Health.

References

1. Fleischmann, R. D., Adams, M. D., White, O., Clayton, R. A., Kirkness, E. F., Kerlavage, A. R., Bult, C. J., Tomb, J.-F., Dougherty, B. A., Merrick, J. M., McKenney, K., Sutton, G., FitzHugh, W., Fields, C., Gocayne, J. D., Scott, J., Shirley, R., Liu, L.-I., Glodek, A., Kelley, J. M., Weidman, J. F., Phillips, C. A., Spriggs, T., Hedblom, E., Cotton, M. D., Utterback, T. R., Hanna, M. C., Nguyen, D. T., Saudek, D. M., Brandon, R. C., Fine, L. D., Fritchman, J. L., Fuhrmann, J. L., Geoghagen, N. S. M., Gnehm, C. L., McDonald, I. A., Small, K. V., Fraser, C. M., Smith, H. O., and Venter, J. C. (1995) Whole-genome random sequencing and assembly of *Haemophilus influenzae* Rd. *Science* **269,** 496–512.
2. Fraser, C. M., Gocayne, J. D., White, O., Adams, M. D., Clayton, R. A., Fleischmann, R. D., Bult, C. J., Kerlavage, A. R., Sutton, G., Kelley, J. M., Fritchman, J. L., Wiedman, J. F., Small, K. V., Sandusky, M., Fuhrmann, J., Nguyen, D., Utterback, T. R., Saudek, D. M., Phillips, C. A., Merrick, J. M., Tomb, J.-F., Dougherty, B. A., Bott, K. F., Hu, P.-C., Lucier, T. S., Peterson, S. N., Smith, H. O., Hutchison, C. A., III, and Venter, J. C. (1995) The minimal gene complement of *Mycoplasma genitalium*. *Science* **270,** 397–403.

3. Tryon, V. V. and Pollack, J. D. (1984) Purine metabolism in *Acholeplasma laidlawii* B: Novel PPi-dependent nucleoside kinase activity. *J. Bacteriol.* **159,** 265–270.
4. Tryon, V. V., and Pollack, J. D. (1985) Distinctions in Mollicutes purine metabolism: pyrophosphate-dependent nucleoside kinase and dependence on guanylate salvage. *Int. J. Syst. Bacteriol.* **35,** 497–501.
5. DeSantis, D., Tryon, V. V., and Pollack, J. D. (1989) Metabolism of Mollicutes: the Embden-Meyerhof-Parnas pathway and the hexose monophosphate shunt. *J. Gen. Microbiol.* **135,** 683–691.
6. Manolukas, J., Barile, M. F., Chandler, D. K. F., and Pollack, J. D. (1988) Presence of anaplerotic reactions and transamination, and the absence of the tricarboxylic acid cycle in Mollicutes. *J. Gen. Microbiol.* **134,** 791–800.
7. Petzel, J., McElwain, M. C., DeSantis, D., Manolukas, J., Williams, M. V., Hartman, P. A., Allison, M. J., and Pollack, J. D. (1989) Enzyme activities of carbohydrate, purine, and pyrimidine metabolism in the Anaeroplasmataceae (class *Mollicutes*). *Arch. Microbiol.* **152,** 309–316.
8. Herrmann, R. personal communication
9. Cordwell, S., D., Basseal, J., Pollack, J. D., and Humphery-Smith, I. (1997) Malate/lactate dehydrogenase in Mollicutes: evidence for a multienzyme protein. *Gene,* **195,** 113–120.
10. Pollack, J. D., unpublished data
11. Pollack, J. D. (1995) Methods for testing metabolic activities in Mollicutes, in *Molecular and Diagnostic Procedures in Mycoplasmology*, vol. 1 (Razin, S. and Tully, J. G., eds.), Academic, San Diego, CA, pp. 277–286.
12. Passonneau, J. V. and O. H. Lowry. (1993) *Enzymatic Analysis*. Humana Press, Totowa, NJ, pp. 403.
13. Pollack, J. D., Banzon, J., Donelson, K., Tully, J. G., Davis, Jr., J. W., Hackett, K. J., Agbanyim, C., and Miles, R. (1996) Reduction of benzyl viologen distinguishes genera of the class *Mollicutes*. *Int. J. Syst. Bacteriol.* **46,** 881–884.
14. Beaman, K. D. and Pollack, J. D. (1983) Synthesis of adenylate nucleotides by Mollicutes (mycoplasmas). *J. Gen. Microbiol.* **129,** 3103–3110.
15. Barber, M. D., Gamblin, S. J., Watson, H. C., and Littlefield, J. A. (1993) Site-directed mutagenesis of yeast phosphoglycerate kinase. *FEBS Lett.* **320,** 193–197.
16. Chevalier, C., Saillard, C., and Bove, J. M. (1990) Organization and nucleotide sequences of the *Spiroplasma citri* genes for ribosomal protein S2, elongation factor Ts, spiralin, phosphofructokinase, pyruvate kinase, and an unidentified protein. *J. Bacteriol.* **172,** 2693–2703.
17. Sakai, H. and Ohta, T. (1993) Molecular cloning and nucleotide sequence of the gene for pyruvate kinase of *Bacillus stearothermophilus* and the production of the enzyme in *Escherichia coli*. Evidence that the genes for phosphofructokinase and pyruvate kinase constitute an operon. *Eur. J. Biochem.* **211,** 851–859.
18. Bucher, T. and Pfleiderer, G. (1955) Pyruvate kinase from muscle. *Methods Enzymol.* **1,** 435–440.
19. Engel, P. C. (ed.). (1996) *Enzymology*. LabFax. Bios Scientific, Oxford, UK and Academic, San Diego, CA.

20. Scopes, R. K. (1993) *Protein Purification,* 3rd ed. Springer-Verlag, New York, pp. 380.
21. Suelter, C. H. (1985) *A Practical Guide to Enzymology*. John Wiley, New York, pp. 288.
22. Pollack, J. D., Razin, S., Pollack, M. E., and Cleverdon, R. C. (1965) Fractionation of Mycoplasma cells for enzyme localization. *Life Sci.* **4,** 973–977.
23. Pollack, J. D., Williams, M. V., Banzon, J., Jones, M. A., Harvey, L., and Tully, J. G. (1996) Comparative metabolism of *Mesoplasma, Entomoplasma, Mycoplasma,* and *Acholeplasma*. *Int. J. Syst. Bacteriol.* **46,** 885–890.
24. Rottem, S., Stein, O., and Razin, S. (1968) Reassembly of mycoplasma membranes disaggregated by detergents. *Arch. Biochem. Biophys.* **125,** 46–56
25. Rottem, S. (1972) Isolation of mycoplasma membranes by digitonin. *J. Bacteriol.* **110,** 699–705.
26. Shirvan, M. H., Rottem, S., Neíeman, Z., and Bittman, R. (1982) Isolation of mycoplasma membranes by dicyclohexylcarbodiimide-induced lysis. *J. Bacteriol.* **149,** 1124–1128.
27. Razin, S. (1963) Osmotic lysis of mycoplasma. *J. Gen. Microbiol.* **33,** 471–475.
28. Razin, S. (1983) Cell lysis and isolation of membranes, in *Methods in Mycoplasmology*, vol. 1 (Razin, S. and Tully, J. G., eds.), Academic, New York, pp. 225–233.
29. Pollack, J. D., Somerson, N. L., and Senterfit, L. B. (1970) Isolation, characterization, and immunogenicity of *Mycoplasma pneumoniae* membranes. *Infect. Immunol.* **2,** 326–339.
30. Breiter, D. R., Resnik, E., and Banaszak, L. J. (1994) Engineering the quatenary structure of an enzyme: Construction and analysis of a monomeric form of malate dehydrogenase from *Escherichia coli*. *Protein Sci.* **3,** 2023–2032.
31. Wynne, S. A., Nicholls, D. J., Stevens, M. D., and Sundaram, T. K. (1996) Tetrameric malate dehydrogenase from a thermophilic *Bacillus*: cloning, sequence and over expression of the gene encoding the enzyme and isolation and characterization of the recombinant enzyme. *Biochem. J.* **317,** 235–245.

11

Determination of Substrate Utilization Rates by Mycoplasmas

Roger J. Miles and Calista Agbanyim

1. Introduction

Mollicutes have restricted metabolic activities, and catabolism is primarily associated with ATP generation rather than the synthesis of metabolic precursors for anabolic metabolism *(1)*. Nevertheless, the pathways of energy substrate metabolism and the range of substrates used by mollicutes are diverse. In mycoplasmas, energy may be obtained by: the fermentation of sugars (via pyruvate) to lactate; the partial oxidation of organic acids, for example, of lactate or pyruvate, to acetate plus CO_2; and the metabolism of arginine, by the arginine dihydrolase pathway *(2)*, to ornithine, NH_3 and CO_2. Individual *Mycoplasma* species may use one or any combination of these reactions to obtain energy, enabling the subdivision of the genus into major physiological groups *(3)*. Within these groups, patterns and rates of substrate utilization have been shown to distinguish certain species and subspecific taxa *(4–6)*, and may be applied to the biochemical characterization and identification of isolates. In addition, knowledge of the substrates used by mycoplasmas and other *Mollicutes* and their kinetics of utilization, may improve understanding of pathogenicity. Kinetic data are important in allowing assessment of the likely significance of substrate metabolism at the concentrations found in host tissues. The utilization of energy substrates at high rates may reduce substrate availability to host cells and result in the formation of toxic products, particularly hydrogen peroxide from carbohydrate metabolism *(4)*. For example, it has recently been shown that European isolates of *Mycoplasma mycoides* subsp. *mycoides* small colony (SC) differ from African and Australian isolates in lacking the ability to oxidize glycerol *(7)*. This may provide an explanation for the reduced virulence of European strains, in causing contagious bovine pleuropneumonia,

From: *Methods in Molecular Biology, Vol. 104: Mycoplasma Protocols*
Edited by: R. J. Miles and R. A. J. Nicholas © Humana Press Inc., Totowa, NJ

since glycerol oxidation leads to the production of large quantities of hydrogen peroxide.

In certain cases, energy sources may be identified in growth studies; for example, the ability to produce acid from glucose metabolism and ammonium from arginine hydrolysis are both widely used in mollicute identification (*see* Chapter 9). However, the techniques used are not quantitative, and routine detection of the metabolism of other substrates during mollicute growth is often not feasible because of low cell yields and the difficulties in detecting the metabolism of specific substrates in complex media containing a wide range of alternative substrates. In addition, the utilization of di- and polysaccharides of glucose cannot be determined in serum-containing medium, since they are rapidly hydrolysed by serum enzymes. The inability to utilize the disaccharides maltose and trehalose was an important feature enabling the differentiation of *M. mycoides* SC strains from closely related strains *(5)*.

The methods described here to determine substrate metabolism are quantitative and use cells suspended in inorganic salts solutions. The utilization of added substrates is monitored by: oxygen uptake, measured as a reduction in the dissolved oxygen tension (DOT) of cell suspensions; acid or alkali production, measured by change in pH; or reduction of 2,2'-di-*p*-nitrophenyl-5,5'-diphenyl-3,3'-[3,3'-dimethoxy-4,4'-diphenylene]-ditetrazolium chloride (nitroblue tetrazolium or NBT), determined spectrophotometrically. In the latter case, NBT (colorless at the concentration used) is reduced to a deep blue formazan dye. All of the methods require only relatively small quantities of cells and are rapid. An important feature in the preparation of cell suspensions is the total time required. The procedure described aims to complete centrifugation, washing, and resuspension procedures rapidly, within about 20 min, in order to retain maximum metabolic activity of cells. In our laboratory, the methods have been successfully used with a wide range of *Acholeplasma* and *Mycoplasma* species.

2. Materials

2.1. Preparation of Cell Suspensions and Test Substrates

1. Liquid (broth) cultures of test strains (*see* **Note 1**).
2. Microcentrifuge and microcentrifuge tubes.
3. Suspension media: 18 g/L HEPES and 160 U/mL catalase (Sigma, Poole, Dorset, UK; product C-10) in one-quarter strength Ringer solution, pH 7.6 (RH solution) or normal saline (0.85% w/v NaCl) plus catalase (160 U/mL), adjusted to pH 7.6 with NaOH/HCl. Sterilize by membrane filtration (0.22-µm filter) (*see* **Notes 2** and **3**).
4. Test substrates, e.g., sugars, amino sugars, organic acids, amino acids, alcohols (5–500 m*M*). Filter-sterilize or prepare using sterile distilled water and store as aliquots at –20°C.

2.2. Detection of Metabolism by Oxygen Uptake

1. Water-jacketed, magnetically stirred reaction vessel(s) for use with oxygen electrode (*see* **Notes 4** and **5**).
2. Clark-type oxygen electrode (*see* **Notes 4** and **5**), fine tissue paper, Teflon membrane, 3*M* KCl solution, and sodium dithionite.
3. Thermocirculator.
4. Chart recorder (*see* **Note 5**).
5. Microsyringe(s) for accurate delivery of 5–50 µL substrate solution.

2.3. Detection of Metabolism by pH Change

1. Water-jacketed, magnetically stirred reaction vessel(s), internal diameter approx 10 mm.
2. pH meter, resolution 0.001 pH units, with 5-mm diameter electrode (*see* **Note 5**).
3. Thermocirculator.
4. Chart recorder (*see* **Note 5**).
5. Microsyringe(s) for accurate delivery of 5–50 µL substrate solution.

2.4. Detection of Metabolism by NBT Reduction

1. NBT (2,2'-di-*p*-nitrophenyl-5,5'-diphenyl-3,3'-[3,3'-dimethoxy-4,4'-diphenylene]-ditetrazolium chloride): Dissolve 10 mg in 10 mL distilled water to give a 1.2-m*M* stock solution. Sterilize by membrane filtration, and store as aliquots in the dark at 4°C.
2. Mineral oil sterilized by autoclaving at 121°C for 15 min.
3. Parafilm.
4. Spectrophotometer and disposable cuvets.

3. Methods

3.1. Preparation of Cell Suspensions

1. Harvest broth cultures towards the end of the exponential growth phase to achieve a maximum yield of metabolically active cells.
2. Dispense culture in 1.5-mL microcentrifuge tubes, and centrifuge in a microcentrifuge at 13,000*g* for 3 min (*see* **Note 6**). Carefully remove the supernatant and resuspend the cell pellet in RH buffer (oxygen uptake and NBT reduction methods) or saline (pH change method) using a Pasteur pipet.
3. Using a similar procedure, wash the cells twice and resuspend in RH buffer or saline at approx 10^9 colony-forming units (CFU) mL. The volume of cell suspension required is 1–2 mL/reaction vessel for oxygen uptake and pH change experiments. In NBT reduction experiments, each 1 mL of cell suspension is sufficient for 10 tests. Keep the total time for the preparation of suspensions to a minimum, i.e., approx 20 min. Use cell suspensions immediately, since activity declines during storage, even at 0–10°C (*see* **Note 7**).

3.2. Determination of Substrate Metabolism by Measurement of Oxygen Uptake

1. The detailed procedure for assembling the oxygen electrode system will vary according to the design of the apparatus, and manufacturer's instructions should be consulted. In the Rank Brothers (Cambridge, UK) instrument, the electrodes (a platinum working electrode and a Ag-AgCl reference electrode) are mounted beneath the cylindrical reaction vessel (maximum volume 5 mL; diameter 15 mm) and are separated from the contents of the vessel by a semipermeable Teflon membrane. Fine tissue (e.g., lens tissue), saturated with 3*M* KCl, is used to maintain contact between the electrodes, and the membrane is held in place using a rubber washer. The tissue has a small central hole that is placed over the central platinum electrode to enable direct contact with the membrane. In instruments of this type, briefly rinse and dry the electrodes, cut and fit a new piece of tissue, add sufficient KCl (3*M*) to wet the electrodes, and cover with a new piece of membrane. Thoroughly clean and wash the reaction vessel with sterile distilled water and assemble the system.
2. Connect the thermocirculator to the water jacket of the reaction vessel and the output of the oxygen electrode to a chart recorder. Set the polarizing voltage of the platinum electrode (relative to the reference electrode) to –0.6 V; at this voltage, current is directly proportional to DOT in the reaction vessel ***(8)***.
3. Calibrate the electrode with air-saturated water (210 nmol dissolved O_2/mL at 37°C; for other temperatures, *see* **ref. *9***) obtained by simply adding sterile distilled water (that has not been shaken, thus avoiding supersaturation) to the reaction vessel and magnetically stirring. The position of the baseline (zero current) may be confirmed by adding a few crystals of sodium dithionite, which will reduce DOT to zero within a few seconds. It is convenient to set up the chart recorder so that air-saturated water gives a 90% deflection.
4. Remove water from the reaction vessel using a vacuum line or Pasteur pipet, taking care not to damage the membrane. Wash the vessel if sodium dithionite has been used.
5. Add 1–2 mL of cell suspension to the oxygen electrode vessel. Set the magnetic stirrer speed in order to give adequate mixing with a high, but stable speed. Do not alter the speed during the course of DOT measurement. Allow up to approx 10 min for temperature equilibration and for the cell suspension to become saturated with oxygen, and then isolate from air. In the Rank Brothers instrument, this is achieved using a cylindrical plug with a fine central pore (1-mm diameter). Adjust the level of the plug, so that the meniscus of the cell suspension is within the central pore, thus essentially eliminating oxygen transfer to the suspension.
6. Allow approx 5 min for the recorded DOT value to become stable (*see* **Note 8**), and add test substrate (*see* **Note 9**), via the central pore in the plug, using a microsyringe. In experiments where there is complete substrate utilization, i.e., after a period of oxygen uptake DOT values become stable, the metabolism of a number of substrates can be followed sequentially (*see* **Note 10**). Using cell populations of 10^9 CFU/mL and initial substrate concentrations of 25–100 μ*M* (*see*

Note 9), the utilization of rapidly metabolized substrates (e.g., glucose) is typically complete within 10–15 min (e.g., **ref. *6***). The contents of the electrode vessel may be reoxygenated as required by raising the plug and allowing air to enter (*see* **Note 11**).

7. Calculation of kinetic parameters of substrate oxidation. Polarographic measurement of DOT to monitor oxygen uptake has been widely used in biological systems and the general principles of the method and the derivation of rates and quantities of oxygen uptake have been previously described (e.g., **ref. *8***). Reaction stoichiometries are determined from the total reduction in DOT and rates of substrate utilization from the rate of decline in DOT (*see* **Note 12**).

   In general, it appears that the relationship between the rate of substrate metabolism (v) and substrate concentration (s) by mollicute cells follows Michaelis kinetics ***(4–6,10)***. Thus, the saturation constant (K_m value) and maximum rate of substrate metabolism (v_{max}) may be estimated from plots of 1/s against $1/v$, in which the intercepts on the ordinate axis and abscissa are $1/v_{max}$ and $-1/K_m$, respectively.

   Where K_m values are high, determine rates of metabolism (proportional to rate of change in DOT) at different values of s (gradually increase substrate concentration). This will enable estimation of K_m values. Values of v_{max} may also be estimated where the relationship between the quantity of O_2 taken up and the amount of substrate oxidized is known; for example, in *Mycoplasma* species where the products of sugar metabolism are acetate plus CO_2, 2 mol of O_2 are consumed in the metabolism of 1 mol of sugar ***(1)***.

   Where K_m values are low, kinetic data are estimated from the analysis of single curves representing the complete utilization of substrate (initial concentration 25–100 μ*M*). Assume that the change in DOT at any time (t) after substrate addition is proportional to the substrate used. Thus, substrate concentration at time t is:

$$(\text{Initial substrate concentration}) \times (a - b)/a \qquad (1)$$

   where a is the total change in DOT following substrate addition and b the change at time t. Rates of substrate metabolism are determined from the gradient (nmol O_2 taken up/min) of the chart recorder curve of DOT vs time. The relationship between the quantity of O_2 taken up and the amount of substrate oxidized is determined from the total change in DOT following substrate addition ***(8)***.

3.3. Determination of Substrate Metabolism by Measurement of pH Change

1. Connect a water-jacketed reaction vessel to a thermocirculator and equilibrate at 37°C (or other temperature). Add the cell suspension in saline (approx 10^9 CFU/mL) to the vessel, and stir using a magnetic stirrer. Clamp the pH electrode in position and record the output of the pH meter continuously using a chart recorder. Using a 10-mm diameter vessel, with a 5-mm diameter pH electrode, 1.5 mL of cell suspension is adequate.

2. Wait until the chart recorder tracing shows that a stable pH has been reached (approx 20 min; *see* **Note 8**); typically, pH is slightly lower than the initial value (pH 7.6) of the saline solution at this time.
3. Add test substrates, in a volume of a few microliters, using a microsyringe (*see* **Note 9**). The fermentation of sugars will cause a decline in pH, whereas arginine hydrolysis will result in an increase in pH. Using cell populations of 10^9 CFU/mL and initial substrate concentrations of 10 μ*M*, the utilization of rapidly-metabolized substrates (e.g., arginine, glucose) is typically completed within 2–10 min (e.g., **ref. *6***). Thus, it may be possible to determine the utilization of a number of substrates using a single-cell suspension (*see* **Note 10**).
4. Calculation of kinetic parameters of substrate utilization (*see* **Note 13**). Where there is a high affinity of the test mollicute for a particular substrate, analyze pH vs time chart recordings, representing the complete utilization of the substrate (initial substrate concentration approx 10 μ*M*). Use the total pH change to determine the pH change/nmol substrate metabolized. This value will depend upon the exact pH at which the test substrate was added and will vary with the cell preparation used. Then, using this value, determine the amount of substrate utilized and rate of substrate utilization (*see* **Note 12**) from the pH change and rate of pH change, respectively, at various times after substrate addition. Calculate V_{max} and K_m values as described in **Subheading 3.2, step 7**. To estimate K_m values, where the affinity for substrate is low, determine rates of pH change for a single-cell suspension at increasing concentrations of substrate. The K_m value is obtained from the double-reciprocal plot of rate of pH change (a measure of rate of substrate metabolism) vs substrate concentration (*see* **Subheading 3.2, step** 7) (*see* **Note 14**).

3.4. Determination of Substrate Metabolism by NBT Reduction

1. Set up a series of cuvets containing 0.1 mL cell suspension, 0.9 mL RH buffer, and 20 μL NBT solution (*see* **Notes 15–17**). Equilibrate at 37°C (or other appropriate temperature) for 10 min.
2. Add test substrate (50 μL/cuvet; *see* **Note 9**) or distilled water (control), and using parafilm to seal the top of the cuvets, quickly mix. Immediately, overlay with 1 mL of mineral oil, sufficient to cover the cell suspension to a minimum depth of 1 cm (see **Note 18**), and continue to incubate at 37°C or other temperature.
3. Record absorbance at 570 nm at regular time intervals (5–15 min). Absorbance is determined against controls without substrate.
4. Determine rates of NBT reduction from plots of absorbance versus time (*see* **Note 19**). Values of *K*m are estimated from double-reciprocal plots of substrate concentration against rate of NBT reduction (*see* **Subheading 3.2**, **step** 7). The rate of reduction is related to the rate of substrate utilization (*see* **Note 20**).

4. Notes

1. The broth medium should be selected to give a high yield of the test strains. Media formulations are described in Chapters 4–7.

2. Catalase is necessary in suspension media, since many mollicute species produce substantial quantities of H_2O_2, leading to reduced metabolic activity and/or cell viability. The presence of catalase also simplifies data analysis using the O_2 uptake method, since in the absence of catalase, oxygen might be reduced to either H_2O or H_2O_2 *(4)*.
3. The suspension medium (RH) for the oxygen uptake and NBT-reduction methods is buffered to eliminate effects of pH change (caused by metabolism) on metabolic activity. In the pH-change method, unbuffered saline is used to maximize pH change resulting from production of acid or alkali products.
4. The oxygen electrode system we use is made by Rank Brothers Ltd. (Cambridge, UK), and incorporates a water-jacketed glass or perspex reaction vessel.
5. In the experimental arrangement we use, two oxygen electrode or pH meter systems are linked to a dual-channel chart recorder. This allows experiments to be run simultaneously with appropriate controls and simplifies comparison of recorder traces.
6. The metabolic activity of mollicute cells declines rapidly if the time taken for washing and resuspending of cells is prolonged. It is for this reason that we recommend the use of a microcentrifuge. Typically, we harvest the growth from 10–15 mL of culture.
7. Liquid nitrogen-stored cells have been used successfully in microcalorimetric studies of metabolism ***(10)***. However, cell viability and metabolic activity are critically dependent on the rates of freezing and thawing and the presence of osmoprotectants.
8. The *Mycoplasma* and *Acholeplasma* strains we have used do not appear to possess a significant endogenous metabolism. Detectable O_2 uptake, NBT reduction, or acid/alkali production was dependent upon addition of metabolizable substrate.
9. The substrate concentrations necessary to give detectable metabolism and allow accurate determination of rates of metabolism will depend on the mollicute strain, the selected substrate, and the method. K_m values, giving half of the maximum rate of substrate utilization, vary from <5 μ*M* (glucose and arginine metabolism by most *Mycoplasma* strains) to >1 m*M* (***4–6,10***). The pH-change method is the most sensitive and for substrates for which there is a high affinity, initial substrate concentrations as low as 10 μ*M* give pH–time curves suitable for kinetic analysis. In the oxygen uptake method, higher initial concentrations are required. We use minimum concentrations of 25 μ*M* for sugars, which are metabolized with the consumption of 2 mol O_2/mol sugar, and 100 μ*M* for organic acids, such as pyruvate, which are metabolized with the consumption of only 0.5 mol O_2 per mol. In the NBT-reduction method, the lowest sugar concentration giving detectable metabolism is approx 250 μ*M*. In this method, in cell suspensions overlaid with mineral oil, dissolved oxygen (210 μ*M* at 37°C) may initially be reduced in preference to NBT. In addition, any metabolism of added sugar to lactate (rather than to acetate plus CO_2) will not lead to NBT reduction (*see* **Note 20**).
10. The maximum time for which cell suspensions can be used to generate reliable data is approx 1.5 h; after this time, the metabolic activity of the cells declines.

11. In those acholeplasmas and mycoplasmas that we have studied, substrate metabolism is independent of DOT in the range 5–100 % of saturation.
12. The metabolic activity of cell suspensions is usually expressed in terms of viable count or cell protein. However, the specific activities of washed cell suspensions prepared on different days will vary, unless the preparation of growth media and inocula, and cell culture and harvesting procedures, are rigorously controlled. In experiments comparing the rates of utilization of different substrates, the effect of variability in the activity of cell suspensions prepared at different times may be minimized by initiating all individual experiments with the addition of a low concentration of a rapidly metabolized "control" substrate, e.g., glucose ***(5,6)***. Rates of metabolism of substrates added subsequently may then be expressed relative to the rate for the control substrate. The control substrate may also be added at the end of experiments to confirm that metabolic activity has not declined.
13. The relationship between pH change and acid or alkali production is not linear, since pH is logarithmically related to $[H^+]$ and the extent to which organic acids are dissociated is pH-dependent. CO_2 produced by substrate metabolism will also lead to increased carbonic acid concentrations. However, the experimental system described has been successfully used to determine saturating levels of substrate (i.e., those causing maximal rate of pH change) and to compare initial rates of pH change following substrate addition ***(4,6,11)***. We have also derived apparent K_m and V_{max} values for substrate metabolism, using pH change and rate of change in pH as measures of substrate utilized and rate of utilization, respectively. For *M. mycoides*, the kinetic data estimated in this way are similar to those obtained by measuring oxygen uptake. This suggests that over narrow pH ranges, there is an essentially linear relationship between acid or alkali produced and pH. In the procedure described, the total pH change during substrate utilization is small, e.g., <0.15 pH units for metabolism of 10 μ*M* glucose and arginine ***(6)***.
14. To derive values of V_{max}, the relationship between substrate utilized and pH change is required. Where the affinity for substrate is low, this cannot readily be determined, since as substrate concentration declines to a low level, metabolism effectively ceases. However, it would appear possible to estimate the relationship between substrate utilized and pH change using data for similar substrates (e.g., sugars) with identical cell preparations.
15. NBT is the preferred tetrazolium salt, although high concentrations of the reduced product (a formazan) give a precipitate. The formazan of MTT (3-[4,5-dimethyl thiazol-2-yl]-2,5-diphenyltetrazolium bromide), which is purple in color, also gives a precipitate. XTT (2,3-*bis*[2-methoxy-4-nitro-5-sulfophenyl]-2H-tetrazolium-5-carboxanilide inner salt) gives a highly soluble orange product, but is relatively expensive and needs to be freshly prepared each day.
16. The rate of NBT reduction increases with increasing NBT concentration up to approx 24 μ*M*, i.e., the concentration used in the experimental system described.
17. We have used the NBT reduction method to compare rates of substrate metabolism in *Acholeplasma* species and those *Mycoplasma* species able to oxidize sugar and organic acids to acetate plus CO_2.

18. Overlaying *M. mycoides* cell suspensions with mineral oil increased the rate of NBT reduction by up to fivefold.
19. The relationship between absorbance at 570 nm (A_{570}) and reduced NBT (formazan) concentration may be readily determined using sodium dithionite-reduced NBT. We observed an approximately linear relationship at concentrations of reduced NBT (formazan) up to 10 μM ($A_{570} = 0.7$).
20. In glucose-fermenting species, the quantity of NBT reduced may be less than that required for the oxidation of sugars to acetate plus CO_2. In *M. mycoides*, it was evident that under the experimental conditions used, a significant proportion of glucose added was metabolized to lactate.

References

1. Miles, R. J. (1992) Catabolism in mollicutes. *J. Gen. Microbiol.* **138,** 1773–1783.
2. Pollack, J. D. (1992) Carbohydrate metabolism and energy conservation in *Mycoplasmas: Molecular Biology and Pathogenesis* (Maniloff, J., McElhaney, R. N., Finch, L. R., and Baseman, J. B., eds.) American Society for Microbiology, Washington, DC, pp. 181–200.
3. Miles, R. J., Taylor, R. R., Abu-Groun, E. A. M., and Alimohammadi, A. (1994) Diversity of energy-yielding metabolism in *Mycoplasma* spp. *IOM Lett.* **3,** 165–166.
4. Miles, R. J., Taylor, R. R., and Varsani, H. (1991) Oxygen uptake and H_2O_2 production by fermentative *Mycoplasma* spp. *J. Med. Microbiol.* **34,** 219–223.
5. Abu-Groun, E. A., Taylor, R. R., Varsani, H., Wadher, B. J., Leach, R. H., and Miles, R. J. (1994). Biochemical diversity within the '*Mycoplasma mycoides* cluster.' *Microbiology* **140,** 2033–2042.
6. Taylor, R. R., Mohan, K., and Miles, R. J. (1996) Diversity of energy-yielding substrates and metabolism in avian mycoplasmas. *Vet. Microbiol.* **51**, 291–304.
7. Houshaymi, B. M., Miles, R. J., and Nicholas, R. A. J. (1997) Oxidation of glycerol differentiates African from European isolates of *Mycoplasma mycoides* subspecies *mycoides* SC (small colony). *Vet. Rec.* **140,** 182–183.
8. Estabrook, R. W. (1967) Mitochondrial respiratory control and the polarographic measurement of ADP:O ratios, in *Methods in Enzymology*, vol. 10 (Eastabrook, R. W. and Pullman, M. E., eds.) Academic, New York, pp. 41–47.
9. Truesdale, G. A., Downing, A. L., and Lowden, G. F. (1955) The solubility of oxygen in pure water and sea water. *J. Appl. Chem.* **5,** 53–62.
10. Miles, R. J., Beezer, A. E., and Lee, D. H. (1985) Kinetics of the utilisation of organic substrates by *Mycoplasma mycoides* subsp. *mycoides*: a flow microcalorimetric study. *J. Gen. Microbiol.* **131,** 1845–1852.
11. Taylor, R. R., Varsani, H., and Miles, R. J. (1994) Alternatives to arginine as energy sources for the non-fermentative *Mycoplasma gallinarum*. *FEMS Microbiol. Lett.* **115,** 163–168.

12

Serological Identification of Mycoplasmas by Growth and Metabolic Inhibition Tests

José B. Poveda and Robin Nicholas

1. Introduction

The lack of a cell wall means that the growth of mycoplasmas can be inhibited by specific antiserum and provides the basis of a simple, economic, and objective means of species identification. In the diagnostic laboratory, unknown mycoplasma isolates are inoculated onto solid media and examined for growth inhibition in the presence of antiserum raised against known mycoplasma species. Zones of inhibition of >2 mm are usually considered significant evidence that the isolate belongs to the species represented by the inhibitory antiserum. Conversely, the absence of inhibition indicates that the isolate is unrelated unless, as is common with many porcine and avian mycoplasmas, there is significant strain heterogeneity. Where there is considerable variation, other tests (i.e., immunofluorescence or metabolic inhibition) are used. Inhibition may also not be seen where the inoculum size of the unknown isolate is too large or the growth rate is too strong; here, these effects will overwhelm the potency of the antiserum, giving false-negative results. Furthermore, the possibility of the isolate being composed of a number of different mycoplasmas ("mixed" cultures) necessitates the cloning of colonies, a laborious and time-consuming occupation. The quality of the antiserum, usually produced in rabbits, which for all intents and purposes are mycoplasma-free, cannot be overstated and is covered elsewhere.

As a result of lipolytic activity, some mycoplasmas can produce a finely mat, slightly iridescent film or pellicle of crystalline appearance under magnification on media, such as Eaton's medium containing horse sera or on solid media containing egg yolk. Film producers that are routinely met in diagnostic veterinary laboratories include: *Mycoplasma bovis*, *Mycoplasma agalactiae*,

From: *Methods in Molecular Biology, Vol. 104: Mycoplasma Protocols*
Edited by: R. J. Miles and R. A. J. Nicholas © Humana Press Inc., Totowa, NJ

Mycoplasma bovigenitialium, *Mycoplasma gallinarum*, *Mycoplasma hyosynoviae*, *Mycoplasma columbinum*, and *Mycoplasma lipofaciens*. The appearance of this film can take between 2 and 3 d for *M. bovis* and 5 and 7 d for *M. bovigenitialium*. Sera from animals actively infected with *M. bovis* and other film producers often inhibit the production of film by a strain of that species without necessarily inhibiting growth. This inhibition is species-specific, not strain-specific, as are some other serological tests; it can therefore be used either for serodiagnosis or for confirmation of a presumptive identification.

The metabolic inhibition (MI) test is a highly specific serological method for identifying species of mollicutes. It is an alternative procedure to the growth inhibition test, but it can also be carried out in liquid media and is especially useful where the organisms under investigation have not grown well on agar plates ***(1)***. This procedure has been used successfully for serotyping strains of *Ureaplasma urealyticum* ***(2)*** and *Spiroplasma* species ***(3)***. The principle is simple: byproducts of the growth of mycoplasmas in liquid medium containing specific substrates alter the pH of the medium, which can be detected by color change. The inhibitory activity of specific antibodies to the mycoplasma under investigation decreases the metabolism, and so prevents the color change. This test can be used to identify isolates with known high-titered antisera or to evaluate the potency of a test serum with known mycoplasma cultures. The tetrazolium reduction test is a modification of the MI test and is based on the observation that some mycoplasmas reduce colorless 2, 3, 5-triphenyltetrazolium chloride to formazan, which is brick red. The inhibition of tetrazolium reduction results in no color change in the medium. This test is particularly useful where the mycoplasma does not hydrolyze arginine or ferment glucose. The procedure for these tests was standardized by Taylor-Robinson ***(4)*** and constitutes the basis for the protocol described here.

The expanding list of mycoplasma species is creating problems for diagnostic laboratories, which must prepare and test an increasingly large bank of specific antisera. Some selection is possible if the species of animal, the site of isolation, and nature of disease, if any, are known. The biochemical characteristics of the mycoplasma are also extremely useful, although these may not always be consistent. A further problem can occur where the mycoplasma is new to or not commonly found in a particular host. The use of 16S rRNA sequencing (*see* Chapter 18), though expensive and time-consuming, can provide an invaluable aid in identification, as seen recently with the isolation of a human mycoplasma from sheep ***(5)***.

2. Materials

1. Heart infusion broth: Dissolve 25 g of dehydrated HIB (Difco, West Molesey, UK) in 1000 mL of double-distilled water. Adjust the pH to 7.6, and autoclave. Store at 4°C.

2. Horse serum: 250 mL. Sterilize by filtration, and inactivate at 56°C for 30 min. Store at –20°C.
3. Yeast extract stock solution 10% (w/v): Dissolve 20 g of dehydrated yeast extract (Oxoid) in 200 mL of double-distilled water. Adjust pH to 7.6, sterilize by filtration, and dispense 10-mL vol into appropriate screw-cap tubes. Store at –20°C.
4. DNA solution 0.2% (w/v): Dissolve 0.1 g of deoxyribonucleic acid (Sigma, Poole, UK) in 50 mL of double-distilled water. Adjust the pH to 7.6, sterilize by filtration, and dispense aseptically 1-mL vol into sterile small aliquots. Store at –20°C.
5. Urea-MI test medium: 114 mL of HIB, 2 mL of DNA solution, 40 mL of horse serum (inactivated), 2 mL of fresh guinea pig serum, 20 mL of yeast extract solution, 20 mL of urea solution 1 % (w/v), and 2 mL of phenol red solution 0.2% (w/v). Adjust the pH to 7.0, and sterilize by filtration. Aseptically dispense 25 mL into appropriate screw-cap flasks. Store at –80°C.
6. Glucose-MI test medium: 114 mL of HIB, 2 mL of DNA solution, 40 mL of horse serum (inactivated), 2 mL of fresh guinea pig serum, 20 mL of yeast extract solution, 20 mL of glucose solution 1% (w/v), and 2 mL of phenol red solution 0.2% (w/v). Adjust the pH to 7.6, and sterilize by filtration. Aseptically dispense 25 mL into appropriate screw-cap flasks. Store at –80°C.
7. Arginine-MI test medium: 114 mL of HIB, 2 mL of DNA solution, 40 mL of horse serum (inactivated), 2 mL of fresh guinea pig serum, 20 mL of yeast extract solution, 20 mL of arginine solution 5% (w/v), and 2 mL of phenol red solution 0.2% (w/v). Adjust the pH to 7.0, and sterilize by filtration. Aseptically dispense 25 mL into appropriate screw-cap flasks. Store at –80°C.
8. Tetrazolium-MI test medium: 134 mL of HIB, 2 mL of DNA solution, 40 mL of horse serum (inactivated), 2 mL of fresh guinea pig serum, 20 mL of yeast extract solution, and 2 mL of tetrazolium solution 2% (w/v). Adjust the pH to 7.6, and sterilize by filtration. Aseptically dispense 25 mL into appropriate screw-cap flasks. Store at –80°C.
9. Liquid standard medium: 296 mL of sterile HIB, 4 mL of sterile DNA solution, 80 mL of sterile horse serum, and 20 mL of sterile yeast extract solution. Aseptically dispense 25 mL into appropriate screw-cap flasks. Store at –20°C.
10. Pure culture of the strains under investigation obtained by triple-cloning procedure.
11. Sterile microtiter plates (96 flat-bottom wells) with cover.
12. Micropipets calibrated to deliver 25, 50, and 200 µL.
13. Multichannel pipets calibrated to deliver 25, 125, 150, and 175 µL.
14. Sterile reagent reservoirs for multichannel pipets.
15. Sterile racks with pipet tips for above.
16. Vortex mixer.
17. Specific antisera (sterilized by filtration and inactivated at 56°C for 30 min).
18. Healthy rabbit serum (sterilized by filtration and inactivated at 56°C for 30 min).
19. Cork borer.
20. 70% Alcohol.
21. Sterile filter disk papers, 6-mm diameter.

3. Methods

3.1. Growth Inhibition Test

1. Triple-clone mycoplasmas under test by selecting well-separated colonies, and inoculate onto fresh agar. Filter through membrane filters (*see* **Note 1**).
2. Prepare an overnight or 48-h broth culture of test mycoplasma.
3. Inoculate at least two dilutions of broth culture (10^{-1} and 10^{-3}) onto predried agar plates by allowing 50 µL of each culture to run down a tilted plate using the "running drop" technique (*see* **Note 2**) and allow to dry. It is possible to apply two or three well-separated running drops to each 9-cm diameter plate.
4. Cut a well using a 6-mm cork borer on the running drop, and remove the agar cylinder.
5. Carefully fill the well with about 60 µL of undiluted antiserum using a pipet. Make sure the well is not overfilled. Incubate as appropriate for the organism under test (*see* **Note 3**).

3.2. Disk Film Inhibition Test

For the identification of mycoplasmas that produce film, such as *M. bovis* and *M. agalactiae*, the disk method can be applied.

1. Prepare antiserum disks by saturating with 30 µL of chosen antiserum in a sterile Petri dish.
2. Prepare a running drop culture as above, and place the impregnated disks over the culture. Incubate as above.

3.3. Reading Plates

1. After 24 h, examine plates for growth. Look for a zone of growth or film inhibition around the well by eye. The width of a zone of inhibition, as measured from the lawn of mycoplasma growth to the disk, should be >2 mm, preferably >5 mm. If a zone of inhibition is not visible by eye, examine under a plate microscope for signs of inhibition, or a gradual thinning out of the number of colonies or reduction in the size of colonies around the well. Examine daily for 7 d.
2. If no inhibition is seen with the culture with normal range of sera, try epi-immunofluorescence test (Chapter 14).

3.4. Metabolic Inhibition Test

3.4.1. Cell Adaptation to Test Media

The mycoplasmas must be able to grow in the test medium producing a satisfactory pH change or reduction of tetrazolium (*see* **Note 4**). It may be necessary to adapt the organisms to each medium by subculturing two or three times. Mycoplasmas with special nutritional requirements can be tested on media containing fresh guinea pig serum (*see* **Note 5**).

1. Mix 1 mL of each culture with 1 mL of the suitable test medium, and incubate at an appropriate temperature for 24–96 h.

2. Transfer the cultures every 24–72 h in the test medium (200 µL into 2 mL) until there is complete adaptation to the medium.
3. Inoculate a batch of test medium with the last subculture, and incubate until the logarithmic phase of growth. Divide the broth culture into 1-mL aliquots, and store at –80°C. One aliquot can be used to determine the number of color-changing units (CCU), and the others for the subsequent metabolic inhibition test.

3.4.2. Determination of Numbers of CCU

1. Prepare serial 10-fold dilutions using 10 tubes with 1.8 mL of test medium. Add 0.2 mL from a vial with broth culture to the first tube, and continue the serial dilutions until 10^{-10}. Use fresh pipet tips for each step, and mix the dilutions on a vortex mixer.
2. Add 50 µL of each dilution to the wells in the microtiter plate and mix with 150 µL of the test medium to produce a total volume of 0.2 mL. Also add 200 µL of test medium to four uninoculated wells as a control.
3. Cover the plate, and incubate at an appropriate temperature (*see* **Note 6**).
4. Read the results daily for the appearance of a red precipitate or for the change of color in the medium. The highest dilution of antigen at which color change is detected contains one CCU (CCU/50 µL).

3.4.3. Metabolic Inhibition Test

1. Prepare the antigen dilution containing 100–1000 CCU/50 µL.
2. Add 25 µL of test medium to each well of the microtiter plate using a multichannel pipet.
3. Add 25 µL of each serum to be tested to the first wells (A1-G1) and 25 µL of healthy rabbit serum to the well H1 (serum negative control).
4. Prepare twofold serial dilutions of sera with the multichannel pipet, transferring 25 µL into all 10 wells in the series.
5. Add 50 µL of the appropriate dilution of antigen with a multichannel pipet up to the 11th well in each series. (These wells serve as antigen controls.)
6. Add 125 µL of test medium with the multichannel pipet up to the 11th well in each series.
7. Add 175 µL of test medium with the multichannel pipet up to the 12th well in each series. (These wells serve as medium controls.)
8. Cover the plate, and incubate at an appropriate temperature.
9. Read the test when the media contained in the wells that serve as antigen controls have changed approx 0.5 pH units, or a red precipitate can be observed on the bottom of the wells. The titer of the serum is recorded as the highest dilution that prevents a color change in the medium.

4. Notes

1. There are several variations on the basic test. The one described here involves placing antiserum in wells cut in the agar, which has the advantage of producing

larger zones of inhibition and enables refilling; a process sometimes necessary with poorly immunogenic isolates. However, it is rather generous with the sera. A more economical approach is to apply approx 30 µL of sera to sterile filter paper disks of 6 mm in diameter in an empty Petri dish; after drying, the disks can be stored in sterile containers at 4°C for many months. An alternative technique to the "running drop" is to flood the whole plate with the culture and then removing the excess. The disks can then be applied to the dried plates allowing 2 cm^2 of surface area for each disk.

2. The growth inhibition test is the preferred serological test. However, because of strain variation, the epi-immunofluorescence test is more often used with avian and porcine mycoplasmas. Another disadvantage of the growth inhibition test is the need to adapt the cultures to growth on solid media.
3. Rapidly growing cultures can sometimes overwhelm the inhibitory effects of the antiserum. Using higher dilutions or incubating cultures at lower temperatures (e.g., 22–30°C) may be useful.
4. In cases where the test will be carried out in tetrazolium-MI test medium, the mycoplasma must be able to reduce tetrazolium aerobically. This reaction can be intensified by including 0.1% sodium thyoglycolate in the test medium. It is also important to culture the mycoplasmas under investigation in liquid standard medium without tetrazolium to prevent the appearance of precipitates than can interfere with the reading of the test.
5. Mycoplasmas with special nutritional requirements can be tested on media containing fresh guinea pig serum (1% final concentration) with the appropriate substrates: glucose (0.1%), arginine (0.5%), urea (0.1%), or tetrazolium (0.02%) incorporating phenol red (0.002%). Phenol red is not added to tetrazolium medium. This procedure can be followed successfully for: spiroplasmas in modified SP4 medium, porcine mycoplasmas in modified Friis medium, and *Mycoplasma synoviae* in modified FM4 medium.
6. Although for normal culture the plates are usually incubated in a moist atmosphere at 37°C, best results for the metabolic inhibition test are achieved when the incubation temperature is reduced to 30°C especially for mycoplasmas, such as *Mycoplasma mycoides* subsp. *mycoides* LC type or *Mycoplasma capricolum* subsp. *capricolum*. Other investigators reduce the incubation temperature to 32°C for spiroplasmas and 30°C for acholeplasmas.

References

1. Subcommittee on the Taxonomy of *Mollicutes*. (1995) Revised minimum standards for description of new species of the class *Mollicutes* (Division *Tenericutes*). *Int. J. Syst. Bacteriol.* **45,** 605–612.
2. Robertson, J. A. and Stenke, G. W. (1979) Modified metabolic inhibition test for serotyping strains of *Ureaplasma urealyticum. J. Clin. Microbiol.* **9,** 673–676.
3. Williamson, D. L., Tully, J. G., and Whitcomb, R. F. (1979) Serological relationships of spiroplasmas as shown by combined deformation and metabolic inhibition tests. Int. J. Syst. Bacteriol. 29, 345–351.

4. Taylor-Robinson, D. (1983) Metabolism inhibition tests. *Methods Mycoplasmol.* **1,** 411–417.
5. Nicholas, R. A. J., Greig, A., Baker, S., Ayling, R. D., Heldtander, M., Johansson, K. E., et al. (1998) Isolation of *Mycoplasma fermentans* from a sheep. *Vet. Rec.* **142,** 220,221.

13

Identification of Mycoplasmas by Dot Immunobinding on Membrane Filtration (MF Dot)

François Poumarat

1. Introduction

The definitive identification of mycoplasmas is usually based on serological procedures, including growth inhibition *(1,2)*, metabolic inhibition *(3,4)*, immunofluorescence *(5–7)*, and immunobinding assays *(8–14)*. The technique most commonly adopted for the routine identification of mycoplasma species isolated from clinical material is, at present, the immunobinding assay involving either broth culture *(12,14)*, mycoplasma colonies on agar plates, or the imprints of colonies *(8–13)*. Several procedures with polyclonal or monoclonal antibodies (MAb) have been described. All these assay systems are based on the detection of mycoplasma surface antigens, which are believed to be highly specific.

At present, immunobinding assays are the most reliable tests for mycoplasma identification, but specificity and sensitivity can be affected in certain circumstances. Shared antigens between some species can lead to cross-reactions, although the use of specific MAbs can greatly improve specificity. However, recently it has become apparent that many mycoplasma species are able to undergo high-frequency surface antigenic variation *(15–18)*. In practical terms, this peculiarity has two main consequences: first, many mycoplasma species may be antigenically highly heterogeneous, so that the selection of reagents, including MAb, which are simultaneously specific and representative of all antigenic variants within the species, is difficult.; second, the usual laboratory practice involving filter cloning and propagation by subcultivation of randomly selected agar-grown subpopulations may result in rapid antigenic drift of the reference strains, as has already been proven.

From: *Methods in Molecular Biology, Vol. 104: Mycoplasma Protocols*
Edited by: R. J. Miles and R. A. J. Nicholas © Humana Press Inc., Totowa, NJ

Since the number of mycoplasma species is increasing, each isolate has to be tested with several sera for complete identification and immunobinding assays involving mycoplasma colonies, or imprints of colonies, are becoming highly laborious. Immunobinding assays using polyclonal antibodies with broth culture bound onto nitrocellulose paper, on the other hand, can be affected by a high level of background staining. The technique of dot immunobinding on membrane filtration in microplates (MF dot) ***(14)*** eliminates background staining problems and offers further advantages over other tests, such as practicality, rapidity, ready standardization, and the possibility of treating many samples against several sera simultaneously. MF dot immunobinding test is described here.

MF dot immunobinding is performed with special 96-well microplates whose well bottoms are made of a durapore 0.22-μm membrane filter (low-protein affinity). These plates allow the removal of well fluids by vacuum filtration. In this way, mycoplasmas are separated from broth media by trapping them on the filtration membranes, and the broth proteins that do not bind to the membrane are easily removed. The membranes are then incubated with hyperimmune rabbit sera. The unbound immunoglobulins (IgG) are removed by filtration as above, and the bound antibodies are detected by means of an enzyme-conjuguated antirabbit IgG. MF dot immunobinding is an easy and reliable test for the identification of mycoplasmas, but in light of the high rate of surface antigenic variability occurring in many mycoplasmas, criteria must urgently be defined for the standardization of mycoplasma strains and diagnostic antisera to ensure that reproducible results can be obtained in different laboratories.

2. Materials

1. 96-Well plates (millititer GV 0.22-μm durapore, Millipore).
2. Vacuum holder (Millititer, Millipore).
3. Vacuum pump with manometer.
4. TBS: 6.057 g Tris, 11.688 g NaCl, in 1 L distilled H_2O adjusted to pH 7.4 with HCl. Store at +4°C, and use within 15 d.
5. Washing buffer (TBS-T): Freshly prepared TBS containing 0.05% of Tween 20.
6. Blocking solution (TBS-B): Freshly prepared solution of 10% normal horse serum in TBS filtered through 0.45 μm.
7. Rabbit hyperimmune sera for the various mycoplasma species: Store lyophilized for long-term use and in 10 μL–20 μL aliquots at –20°C for short-term experiments. Do not freeze and thaw each aliquot more than four times.
8. Conjugate: affinity-isolated swine antirabbit IgG conjugated to horseradish peroxidase (HRP) diluted in TBS-B to a predetermined optimum concentration.
9. Color reagent: 3,3' diaminobenzidine tetrachloride (DAB) in powder ($C_{12}H_{14}N_4$, 4 HCl). DAB is unstable and light-sensitive, store dry at –20°C, and replace frequently (harmful by inhalation and contact with skin; a possible carcinogen).

10. Enzyme substrate: H_2O_2 30%, store at –4°C in a large volume, light-sensitive.
11. Substrate solution: dissolve 25 mg DAB in 50 mL TBS and then add 50 µL H_2O_2. Freshly prepare.

3. Methods

1. Before use, wash the plates once with TBS-T and then three times with TBS without incubation. After each wash, remove the fluids by applying a vacuum, and after the last rinse, remove any drops on the plate bottom with disposable tissues.
2. Use broth mycoplasma cultures without any preparation. Pipet the mycoplasma cultures to be identified and the reference cultures in 200-µL aliquots/well (*see* **Notes 1** and **2**). Usually, 12 cultures/plate and 1/column are tested. Filter the well contents by vacuum suction, and remove any drops on the plate bottom with tissue.
3. Add blocking solution in 200-µL aliquots/well and leave for 30 min of incubation with slow agitation. Filter the well contents as in **step 2**, and remove any drops on the plate bottom.
4. Dispense rabbit hyperimmune sera, and diluted in TBS-B, in 200-µL aliquots/well (*see* **Note 3**). Cultures to be identified are usually tested againts eight sera, 1/line. After 30 min of incubation with slow agitation, remove the fluids by vacuum filtration. Wash the wells by filtration, three times with TBS-T and once with TBS. Each wash lasts 5 min. Remove any drops on the plate bottom.
5. Dispense HRP labeled antirabbit IgG diluted in TBS-B in 200-µL aliquots/well. After 30 min of incubation with slow agitation, remove the well contents by filtration. Wash the wells by filtration, three times with TBS-T and once with TBS. Each wash lasts 5 min. Remove any drops on the plate bottom.
6. Add the developing solution in 200 µL aliquots/well. A reddish coloration appearing on the membrane filter within 1 min is the sign of a positive reaction. When the reaction is complete, wash the plate with distilled water (without filtration), and examine before drying (*see* **Note 4**).

4. Notes

1. MF dot is specific, but not very sensitive. Sensitivity varies from 10^4–10^7 mycoplasmas/well depending on the hyperimmune serum used and mycoplasma species tested ***(14)***. To avoid false-negative reactions, only cultures in which growth turbidity can be visually detected should be used. A 0.1-µm durapore filtration membrane must be used for ureaplasma serotyping because of the smaller cell size of this species.
2. The blocking of well filters occasionally occurs during the first step of vacuum filtration. There are two main causes: (a) the high density of the cultures of certain fast growing mollicutes (e.g., Acholeplasma laidlawii). These cultures should be diluted before use from 1/2 to 1/10; (b) precipitate in the broth medium or cell remains from the sample. These must to be eliminated by filtration through a 0.80-µm filter before use. A clear broth medium should always be used for cul-

ture, and vacuum depression should never exceed –40 to –60 kPa for more than 2 min, or the filtration membranes may be distorted.

3. The quality of the hyperimmune sera strongly affects the results. Many preparative techniques have been proposed, and the the following procedure has proven to be reliable. Hyperimmune sera are produced in rabbits, because the absence of a natural mycoplasma flora makes the rabbit highly suitable for the production of specific mycoplasma antisera. The mycoplasma culture in late log phase is centrifuged at 10,000*g* for 45 min, washed, and sedimented three times in PBS, pH 7.4, and resuspended in PBS to obtain a final concentration close to 10^{10} mycoplasmas/mL. Aluminum hydroxide gel is used as adjuvant. Immunization is performed as follows: at d 0 and 2, 10^5 mycoplasmas are inoculated intravenously and 10^5 intraperitonealy with adjuvant; at d 4, 10^8 mycoplasmas are inoculated subcutaneously with adjuvant; at d 6 and 8, 10^{10} are inoculated intramuscularly with adjuvant at six sites. Blood samples are obtained regularly from d 15 onward to test specificity and sensitivity of the sera by MF dot. As soon as working titres of 1/1500 to 1/2500 are obtained, the rabbits are bled. Higher titers have to be avoided, since problems of background staining may occur beyond dilutions of 1/5000.
4. Interpretation: always include a reference strain as a technical control. Deterioration of the color reagents or enzyme substrate is the most common cause of failure (*see* conservation of these solutions). MF dot immunobinding, like all serological identification tests, is only qualitative. Owing to the high rate of variability of the surface antigens in many species, reaction intensity varies from strain to strain, even with equal numbers of mycoplasmas in the broths. A positive reaction toward two or more hyperimmune sera may occur with some field isolates. In most cases, this does not result from technical problems, but from crossreactions or species mixtures. The causes may be:
   a. "Classical" crossreactions between reference strains. The homologous reaction is usually stronger than the heterologous one; in this case, MAb must be used;
   b. "Occasional" crossreactions. A few field strains appear to be antigenically intermediate between different reference strains. This presents a real problem in certain mycoplasma groups, such as in the "*Mycoplasma mycoides* cluster" ***(19)***;
   c. Mixed cultures of mycoplasma species frequently occur, especially in samples from respiratory tract. Mixed cultures cannot be distinguished from crossreactions in theory. In practice, however, crossreactions and mixtures do not involve the same species, but only immunobinding on colonies obtained according to appropriate procedures ***(14)*** will allow definitive conclusions to be drawn.

References

1. Clyde, W. A. (1964) *Mycoplasma* species. Identification based upon growth inhibition by specific antiserum. *J. Immunol.* **92,** 958–965.

2. World Health Organization (1976) "The Growth Inhibition Test" Working Document. VHP/MIV/76.7. Working Group of the FAO/SHO. Programme on comparative mycoplasmology WHO, Geneva, pp. 1–11.
3. Taylor-Robinson, D., Purcell, R. H., Wong, D. G., and Chanock, R. M. (1966) A color test for the measurement of antibody to certain mycoplasma species based on the inhibition of acid production. *J. Hyg.* **64,** 91–104.
4. World Health Organization (1975) "The Metabolism Inhibition Test" Working Document VHP/MIC/75.6. Working Group of the FAO/WHO Programme on comparative mycoplasmology WHO, Geneva, pp. 1–10.
5. Del Giuidice, R. A., Robillard, N. F., and Carski, T. R. (1967) Immunofluorescence identification of mycoplasma on agar by use of incident illumination. *J. Bacteriol.* **93,** 1205–1209.
6. Gardella, R. S., Del Guidice, R. A., and Tully, J. G. (1983) Immunofluorescence, in *Methods in Mycoplasmology,* vol. I. *Mycoplasma Characterization* (Razin, S. and Tully, J. G., ed.), Academic, New York, pp. 431–439.
7. Rosendal, S. and Black, F. T. (1972). Direct and indirect immunofluorescence of unfixed and fixed mycoplasma colonies. *Acta. Pathol. Microbiol. Scand.* **80,** 615–622.
8. Bencina, D. and Bradbury, J. M. (1992) Combination of immunofluorescence and immunoperoxidase for serotyping mixtures of *Mycoplasma* species. *J. Clin. Microbiol.* **30,** 407–410.
9. Bradbury J. M. and Mac Clenaghan, M. (1982) Detection of mixed *Mycoplasma* species. *J. Clin. Microbiol.* **16,** 314–318.
10. Brown, M. B., Gionet, P., and Senior, D. F. (1990) Identification of *Mycoplasma felis* and *Mycoplasma gatae* by an immunobinding assay. *J. Clin. Microbiol.* **28,** 1870–1873.
11. Imada, Y., Nonomura, I., Hayashi, S., and Tsurubuchi, S. (1979). Immunoperoxidase technique for identification of *Mycoplasma gallisepticum* and *M. synoviae. Natl. Inst. Anim. Health Q.* **19,** 40–46.
12. Kotani, H. and MacGarrity, G. J. (1986) Identification of mycoplasma colonies by immunobinding. *J. Clin. Microbiol.* **23,** 783–785.
13. Polak-Vogelzang, A. A., Hagenaars, R., and Nagel, J. (1978) Evaluation of an indirect immunoperoxidase test for identification of *Acholeplasma* and *Mycoplasma. J. Gen. Microbiol.* **106,** 241–249.
14. Poumarat F., Perrin B., and Longchambon, D. (1991) Identification of ruminant mycoplasmas by dot immunobinding on membrane filtration (MF dot). *Vet. Microbiol.* **29,** 329–338.
15. Rosengarten, R. and Wise, K. S. (1990) Phenotypic switching in mycoplasmas: phase variation of diverse surface lipoproteins. *Science* **247,** 315–318.
16. Rosengarten, R. and Yogev, D. (1996) Variant colony surface antigenic phenotypes within mycoplama strain populations: implications for species identification and strain standardization. *J. Clin. Microbiol.* **34,** 149–158.
17. Wise, K. S., Yogev, D., and Rosengarten, R. (1992) Antigenic variation, in Mycoplasmas: *Molecular Biology and Pathogenesis* (Maniloff, J., McElhaney, R. N., Finch, L. R., and Baseman, J. B., eds.), American Society for Microbiology, Washington, DC, pp. 473–490.

18. Wise, K. S. (1993) Adaptative surface variation in mycoplasmas. *Trends Microbiol.* **1,** 59–63.
19. Poumarat, F., Longchambon, D., and Martel, J. L. (1992) Application of dot immunobinding on membrane filtration (MF dot) to the study of relationships within "*M. mycoides* cluster" and within "glucose and arginine-negative cluster" of ruminant mycoplasmas. *Vet. Microbiol.* **32,** 375–390.

14

Identification of Mycoplasmas by Immunofluorescence

Janet M. Bradbury

1. Introduction

Immunofluorescence has been used as a diagnostic tool for identification of mycoplasmas cultured on artificial medium and also for detection of the organisms *in situ* in infected hosts and for detection of contaminated cell cultures. The technique has been used in research investigations to locate mycoplasmas in pathogenicity studies in both animals and in organ cultures, and it is also one of the recommended serological tests for the characterization of new species of *Mollicutes* *(1)*. Another diagnostic application of immunofluorescence is the detection of mycoplasma antibodies *(2)*, but the topic falls outside the scope of this chapter.

Immunofluorescence offers an excellent and reliable means of identifying mycoplasmas, provided that good reagents and appropriate controls are used, and the operator has some experience in interpretation. A primary requirement is a supply of high-quality polyclonal antisera specific for each mycoplasma species of interest. The surface antigens of the organisms are generally species-specific and are the main targets for immunofluorescence *(3,4)*. However, recent work has suggested that some mycoplasmas exhibit variation in expression of their surface epitopes *(5)*. Therefore, this suggests that monoclonal antibodies (MAb) should be used with caution for species identification, unless there is good evidence that the epitope in question is permanently expressed in all strains. The complex relationships between the organisms in the so-called *Mycoplasma mycoides* cluster *(6)* can present special difficulties in identification by routine methods such as immunofluorescence, and in such cases, the use of MAb may offer an advantage *(7)*.

Immunofluorescence tests can be carried out by the direct or indirect method. For the direct method, antiglobulins prepared against each mycoplasma species

From: *Methods in Molecular Biology, Vol. 104: Mycoplasma Protocols*
Edited by: R. J. Miles and R. A. J. Nicholas © Humana Press Inc., Totowa, NJ

of interest are first conjugated with a suitable fluorochrome. The conjugates are applied, appropriately diluted, to the unidentified mycoplasma (antigen) and, after suitable incubation and washing procedures, the reaction is examined by microscope under UV illumination. A specific antigen–antibody reaction is revealed by fluorescence of the sample owing to the presence of attached antibody with fluorochrome. Fluorochromes, such as fluorescein isothiocyanate and tetramethylrhodamine isothiocyanate, which show different colors on excitation, can allow detection of two different target species in one preparation. Moreover, the test can be combined with immunoperoxidase staining for detection of up to three components ***(8)***. Detailed descriptions of the direct test ***(9)***, and of the conjugation procedure for antisera can be found in the literature ***(9,10)***.

For the indirect immunofluorescence test, antisera prepared against the mycoplasma species in question are first applied to samples of the specimen, and any bound antibodies are then detected by adding a conjugated antiglobulin to the host providing the antiserum. Although it involves an extra step, the indirect test is more convenient than the direct test for general diagnostic use because only one fluorescent conjugate is required. A suitable conjugate (e.g., goat antirabbit immunoglobin (IgG) conjugated with fluorescein isothiocyanate) can be purchased without difficulty. The indirect test for identifying mycoplasma colonies *in situ* is the method described in detail in this chapter.

Specimens from direct or indirect tests can be viewed by either incident or transmitted UV light, but the former (also called "epiillumination") tends to give brighter fluorescence with less background reaction.

When identifying clinical isolates by immunofluorescence, it is often possible to detect pathogenic mycoplasma species, even when they are mixed with faster-growing nonpathogens ***(11)***. A similar technique can be used for immunoperoxidase staining, and this obviates the need for an expensive fluorescence microscope, but it is generally easier to elucidate the components of mixed cultures using immunofluorescence. These tests offer an advantage over growth inhibition or metabolism inhibition tests that require a pure culture and that may fail to detect the pathogen if the wrong colony is selected from a mixture. Another advantage of immunofluorescence and immunoperoxidase tests is that results can be obtained in half a day instead of the several days needed for the inhibition tests.

In addition to a description of indirect immunofluorescence for identifying mycoplasma colonies, a method is given below for detecting mycoplasmas in infected tissues after cryostat sectioning. In the absence of a cryostat, a paraffin-embedding method using ethanol as a fixative can be tried ***(12)***, but its suitability for detecting the mycoplasma antigens of interest needs to be established.

The use of immunofluorescence for detection of mycoplasma contaminants in cell culture is described in Chapter 24. Useful basic information on fluorescent protein tracing techniques and preparation of appropriate reagents can be found in a number of publications ***(13,14)***.

2. Materials

1. Test samples: mycoplasma isolates are preferably colonies on mycoplasma agar ***(11,15)***, although colony impressions on microscope slides have also been used ***(16)***, as have smears of centrifuged broth cultures (*see* **Note 1**). Tissues may be examined by cryostat sectioning of small snap-frozen samples or by staining of impression smears (*see* **Note 2**).
2. Control cultures: A reference strain (*see* **Note 3**) of each mycoplasma species under test should be prepared in the same manner as the test sample. An unrelated species should be prepared as a negative control culture.
3. Antisera: High-titered polyclonal antimycoplasma sera prepared in a suitable host (*see* **Note 4**) should be aliquoted and stored at –20°C. Sera at their appropriate working dilutions in phosphate-buffered saline (PBS) can be stored at –20°C for approx 6 mo.
4. Conjugate: Fluorescein-conjugated antiglobulin to the host species providing the antiserum. Affinity-purified products can be purchased.
5. "Normal" sera: These should be sera from an uninfected host of the species providing the antiserum.
6. PBS: 8.5 g/L NaCl, 0.01*M* phosphate, pH 7.1, but made up as a 10X concentrate and diluted with distilled water as required.
7. Fluorescence microscope, preferably equipped with an epiillumination system, and fitted with suitable excitation and barrier filters for the fluorochrome in use ***(9,13)***.
8. Tube mixer (e.g., Rotator Drive STR4, Stuart Scientific Co. Ltd, Redhill, UK).
9. For making cryostat tissue sections use embedding medium (OCT compound, Lab-Tek Products, Miles Lab. Inc., Naperville, IL).
10. Freezing mixture: for example, liquid nitrogen, or dry ice and isopentane, in a Dewar flask.
11. Cryostat.
12. For washing microscope slides: slide rack, beaker, and magnetic stirrer.
13. Nonfade mountant: for samples prepared on slides and mounted under coverslips, this mountant will enhance and extend the fluorescence ***(17)***. Prepare a solution of *p*-phenylenediamine containing 100 mg in 10 mL PBS. Add this to 90 mL glycerol and adjust pH to 8.0 with carbonate-bicarbonate buffer (0.5*M*, pH 9.0). Store at –20°C in the dark.

3. Methods

3.1. Identification of Mycoplasma Colonies on Agar by the Indirect Fluorescent Antibody Test

1. The method described here is that of Rosendal and Black ***(15)*** with slight modifications. Select areas of the agar that show plentiful small, discrete colonies (*see*

Note 5), and cut into rectangular blocks (approx 10 × 5 mm) using a sterile scalpel blade. Cut the bottom right-hand corner off the block to help with subsequent orientation (*see* **Note 6**). Place the blocks colony-side-up on appropriately labeled microscope slides.

2. The species to be targeted depends on the host from which the sample was taken. Always include a colony-bearing block of a reference strain of each target species as a positive control (to be tested with the homologous antiserum). Select also an unrelated species for use as a negative control with this antiserum. Blocks bearing known positive, known negative, and the test colonies can be placed on one microscope slide so that the same antiserum is used for all. Place a further block of the test culture on a separate slide to act as a "normal" serum control.
3. Add 20–25 μL of the appropriately diluted antiserum (*see* **Note 7**) to each block of the known positive, known negative, and test colonies. Add normal serum, similarly diluted, to the other block of the test colonies. This will serve as a test for any autofluorescence of the test colonies. Incubate slides in a humid chamber for 30 min at room temperature (*see* **Note 8**).
4. During the incubation, prepare and label a 12-mL test tube and rubber stopper for each agar block. Dispense PBS to three-quarters of the capacity of each tube.
5. Gently push each agar block into its appropriate tube, and wash by rotation on a tube mixer at approx 30 revolutions/min for 10 min at room temperature. Drain the PBS from each tube into a beaker of disinfectant. The rubber stopper can be used to prevent the agar block from falling out of the tube. Refill each tube with PBS, and wash for a further 10 min (*see* **Note 9**). Drain again, and replace each block onto its original microscope slide. Allow to dry for 5 min.
6. Add 20–25 μL diluted conjugate (*see* **Note 7**) to every block including all the controls. Incubate, wash twice, and replace blocks on slides as above.
7. Examine the blocks by incident UV light (*see* **Note 10**).

3.2. Identification of Mycoplasmas in Tissues Using the Indirect Fluorescent Antibody Test

1. Prepare the chosen freezing mixture, and have ready a small vessel containing OCT compound for each tissue to be taken (*see* **Note 11**).
2. Submerge a small piece of the selected tissue in the OCT compound, avoiding the production of air bubbles.
3. Freeze the OCT-containing tissue immediately by lowering the vessel slowly into the freezing mixture. For storage, place each frozen sample in an airtight container, such as a self-sealing plastic bag, and hold at –70°C or lower.
4. Make cryostat sections of 4–5 μm by the standard technique. Enough sections should be cut from each tissue to provide replicate samples and suitable controls (*see* **Note 12**).
5. Air-dry the sections, and fix in acetone for 10 min at room temperature.
6. Cover the section with the appropriate antiserum at the predetermined dilution (*see* **Note 7**). Incubate for 30 min at room temperature in a humid chamber (*see* **Note 8**).

7. Rinse the slide in two changes of PBS for 15 min each time.
8. Add diluted conjugate (*see* **Note 7**) to cover the section, incubate, and wash as before.
9. Mount the section in nonfade mountant under a coverslip, and examine on the fluorescence microscope. Incident light illumination generally gives brighter fluorescence than transmitted light.

4. Notes

1. Colonies *in situ* on agar plates are much preferred to colony impressions or broth deposits. Impressions and deposits require fixing with methanol or gentle heat, which may destroy some of the surface antigens. Furthermore, colonies, particularly unreacting ones, are much easier to *see* if they are still intact.
2. For clinical diagnosis, it is often preferable to isolate and identify the mycoplasmas rather than to examine tissues or smears, in which it can be difficult to distinguish the tiny organisms from debris and other artifacts that may also fluoresce under UV light. Furthermore, the necessary uninfected control tissue or smear may not always be available. Possible exceptions are the detection of *M. mycoides* subsp. *mycoides* in the lungs of cattle with contagious bovine pleuropneumonia ***(18)*** and of *Mycoplasma hyopneumoniae* in lungs of pigs with enzootic pneumonia ***(19)***. Successful detection of early *Mycoplasma pneumoniae* infection in human respiratory exudates using indirect immunofluorescence has also been described ***(20)***. In experimental studies, the required control tissues or smears should be available, and the technique can be very useful for locating organisms, for example, on the mucosal surface of the respiratory tract and in organ cultures. Tissue sections are generally preferred to smears.
3. Reference strains of many mycoplasma species are available for purchase through the recognized culture collections.
4. Antisera have been prepared in hosts, such as rabbits, goats, horses, and mules. For the indirect test, the choice of host may depend on the availability of a suitable antiglobulin conjugate for that host. Methods of preparation are available in the literature ***(21)***. Antisera may also be available for purchase from recognized mycoplasma reagent collections.
5. Small discrete colonies are essential for good results, because overcrowded or overlarge colonies can give a very poor reaction. When selecting suitable areas of colonies, it is useful to indicate their location on the base of the Petri dish with a fiber-tip pen.
6. If the block is not rectangular, but square, or the corner is not cut off, there is no way of ensuring that the colonies are uppermost during the rest of the procedure, unless every block is examined under a microscope after each of the washing stages.
7. Appropriate dilutions of antiserum and conjugate must first be determined by checkerboard titration. The optimal dilutions should be determined for the system used (i.e., colonies or sections), since they will not necessarily be the same.
8. A simple humid chamber consists of an upside-down plastic sandwich box with a layer of moist paper towel or sponge placed in the lid.

9. To reduce the background fluorescence that sometimes occurs within the agar, the procedure can be interrupted at this stage and the blocks allowed to stand in the second PBS wash overnight at 4 °C (or for a minimum of 2 h if results are needed urgently).
10. If incident light illumination is not available, it is possible to examine colonies using transmitted UV light, but there may be unacceptably high background fluorescence within the agar. Improved results may be obtained by examining a thin slice off the top of the agar.
11. A suitable small cylindrical cup can be made by moulding an aluminum foil cap around the blunt end of a pencil. For smaller tissues, a piece of card can be used.
12. Suitable controls are an essential part of this technique, because it can be prone to spurious reactions, particularly when polyclonal antisera are used. Controls consist of:
    a. Uninfected (matching) tissue with antiserum and conjugate;
    b. Infected tissue with "normal" serum from same host as the antiserum and conjugate; and
    c. Infected tissue with conjugate.

References

1. Whitcomb, R. F., Tully, J. G., Bové, J. M., Bradbury, J. M., Christiansen, G., Kahane, I., Kirkpatrick, B. C., Laigret, F., Leach, R. H., Neimark, H. C., Pollack, J. D., Razin, S., Sears, B. B., and Taylor-Robinson, D. (1995) Revised minimal standards for description of new species of the class *Mollicutes* (division *Tenericutes*). *Int. J. Syst. Bacteriol.* **45,** 605–612.
2. Taylor-Robinson, D. (1996) Microimmunofluorescence, in *Molecular and Diagnostic Procedures in Mycoplasmology,* vol. II, *Diagnostic Procedures* (Tully, J. G. and Razin, S., eds.), Academic, San Diego, pp. 147–150.
3. Tully, J. G. (1983) Introductory remarks, in *Methods in Mycoplasmology,* vol. I, *Mycoplasma Characterization,* (Razin, S. and Tully, J. G., eds.), Academic, New York, pp. 399, 400.
4. Tully, J. G. (1996) Introductory remarks, in *Molecular and Diagnostic Procedures in Mycoplasmology,* Vol. II, *Diagnostic Procedures* (Tully, J. G. and Razin, S., eds.), Academic, San Diego, pp. 89–91.
5. Rosengarten, R. and Yogev, D. (1996) Variant colony surface antigenic phenotypes within mycoplasma strain populations—implications for species identification and strain standardization. *J. Clin. Microbiol.* **34,** 149–158.
6. Cottew, G. S., Breard, A., DaMassa, A. J., Ernø, H., Leach, R. H., Lefevre, P. C., Rodwell, A. W., and Smith, G. R. (1987) Taxonomy of the *Mycoplasma mycoides* cluster. *Isr. J. Med. Sci.* **23,** 632–635.
7. Belton, D., Leach, R. H., Mitchelmore, D. L., and Rurangirwa, F. R. (1994) Serological specificity of a monoclonal antibody to *Mycoplasma capricolum* strain F38, the agent of contagious caprine pleuropneumonia. *Vet. Rec.* **134,** 643–646.
8. Bencina, D. and Bradbury, J. M. (1992) Combination of immunofluorescence and immunoperoxidase techniques for serotyping mixtures of *Mycoplasma* species. *J. Clin. Microbiol.* **30,** 407–410.

9. Gardella, R. S., DelGiudice, R. A., and Tully, J. G. (1983) Immunofluorescence, in *Methods in Mycoplasmology,* vol. I, *Mycoplasma Characterization* (Razin, S. and Tully, J. G., eds.), Academic, New York, pp. 431–439.
10. Whitford, H. W., Rosenbusch, R. F., and Lauerman, L. H. (eds.) (1994) *Mycoplasmas in Animals: Laboratory Diagnosis.* Iowa State University Press, Ames, pp. 149–151.
11. DelGiudice, R. A., Robillard, N. F., and Carski, T. R. (1967) Immunofluorescence identification of mycoplasma on agar by use of incident light illumination. *J. Bacteriol.* **93,** 1205–1209.
12. Sainte-Marie, G. (1962) A paraffin embedding technique for studies employing immunofluorescence. *J. Histochem. Cytochem.* **10,** 250–256.
13. Nairn, R. C. (1976) *Fluorescent Protein Tracing,* 4th ed. Churchill Livingstone, Edinburgh.
14. Harlow, E., and Lane, D. (1988) *Antibodies: a Laboratory Manual.* Cold Spring Harbor Laboratory, Cold Spring Harbor, NY.
15. Rosendal, S. and Black, F. T. (1972) Direct and indirect immunofluorescence of unfixed and fixed mycoplasma colonies. *Acta Pathol. Microbiol. Scand.* **80,** 615–622.
16. Bradbury, J. M., Oriel, C. A., and Jordan, F. T. W. (1976) Simple method for immunofluorescent identification of mycoplasma colonies. *J. Clin. Microbiol.* **3,** 449–452.
17. Johnson, G. D. and Araujo, G. M. (1981) A simple method of reducing the fading of immunofluorescence during microscopy. *J. Immunol. Methods* 43, 349–350.
18. Trichard, C. J. V., Basson, P. A., Jacobsz, E. P., and van der Lugt, J. J. (1989) An outbreak of contagious bovine pleuropneumonia in the Owambo Mangetti area of South West Africa/Namibia: microbiological, immunofluorescent, pathological and serological findings. *Onderstepoort J. Vet. Res.* **56,** 277–284.
19. Ross, R. F. (1992). Mycoplasmal diseases, in *Diseases of Swine,* 7th ed. (Leman, A. D., Straw, B. E., Mengeling, W. L., D'Allaire, S., and Taylor, D. J., eds.), Wolfe Publishing, Ames, pp. 537–551.
20. Hirai, Y., Shiode, J., Masayoshi, T., and Kanemasa, Y. (1991) Application of an indirect immunofluorescence test for detection of *Mycoplasma pneumoniae* in respiratory exudates. *J. Clin. Microbiol.* **29,** 2007–2012.
21. Senterfit, L. B. (1983) Preparation of antigens and antisera, in *Methods in Mycoplasmology*, vol. I, *Mycoplasma characterization* (Razin, S. and Tully, J. G., eds.), Academic, New York, pp. 401–404.

15

Diagnostic Application of Monoclonal Antibody (MAb)-Based Sandwich ELISAs

Hywel J. Ball and David Finlay

1. Introduction

The advantages of microtiter-based ELISAs in diagnostic techniques can be briefly summarized by the economic use of reagents and by the ease of their application to large numbers of test samples. ELISAs are widely applied to the serological diagnosis of both human and animal bacterial and mycoplasmal disease, but similar assays for the diagnostic detection of antigen rather than antibody are not as commonly used. Although many antigen-capture or sandwich ELISAs have been developed, their application has been largely confined to research. This is mainly owing to their limited sensitivity in comparison with standard culture techniques. This means that pre-enrichment is inevitably necessary before testing to achieve an acceptable sensitivity.

The development of monoclonal antibodies (MAbs) has contributed additional advantages to ELISA methodology, by increasing the reproducibility of individual assays with the availability of an unlimited supply of standardized reagents. In addition, the use of MAbs improves test specificity. For the detection and identification of *Mycoplasma* spp., this has particular advantages, since these organisms, despite well-documented problems with species crossreactivity and interference by nonspecific media components, are largely speciated by serological methods.

The methods described in this chapter have been developed to improve the sensitivity of mycoplasma sandwich ELISAs for the detection of specific antigen for diagnostic purposes. The combination of enrichment with the ELISA capture stage, although it increases the test time, includes the necessary pre-enrichment conveniently in the same assay. The result is a diagnostic procedure, with the ELISA and MAb advantages listed above, with a sensitivity

From: *Methods in Molecular Biology, Vol. 104: Mycoplasma Protocols*
Edited by: R. J. Miles and R. A. J. Nicholas © Humana Press Inc., Totowa, NJ

comparable to culture diagnosis. The approach has been successfully applied to the routine diagnosis of *Mycoplasma bovis* ***(1)***, and its application to strains of the *Mycoplasma mycoides* cluster ***(2,3)*** and *Mycoplasma gallisepticum* (unpublished results) examined.

2. Materials

2.1. Purification of MAb

1. 0.06*M* Acetate buffer: make up a solution of 0.817 g of sodium acetate, adjust pH to 4.3 with 20% (v/v) glacial acetic acid, and make up to 100 mL.
2. Caprylic acid.
3. 0.01*M* phosphate-buffered saline (PBS): 27 mL of solution A (4.8 g NaH_2PO_4 made up to 200 mL) and 73 mL of solution B (14.2 g Na_2HPO_4 made up to 500 mL), plus 17 g NaCl; adjust pH to 7.2, and make up to 2000 mL.

2.2. Biotinylation of MAb

1. 0.1*M* Sodium bicarbonate: 8.4 g $NaHCO_3$ made up to 1000 mL.
2. Biotin: 10 mg/mL (w/v) aminohexanoyl-biotin-*N*-hydroxysuccinimide in dimethyl formamide (Cambridge Bioscience, Cambridge, UK, 004302).
3. 0.01*M* PBS, pH 7.2.

2.3. Sandwich ELISA

1. 0.05*M* Carbonate coating buffer: 10 mL of solution A (2.1 g Na_2CO_3 made up to 100 mL) and 40 mL of solution B (8.4 g $NaHCO_3$ made up to 500 mL); adjust pH to 9.5, and make up to 200 mL. Store at +4°C, and adjust pH before use, if stored for longer than 3 d.
2. 0.01*M* PBS, pH to 7. 2.
3. Dilution buffer (PTN-phosphate buffer, Tween, and NaCl): 200 mL of PBS plus an additional 4 g NaCl and 400 µL 20% (v/v) Tween 80.
4. Wash fluid: 2000 mL of PBS plus 5 mL 20% (v/v) Tween 20.
5. Streptavidin peroxidase (Sigma S5512): Reconstitute at a dilution of 1 mg/5 mL distilled water, and store in small volumes at –70°C.
6. Substrate buffer: 24 mL citric acid solution (1.92 g made up to 100 mL) and 26 mL of solution B of PBS; adjust pH to 5.0, and make up to 100 mL. Store at +4°C, and adjust pH before use, if stored for longer than 3 d.
7. Substrate (per microtiter plate): 10 mL substrate buffer and 100 µL TMB/DMS (10 mg 3,3,5,5-tetramethyl-benzidine in 1 mL dimethyl sulfoxide) and 10 µL H_2O_2. Make up fresh before use.
8. Substrate stopper: 2.5 M H_2SO_4.
9. Mycoplasma medium: Numerous medium formulas are available for the culture of mycoplasmas, the choice dependent to some extent on the species being assayed for. It is essential that antibiotics are included in the medium to limit bacterial growth. This laboratory uses ampicillin (100 µg/mL) and bacitracin (100 µg/mL) (*see* **Note 1**).

10. Test samples: Samples for mycoplasma culture are generally titrated in broth culture to dilute out any nonspecific tissue inhibitors and bacterial contamination. Samples, such as milk and joint fluid, are titrated directly; swabs are broken into mycoplasma transport medium, or broth and tissue samples are homogenized in the same (approx 10% w/v).

3. Methods

3.1. Purification of MAb

The immunoglobin (IgG) in ascites must be purified before use in a sandwich ELISA, to limit the nonspecific effects of the other contaminating serum proteins. A simple effective method of purification is by caprylic acid precipitation *(4)*.

1. Slowly add 2 vol of acetate buffer to 1 vol of ascites while stirring; adjust the pH to 4.6 with 0.1*M* NaOH if necessary.
2. Slowly, over the course of 5 min, add 25 µL caprylic acid/mL of diluted ascites while stirring rapidly.
3. Leave for 30 min, stirring more slowly.
4. Centrifuge at 10,000*g* for 15 min; discard the precipitate and ignore any floating material, which normally settles out after dialysis.
5. Dialyze overnight in 0.01*M* PBS, pH 7.2, and discard any precipitate left after centrifugation at 10,000*g* for 15 min.

3.2. Biotinylation of MAb

Biotinylation of a MAb *(5)* enables mouse MAbs to be used on both sides of a sandwich ELISA; streptavidin conjugated to an enzyme is used to link any of the biotinylated MAb binding to captured antigen, to the substrate.

1. Dialyze caprylic acid-purified IgG overnight against at least 100 volumes of 0.1*M* $NaHCO_3$.
2. Calculate the protein concentration of the dialyzed solution by spectrophotometric measurement at 280 nm, and adjust to between 5 and 10 mg/mL with 0.1*M* $NaHCO_3$.
3. Add 1 vol of biotin to 25 vol of protein solution, and place on a mixing wheel for 1 h at room temperature. If the eventual test sensitivity is poor, a series of biotin:antibody ratios between 1:5 and 1:50 should be used. Low biotinylation can result in reduced sensitivity owing to the presence of nonbiotinylated antibody, and over biotinylation may give a high level of nonspecific binding of the biotinylated antibody.
4. Dialyze the biotinylated antibody against two changes of at least 100 volumes of 0.01*M* PBS.

3.3. Optimization of the Reagents

The MAbs are stored at –20°C, and can withstand freezing and thawing several times without any loss of titer. They are, however, best stored in small volumes to reduce this risk. It is not possible to predict whether a combination of MAbs will be successful in a sandwich ELISA before experimental trial.

1. Using the sandwich ELISA protocol described below, carry out a checkerboard titration, with the purified coating MAb being diluted in one direction and the biotinylated MAb in the other. Reagents should be titrated from a 1:100 dilution for the coating MAb and a 1:500 dilution for biotinylated MAb.
2. Use an antigen that is reactive to both reagents when used as a coating in an ELISA.
3. Use streptavidin peroxidase at a dilution of 1:2000 in PTN until the optimum dilutions of the MAb capture and biotinylated reagents have been established; then titrate it further to determine its optimum dilution.
4. Essential controls to detect nonspecific activity of the ELISA reagents consist of the omission of coating MAb, test sample, and both. The missing reagent should be replaced with a corresponding volume of coating buffer or PTN.
5. Reagent dilutions that give maximum sensitivity with minimum background are selected for further use.

3.4. Sandwich ELISA Protocol

1. All ELISA reagents are used at 100 µL/well except for the final addition of 50 µL of H_2SO_4 to stop the substrate reaction. Positive and negative controls of an ELISA-specific mycoplasma strain and uninoculated mycoplasma broth, respectively, should be included in each microtiter plate used.
2. Coat the microtiter plate wells with the optimum dilution of the purified capture MAb in 0.05*M* carbonate buffer, pH 9.5; this is done either overnight at +4°C or for 1 h at 37°C (*see* **Note 2**).
3. Wash the microtiter wells with six changes of PBS.
4. Divide the microtiter plate lengthways into quarters, so that four groups of eight rows containing three wells each are created (*see* **Note 3**).
5. Add mycoplasma growth medium to the wells, add 10 µL of sample to the first well of each three well row, and titrate further to the remaining two wells of the row. The sample addition and titration can be carried out by using changes of sterile microtiter pipet tips, or using a tip resterilized by washing in boiling water in an adjacent beaker. A multitip microtiter pipet can be conveniently used for the titrations.
6. Cover the microtiter plate with a transparent microplate sealer (Greiner Labortechnik Ltd., Dursley, Gloucester), and incubate at 37°C for 1–3 d (*see* **Note 4**).
7. Wash the microtiter wells with six changes of wash buffer.
8. Add the optimum dilution of biotinylated MAb in PTN buffer, and incubate at 37°C for 1–2 h.
9. Wash the microtiter wells with six changes of wash buffer.
10. Add the optimum dilution of streptavidin-peroxidase in PTN, and incubate at 37°C for 1 h.
11. Wash the microtiter wells with six changes of wash buffer.
12. Add the substrate and incubate for 10–20 min at room temperature or at 37°C.
13. When sufficient substrate reaction has occurred, with reference to the positive and negative controls, add 50 µL 2.5*M* H_2SO_4/well to stop the reaction (*see* **Note 5**).
14. Read the reaction on an ELISA microtiter plate reader at 450 nm.

15. The result is calculated with reference to the negative control; typically, absorbency values of more than three times the average negative control are regarded as positive.

4. Notes

1. The antibiotics are an essential selective inclusion in mycoplasma media used for culture from diagnostic samples to limit contaminating bacterial growth. In particular, it is necessary to prevent the growth of bacteria, like *Staphylococcus aureus,* which can nonspecifically bind immunoglobulins and give rise to false-positive results in sandwich ELISAs.
2. It might be advisable to filter-sterilize the coating buffer before use if contamination problems are encountered during the combined capture/enrichment stage. Do not filter-sterilize the MAb dilution in coating buffer, since this will remove some of the MAb and significantly affect the dilution and capture.
3. The division of the microtiter plate for applying the sample dilutions is arbitrary. The one routinely used in this laboratory is described. It is possible that dilutions using two wells, or even single wells, are sufficient for screening large numbers in a preliminary survey. Little evidence of crosscontamination between samples in adjacent wells has been observed during the extended capture incubation stage in the protocol described here. Care is taken in avoiding the creation of aerosols during the dilution process. In a prolonged trial with the *M. bovis* ELISA, comparing the microtiter sample arrangement described in this chapter with another arrangement that left well spaces between each sample, no significant differences in results were obtained.
4. After incubation for the chosen time, selected wells, such as the lowest sample dilution showing indication of growth by color change or showing least indication of bacterial contamination, can be subcultured onto equivalent agar medium, before development of the ELISA. This step would confirm any ELISA-positive reactions, and enable culture and storage of any ELISA detected isolates. In addition, it would demonstrate the presence of any *Mycoplasma* spp. that were not detectable by the ELISA.
5. The reaction indicated by a color change is dependent on the concentration of streptavidin-peroxidase remaining in each well, which in turn is dependent ultimately on the level of specific antigen captured. This makes the necessary substrate incubation time variable, but if the MAb reagent dilutions have been optimized adequately, this time should be between 10 and 20 min. Since the negative control absorption reading is the reference point of each ELISA microtiter plate, as a general rule, the substrate incubation time should be limited to keep the negative control reading to below an absorption reading of 0.1.

References

1. Ball, H. J., Finlay, D., and Reilly, G. A. C. (1994) Sandwich ELISA detection of *Mycoplasma bovis* in pneumonic calf lungs and nasal swabs. *Vet. Rec.* **135,** 531–532.

2. Rodriguez, F., Ball, H. J., Finlay, D., Campbell, D., and Mackie, D. P. (1996) Detection of *Mycoplasma mycoides* subspecies *mycoides* by monoclonal antibody-based sandwich ELISA. *Vet. Microbiol.* **51,** 69–76
3. Ball, H. J., Finlay, D., Rodriguez, F., and Mackie, D. P. (1996) Diagnostic application of monoclonal antibody-based sandwich ELISAs by combining enrichment with the capture stage. 11th International Congress of the International Organization for Mycoplasmology. *IOM Lett.* **4,** 81.
4. McKinney, M. M. and Parkinson, A. (1987) A simple, non-chromatographic procedure to purify immunoglobulins from serum and ascites fluid. *J. Immunol. Methods* **96,** 271–278.
5. Hofmann, K., Titus, G., Montibeller, J., and Finn, F. M. (1982) Avidin binding of carboxyl-substituted biotin and analogues. *Biochemistry* **21,** 978–984.

16

Immunohistochemical Staining of Fixed Tissues

Eugenio Scanziani

1. Introduction

Immunohistochemistry is a technique in which the specific interaction between an immunoglobulin and its homologous antigen is visualized on histological sections by a microscopically detectable label. Generally, the label consists of an enzyme, such as peroxidase, alkaline phosphatase, or glucose oxidase that reacts with an appropriate substrate-chromogen solution to produce a specific color at the site of reaction. Several immunohistochemical techniques have been developed and the most important are schematically represented in **Fig. 1**. In the direct method, the primary antibody is directly labeled with the enzyme. In the indirect method, an enzyme-labeled secondary antibody is directed against the immunoglobulin type of the animal species in which the primary antibody has been raised. Both methods have a relatively low sensitivity and are therefore not frequently used.

The peroxidase–antiperoxidase (PAP) complex procedure is based on the immunological affinity of antibody and enzyme, and introduces an enzyme–antibody immune complex as a third step of the reaction *(1)*. Numerous enzyme molecules are then linked to each antigenic site, considerably increasing the sensitivity of this test. The avidin–biotin complex (ABC) procedure is based on the high affinity of the egg-white glycoprotein avidin to the vitamin biotin *(2)*. This affinity is considerably higher than that of antibody–antigen linking. Moreover, avidin has four binding sites for biotin that can be easily conjugated with immunoglobulins and enzymes.

From: *Methods in Molecular Biology, Vol. 104: Mycoplasma Protocols*
Edited by: R. J. Miles and R. A. J. Nicholas © Humana Press Inc., Totowa, NJ

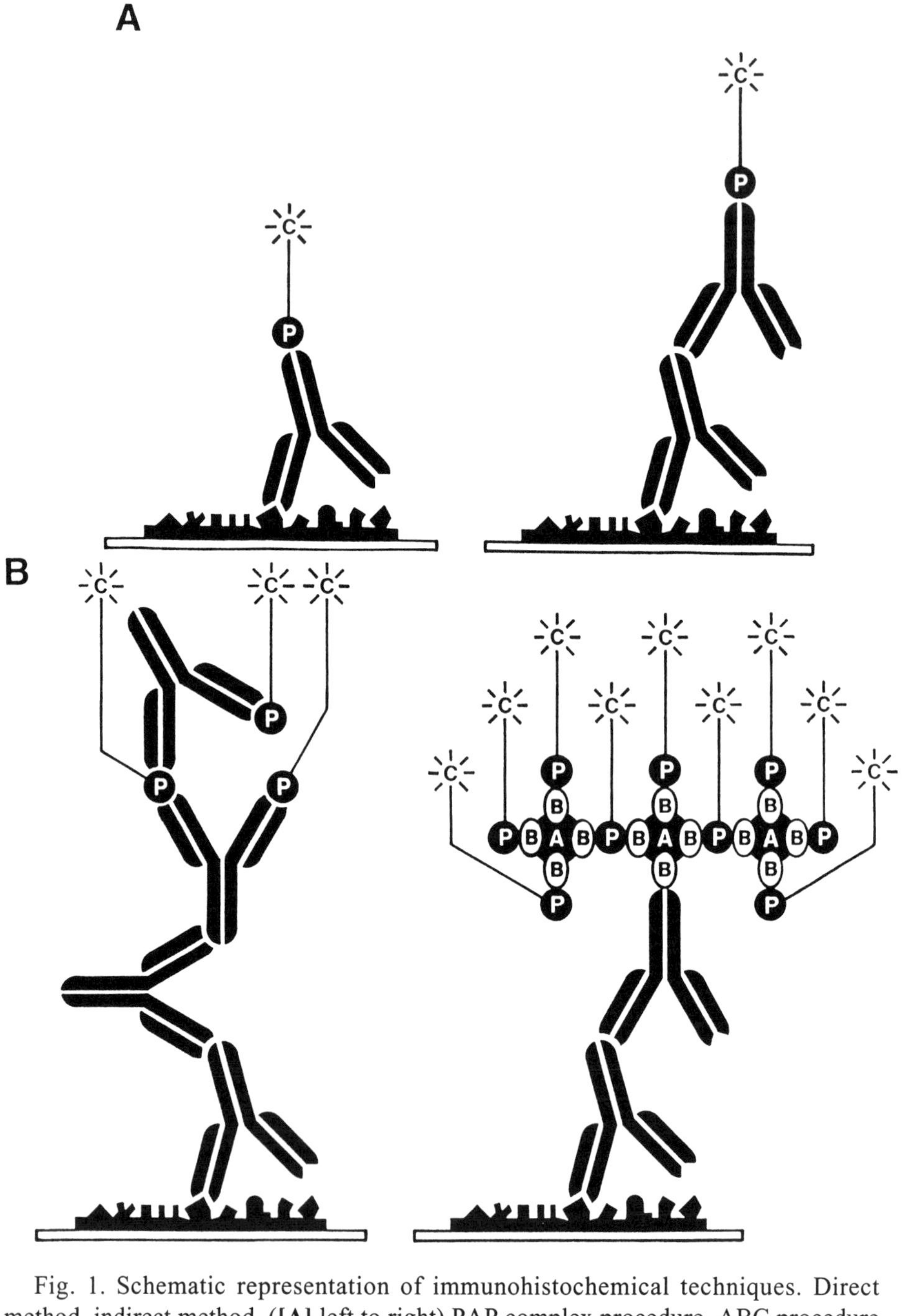

Fig. 1. Schematic representation of immunohistochemical techniques. Direct method, indirect method, (**[A]** left to right) PAP complex procedure, ABC procedure (**[B]** left to right). A = avidin; B = biotin; C = substrate-chromogen; P = peroxidase.

Fixation and embedding procedures routinely used during histological processing can alter or destroy many antigens present in the tissues. For this reason, high sensitivity is the most important feature of an immunohistochemical method to be applied to a section of formalin-fixed, paraffin-embedded tissue. PAP and ABC procedures are the methods of choice for this purpose. At the present time, owing to its sensitivity and versatility, the ABC method is the most commonly used immunohistochemical method and will therefore be described in this chapter.

Immunohistochemical techniques are widely used for the demonstration of various substances, such as immunoglobulins, leukocyte antigens, enzymes, oncodevelopmental antigens, cell proliferatin markers, cell receptors, hormones, tissue specific antigens, and microorganisms ***(3,4)***. It offers numerous advantages in the diagnosis and study of mycoplasma infections as well as other bacterial infections. It allows the simultaneous visualization of mycoplasma and its cellular/tissue localization, enabling detailed pathogenesis studies. The performance of an immunohistochemical test requires only a few days. It can detect mycoplasma antigen in the presence of degraded organisms or when endogenous or exogenous antimicrobial agents, such as antibiotics or antibodies, can inhibit in vitro mycoplasma growth. It is performed on fixed tissue that does not need to be sent to the laboratory immediately and can be handled without specific precautions. The test is not significantly affected by bacterial contamination, which is an important complication in culturing specimen for mycoplasma. Finally, it allows the preparation of permanent record enabling retrospective studies of archival material.

Recent studies have shown the reliability of immunohistochemical techniques for the specific identification of several mycoplasma species in histological sections of fixed tissues in human as well as in veterinary pathology. The occurrence of *Mycoplasma fermentans* infection in humans has been documented in several studies using immunohistochemical techniques ***(5)***. In veterinary pathology, immunohistochemistry has been used in the diagnosis and study of *Mycoplasma hyorhinis* infection in pigs ***(6)*** and in *Mycoplasma gallisepticum, Mycoplasma gallinarum,* and *Mycoplasma gallinaceum* infections in poultry ***(7,8)***.

Immunohistochemical identification of the small colony form of *Mycoplasma mycoides* subspecies. *mycoides (M. m. mycoides* SC), the cause of contagious bovine pleuropneumonia (CBPP), has been carried out in the lungs of naturally infected cattle with polyclonal antibodies ***(9,10)***. The specificity of the method was demonstrated by the absence of crossreactivity in samples in which *Mycoplasma bovis, Mycoplasma arginini,* or other bacteria *(Pasteurella multocida, Pasteurella haemolytica, Actinomyces pyogenes)* were isolated. Immunohistochemistry was found to be more sensitive than the culture of the mycoplasma *(11)*. Moreover, immunohistochemical studies demonstrated that in animals affected with CBPP the presence of *M. mycoides* SC antigen is not restricted to the lungs, but also involves the thoracic lymph nodes, the kidneys, the liver, and the muscular tissue *(12)*. The reliability of a monoclonal antibody- (MAb) based immunohistochemical technique in the specific identification of members of the *M. mycoides* cluster and *M. bovis* in lung samples from cattle and goats has been reported *(13)*. Other workers used a pool of three MAbs in an immunoperoxidase test that has been demonstrated to be reliable, sensitive, and specific in the visualization of *M. bovis* antigen in formalin-fixed, paraffin-embedded lung tissue from field cases of calf pneumonia *(14)*.

2. Materials

1. Neutral buffered formalin, pH 7.0: 100 mL formalin (37–40% formaldehyde solution), 900 mL distilled water, 4 g acid sodium phosphate monohydrate, 6.5 g anhydrous disodium phosphate. Neutral buffered formalin is stable for years at room temperature.
2. Polylisine-coated slides: Place microscopic slides in a plastic rack. Clean in 5% nitric acid for 4 h at room temperature. Wash slides under running tap water for 4 h, and rinse in 10 changes of distilled water. Soak slides in a 0.01 solution of Poly-L-lysine in distilled water for 1 h at room temperature. Rinse slides in two changes of distilled water for 5 min each, and then dry slides overnight at 37°C. Store slides in a dust-free box at room temperature. Poly-L-lysine solution can be reused if stored at –20°C. Ready-to-use polylisine-coated slides are commercially available.
3. Peroxidase inhibition reagent: Add 1 mL of 30% hydrogen peroxide to 100 mL methanol. Reagent should be prepared each time just before use.
4. 0.05*M* Tris-HCl buffer, pH 7.6: Dissolve 6.1 g Tris in 100 mL of distilled water, add 37 mL of 1*N* hydrochloric acid, and add distilled water to a total volume of 1 L. For immunostaining, 2 L of Tris-HCl buffer are required. Tris-HCl buffer should be prepared fresh each week. Store at 4°C.
5. DAB peroxidase substrate solution: Dissolve 5 mg of 3,3 diaminobenzidine tetrahydrochloride (DAB) in 10 mL of 0.05*M* Tris-HCl buffer, pH 7.6. Filter and add 4 μL of 30% hydrogen peroxide. Solution should be prepared each time just before use. **Note:** DAB is a suspected carcinogen. Handle with gloves in a fume cupboard. Avoid contamination, use precautions, and dispose of properly. To avoid aerosol during weighing, use already weighed DAB tablets.
6. Mayer's hematoxylin: Dissolve 1 g hematoxylin and 50 g aluminum potassium sulfate dodecahydrate in 500 mL of distilled water by heating. When the solution is cooled, add 50 mL of distilled water, in which 0.2 g sodium iodate has been dissolved, and 450 mL distilled water, in which 50 g chloralhydrate and 1 g citric acid have been dissolved. Mayer's hematoxylin is stable for years at room temperature. It can be reused several times until the intensity of staining decreases.
7. Primary antibodies: high-titered monospecific antiserum to mycoplasma of interest.
8. Normal serum: serum from healthy animals free of antibodies to a range of relevant mycoplasmas (same animal species in which the biotinylated antibody has been produced).

9. Biotinylated antibodies: Raised against immunoglobulin of animal used for primary antibody production (Vector Laboratories).
10. Avidin-biotin peroxidase complex (Vector Laboratories).
11. 0.01% Triton X-100 in Tris-HCl buffer (as above).

3. Methods

3.1. Preparation of Paraffin Sections

1. Fix tissue samples in neutral buffered formalin for 24–48 h at room temperature (*see* **Note 1**).
2. Process the tissue samples for paraffin embedding as for routine histology. Paraffin blocks can be stored indefinitely at room temperature.
3. Cut 5-μm thick sections with a microtome, and mount on polylisine-coated slides. Dry sections onto slides very thoroughly in a 50°C oven overnight (*see* **Note 2**). Paraffin sections can be stored indefinitely in a dust-free box at room temperature.

3.2. Staining Procedure

Include appropriate controls in each immunostaining test (*see* **Note 3**). Unless otherwise specified, all procedures are carried out at room temperature. **Steps 1**, **2**, **3**, **6**, **8**, **10**, **12**, and **13** should be performed placing the slides in a Coplin jar. **Steps 4**, **5**, **7**, **9**, and **11** should be performed laying the slides flat in a humidity chamber and applying 100–200 μL (depending on the size of the section) of reagent. To prevent mixing of the reagents do not allow slides to touch each other. Do not let the sections dry out at any point.

1. Deparaffinize sections in two changes of xylene for 5 min each. Hydrate sections through the following graded alcohol series: two changes of absolute alcohol for 5 min each, 95% alcohol for 5 min, 70% alcohol for 5 min, distilled water for 5 min.
2. Quench endogenous peroxidase by treating the sections with the inhibition peroxidase reagent for 20 min.
3. Wash slides in Tris-HCl buffer for 5 min (*see* **Note 1**).
4. Remove excess liquid from around the sections with the aid of a disposable tissue, and make a circle around the section on the slide with a diamond pencil. The circle delineates the area in which the section is located. Moreover, it creates a surface tension that keeps the reagents on the section. Apply 100–200 μL of 2% normal serum from the animal species in which the secondary biotinylated antibody has been produced, diluted in Tris-HCl buffer. Incubate for 20 min.
5. Blot the normal serum, and without washing, apply 100–200 μL of an optimal dilution of the primary antibody in Tris-HCl buffer (*see* **Note 4**). Incubate for 18 h at 4°C.
6. Wash slides three times in 0.01% Triton X-100 in Tris-HCl buffer for 3 min each.
7. Remove excess liquid from around the sections. Apply 100–200 μL of 1/200 dilution of secondary biotinylated antibody directed against the immunoglobulins of the species in which the primary antibody has been produced diluted in Tris-HCl buffer. Incubate for 30 min.
8. Wash slides for three times in 0.01% Triton X-100 in Tris-HCl buffer for 3 min each.

9. Remove excess liquid from around the sections. Apply 100–200 μL of the avidin–biotin peroxidase complex reagent. This reagent must be prepared 30 min before use by adding 10 μg avidin and 2.5 μg biotin–peroxidase to 1 mL Tris-HCl buffer. Incubate sections for 30 min.
10. Wash slides three times in Tris-HCl buffer for 3 min each.
11. Remove excess liquid from around the sections. Develop the reaction by applying to the sections 100–200 μL of the DAB peroxidase substrate solution for 1–3 min. Stop the reaction by washing in tap water for 5 min.
12. Counterstain sections with Mayer's hematoxylin for 1 min, and then wash in running tap water for 5 min.
13. Dehydrate sections through a sequence of alcohols starting at 70% alcohol for 5 min, then 95% alcohol for 5 min, and two changes of absolute alcohol for 5 min each. Clear sections by two changes of xylene for 10 min each.
14. Place a drop of mounting media over the section, and gently apply a cover slip (*see* **Note 5**).

4. Notes

1. It is important to fix the tissue as soon as possible after sampling. For an optimal fixation, the tissue samples should not exceed 7 mm in thickness and should be immersed in abundant fixative fluid with a ratio of 10 parts formalin to 1 part tissue. Prolonged fixation times may adversely influence the immunohistochemical reaction by altering or destroying the antigen under investigation. The adverse effect of prolonged fixation is particularly evident when MAbs are used as primary antibody, whereas fixation times of several days generally do not influence the immunohistochemical reactivity of polyclonal antibodies. Mild fixatives, such as B-5, Bouin's, and absolute alchol, can be used instead of formalin to prevent loss of antigenicity. A number of methods have been described to restore antigenicity altered by formalin fixation. These treatments include trypsin or other proteolytic enzymes. More recently, an antigen-retrieval technique has been developed that involves heating sections at high temperatures in a microwave oven or in an autoclave ***(15)***. Proteolytic and antigen-retrieval treatments are performed after rehydration and quenching endogenous peroxidase, but before blocking.
2. Owing to the frequent rinsing of the slides, lifting and subsequent loss of the section are a common problem in immunohistochemical staining. Adequate drying of the sections and the use of polylysine-coated slides can considerably improve the adherence of sections on the slide. Do not use adhesive containing egg albumin because of the presence of avidin.
3. It is necessary to have a negative control for each sample. This control is made on a serial section in which the primary antibody is substituted by the preimmune serum, a nonimmune serum, or another irrelevant antibody (for example, an antibody directed against another mycoplasma). Moreover, it is necessary to introduce in each immunostaining run at least one known negative section in which the mycoplasma antigen under investigation is not present and one known positive section in which the antigen is present.

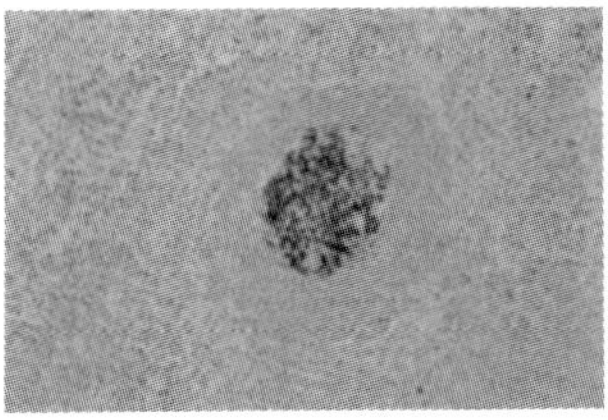

Fig. 2. Posterior mediastinal lymph node of a cow with chronic pulmonary lesions of CBPP. Diffuse positivity is present in the centrofollicular area of a follicle. Immunohistochemical staining for *M. m. mycoides* SC, hematoxylin counterstain, × 200.

4. For long storage, keep aliquots of the primary antibodies at –20°C, and avoid repeated freezing and thawing. For continuous use, a 1/10 dilution of the primary antibodies may be stored at 4°C in 0.05*M* Tris-HCl buffer containing 0.1% sodium azide. When stored in this manner, no loss of reactivity is seen for up to 6 mo. The optimal working dilution of the primary antibody should be determined by titration assay. For this purpose, apply a twofold dilution series of the primary antibody on serial sections of a known positive control. The optimum dilution is where the strongest positive reaction is present in the absence of background staining. The following are suggested dilutions of primary antibodies used in the ABC method: polyclonal antibodies 1/5000–20000, mouse MAbs (ascitic fluid) 1/1000–10000, mouse MAbs (supernatant fluid) 1/5–100.
5. The interpretation of the result is not always straightforward and should be performed by a skilled pathologist. A positive reaction is visualized by a brown color against the blue counterstained tissue (**Fig. 2**). Peroxidase chromogens other than DAB are also available. 3-Amino-9-ethylcarbazole (AEC), which gives a red color at the site of reaction, may be used instead of DAB when endogenous brown pigments, such as melanin or hemosiderin, are present in the sample. Several artifacts can mimic the positive reaction. To avoid false-positive results, check carefully the corresponding negative control section for the absence of positive staining.

References

1. Sternberger, L. A., Hardy, P. H., Cuculis, J. J., and Meyer, H. G. (1970) The unlabelled antibody-enzyme method of immunohistochemistry. Preparation and properties of soluble antigen–antibody complex (horseradish peroxidase-anti-horseradish peroxidase) and its use in identification of spirochetes. *J. Histochem. Cytochem.* **18,** 315–333.
2. Hsu, S. M., Raine, L., and Fanger, H. (1981) Use of avidin-biotin-peroxidase complex (ABC) in immunoperoxidase techniques: a comparison between ABC and unlabeled antibody (PAP) procedures. *J. Histochem. Cytochem.* **29,** 577–580.
3. Mukai, K. and Rosai, J. (1980) Applications of immunoperoxidase techniques in surgical pathology, in *Progress in Surgical Pathology,* vol. 1 (Fenoglio, C. M. and Wolff, M., eds.), Masson, New York, New York, pp. 15–49.

4. Elias, J. M. (1982) *Principles and Techniques in Diagnostic Histopathology.* Noyes Publications, Park Ridge, New Jersey.
5. Lo, S. C., Wear, D. J., Green, S. L., Jones, P. G., and Legier, J. F. (1993) Adult respiratory distress syndrome with or without systemic disease associated with infections due to *Mycoplasma fermentans. Clin. Infect. Dis.* **17 (Suppl. 1),** 259–263.
6. Morita, T., Fukuda, H., Awakura, T., Shimada, A., Umemura, T., Kazama, S., and Yagihashi, T. (1995) Demonstration of *Mycoplasma hyorhinis* as a possible primary pathogen for porcine otitis media. *Vet. Pathol.* **32,** 107–111.
7. De las Mulas, J. M., Fernandez, A., Sierra, M. A., Poveda, J. B., Carranza, J., and De las Mulas, M. (1990) Immunohistochemical demonstration of *Mycoplasma gallinarum* and *Mycoplasma gallinaceum* in naturally infected hen oviducts. *Res. Vet. Sci.* **49,** 339–345.
8. Nunoya, T., Yagihashi, T., Tajima, M., and Nagasawa, Y. (1995) Occurrence of keratoconjunctivitis apparently caused by *Mycoplasma gallisepticum* in layer chickens. *Vet. Pathol.* **32,** 11–18.
9. Ferronha, M. H., Nunes Petisca, J. L., Sousa Ferreira, H., Machado, M., Regalla, J., and Penha Goncalves, A. (1990) Detection of *Mycoplasma mycoides* subsp. *mycoides* immunoreactive sites in pulmonary tissue and sequestra of bovines with contagious pleuropneumonia, in *Contagious Bovine Pleuropneumonia* (Regalla, J., ed.), Commission of the European Communities, Luxembourg, pp. 17–25.
10. Scanziani, E., Paltrinieri, S., and Gelmetti, D. (1991) Identificazione immunoistochimica di *Mycoplasma mycoides* subsp. *mycoides.* Osservazioni preliminari. *Selezione Veterinaria* **32,** 33–40.
11. Scanziani, E., Grieco, V., Boldini, M., Giusti, A. M., and Monaci, C. (1994) Use of immunohistochemistry for diagnosis of contagious bovine pleuropneumonia (CBPP). Proceedings of the 10th International Congress of the International Organization for Mycoplasmology, July 19–26, Bordeaux, France, p. 83.
12. Scanziani, E., Grieco, V., Boldini, M., and Mandelli, G. (1994) Immunohistochemical identification of *Mycoplasma mycoides* subsp. *mycoides* in cases of contagious bovine pleuropneumonia (CBPP). Proceedings of the 12th autumn meeting of the European Society of Veterinary Pathology, September 18–22, Mondovì, Italy, p. 130.
13. Rodriguez, F., Kennedy, S., Bryson, T. D. G., Fernandez, A., and Ball, H. J. (1996) An immunohistochemical method of detecting *Mycoplasma* species antigens by use of monoclonal antibodies on paraffin sections of pneumonic bovine and caprine lungs. *J. Vet. Med.* B **43,** 429–438
14. Adegboye, D. S., Rasberry, U., Halbur, P. G., Andrews, J. J., and Rosenbusch, R. F. (1995) Monoclonal antibody-based immunohistochemical technique for the detection of *Mycoplasma bovis* in formalin-fixed, paraffin-embedded calf lung tissue. *J. Vet. Diagn. Invest.* **7,** 261–265.
15. Shi, S. R., Gu, J., Kalra, K. L., Chen, T., Cote, R. J., and Taylor, C. R. (1995) Antigen retrieval technique: a novel approach to immunohistochemistry on routinely processed tissue sections. *Cell Vision* **2,** 6–22.

17

Extraction of DNA from Mycoplasmas

John B. Bashiruddin

1. Introduction

The manipulation of genetic material for the purpose of diagnosis or analysis almost always requires the preparation of sample to expose genomic nucleic acid or the extraction and purification of DNA. Molecular techniques, such as restriction enzyme analysis, Southern hybridization, random amplified polymorphic DNA (RAPD) analysis, and nucleic acid sequencing, rely on the purity and integrity of DNA for consistent results. Generally, cells are disrupted mechanically or by detersive agents, proteins inactivated by heat denaturation or enzymatic digestion, and cellular material removed from nucleic acids by phase separation in solvents. Many other methods have been described for the preparation of DNA suitable for amplification and restriction enzyme analysis. Some include new reagents that are mixtures of solvents and result in the reduction of handling time, whereas others are special polymers that sequester or chelate materials other than nucleic acids. The method that combines the action of the detergent sodium dodecyl sulfate (SDS), the proteolytic enzyme proteinase K (PK), and the solvent phenol is widely used for DNA extraction from a variety of eukaryotic and prokaryotic sources. With the exception of pulse-field gel electrophoresis (PFGE), which requires unsheared DNA, this method provides DNA suitable for most techniques, including PCR.

A modification of the method that uses SDS, PK, and phenol, and one that uses guanidium thiocyanate, which have provided DNA suitable for analysis by most techniques, is described here for mycoplasmas *(1,2)* (*see* **Note 1**). Extraction and sample preparation methods for the handling of clinical material and bacteriological culture are described in Chapter 16.

From: *Methods in Molecular Biology, Vol. 104: Mycoplasma Protocols*
Edited by: R. J. Miles and R. A. J. Nicholas © Humana Press Inc., Totowa, NJ

2. Materials

2.1. Phenol/Chloroform Method (See Note 2)

1. Washed and pelleted cells from which the DNA will be extracted.
2. Cell resuspension buffer (TNE): 0.01*M* Tris-HCl, pH 8.0, 0.01*M* NaCl, 0.01*M* EDTA.
3. 10% SDS: (w/v) in water.
4. Sarcosine: 10% (w/v) in water.
5. PK: 20 mg/mL in water.
6. DNase-free RNase: 10 mg/mL.
7. Phenol saturated with 0.2*M* Tris-HCl, pH 7.2.
8. Phenol:chloroform:isoamyl alcohol.
9. Sodium acetate: 3.0*M* in water.
10. 100% Ethanol at –20°C.
11. 80% Ethanol at –20°C.
12. TE buffer: 0.01*M* Tris-HCl, pH 8.0, 0.01*M* EDTA.

2.2. Guanidium Thiocyanate Method

1. Washed and pelleted cells from which the DNA will be extracted.
2. GES buffer: 5*M* guanidium thiocyanate, 0.1*M* EDTA, 0.5% sarcosyl.
3. Ammonium acetate: 7.5*M*, pH 7.7 in water.
4. Phenol saturated with 0.01*M* Tris-HCl, pH 7.2.
5. Phenol:chloroform:isoamyl alcohol.
6. 100% 2-Propanol.
7. 80% Ethanol.
8. TE buffer: 0.01*M* Tris-HCl, pH 8.0, 0.01*M* EDTA.

3. Method

3.1. Phenol/Chloroform Method

1. Collect the cells from 25 mL of broth culture by centrifugation at 10,000*g* for 30 min at 4°C. Wash them once in TE at 4°C, and use the pellet immediately or store at –80°C.
2. Resuspend the cell pellet in 0.5 mL of TNE (*see* **Note 3**).
3. Lyse the cells by the addition of 10 µL of 10% SDS and 10 µL of 10% sarcosine.
4. Add 10 µL of 20 mg/mL PK and incubate at 37°C for 2 h.
5. Add DNase-free RNase to 100 µg/mL, and incubate at 37°C for 30 min.
6. Extract the lysate once with 0.5 mL of phenol.
7. Centrifuge the emulsion at 13,000*g* for 10 min, and transfer the aqueous upper phase to another tube.
8. Add 0.5 mL of phenol:chloroform:isoamyl alcohol, extract again, and repeat the extraction with phenol:chloroform:isoamyl alcohol and remove 0.4 mL of the aqeous phase into a new tube.
9. Add 40 µL of sodium acetate and 0.8 mL of 100% ethanol, mix gently, and allow the DNA to precipitate at –20°C for 16 h.

10. Pellet the DNA by centrifugation at 13,000g for 10 min, and discard the supernatant.
11. Add 0.5 mL of 80% ethanol, mix, and pellet the washed DNA by centrifugation at 13,000*g* for 10 min.
12. Discard the supernatant, and dry the DNA in air. Resuspend the DNA in water or TE for immediate use, or keep the DNA pellet at –20°C for long-term storage.

3.2. Guanidium Thiocyanate Method

1. Collect the cells from 25 mL of broth culture by centrifugation at 10,000*g* for 30 min at 4°C. Wash them once in TE at 4°C, and use the pellet immediately or store at –80°C.
2. Resuspend the cell pellet in 2.5 mL of TE.
3. To 100-μL aliquots, add 500 μL of GES buffer, and hold at room temperature for 10 min.
4. Place tubes on ice, and add 250 μL of 7.5*M* ammonium acetate, pH 7.7.
5. Extract the lysate three times with 0.5 mL of phenol:chloroform:isoamyl alcohol.
6. Centrifuge the emulsion at 13,000*g* for 10 min, transfer the aqueous upper phase to another tube, and proceed with the next extraction.
7. To the final aqeous phase, add 600 μL of 2-propanol, and pellet the DNA by centrifugation at 13,000*g* for 15 min.
8. Discard the supernatant, and wash the DNA pellet three times with 80% ethanol by centrifugation at 13,000*g* for 10 min.
9. Discard the supernatant, and dry the DNA in air. Resuspend the DNA in water or TE for immediate use, or keep the DNA pellet at –20°C for long-term storage.

4. Notes

1. These procedures are two of the many that may provide genomic DNA from mycoplasmas. For further general methods for the extraction of DNA from bacterial sources, *see* **refs. *3–5***.
2. Use molecular biology-grade chemicals and solvents, including water, for the preparation of all solutions.
3. DNA from mycoplasmas, in particular *M. m. mycoides,* is delicate and susceptible to rapid degradation by nucleases. Care must be taken to keep the pellets, suspensions, and lysates cold by placing them on ice. This becomes increasingly important when delays are expected in the processing when several suspensions must be treated at the same time.
4. Aliquot all solutions and enzymes into convenient lots. This ensures the freshness of reagents and allows quick refreshment of reagents in case of a failure.
5. Use DNA-grade recrystallized phenol and DNA-grade solvents. Be aware of the health hazards from phenol and guanidium thiocyanate.
6. With the methods described, extractions may be performed in 1.5-mL Eppendorf tubes, but they may be adapted to larger volumes if necessary.
7. The incubation time during the lysis of the cells in **Subheading 3.1.** may need to be optimized; for more delicate mycoplasmas, reduce the time at 37°C to 1 h.

8. The purity of the DNA may be assessed by $OD_{260/280}$ measurements, and the concentration of the resuspended DNA may be quantified and adjusted based on these readings. Agarose electrophoresis of extracted DNA may be used to verify the integrity of DNA.

References

1. Taylor, T. K., Bashiruddin, J. B., and Gould, A. R. (1992) Application of a diagnostic DNA probe for the differentiation of the two types of *Mycoplasma mycoides* subspecies mycoides. *Res. Vet. Sci.* **53,** 154–159.
2. Cheng, X., Nicolet, J., Poumarat, F., Regalla, J., Thiacourt, F., and Frey, J. (1995) Insertion element IS*1296* in *Mycoplasma mycoides* subspecies *mycoides* small colony identifies a European clonal line distinct from African and Australian strains. *Microbiology* **53,** 154–159.
3. Delidow, B. C., Lynch, J. P., Peluso, J. J., and White, B. W. (1993) Polymerase chain reaction: Basic protocols, in *PCR Protocols: Current Methods and Applications* (White, B., ed.), Humana Press, Totowa, NJ, pp. 1–29.
4. Graves, L. M. and Swaminathan, B. (1993) Universal bacterial DNA isolation procedure, in *Diagnostic Molecular Microbiology: Principles and Applications* Persing, D. H., Smith, T. F., Tenover, F. C., and White, T. J., eds.), ASM, Washington, DC, pp. 617–621.
5. Rolfs A., Schuller, I., Finckh, U., and Weber-Rolfs, I. (1992) Isolation of DNA from cells and tissue for PCR, in *PCR: Clinical Diagnosis and Research.* (Rolfs, A., Schuller, I., Finckh, U., and Weber-Rolfs, I., eds.) Springer-Verlag, Berlin, pp. 79–89.

18

Characterization of Mycoplasmas by PCR and Sequence Analysis with Universal 16S rDNA Primers

Karl-Erik Johansson, Malin U. K. Heldtander, and Bertil Pettersson

1. Introduction

Ribosomes are present in all self-replicating cells and constitute their protein-synthesizing machinery. The ribosomes are composed of ribosomal proteins and ribosomal RNA (rRNA). Bacteria have three kinds of rRNA (5S, 16S, and 23S rRNA), and the genetic information of these molecules is organized in the genome in the form of rRNA operons. The nucleotide sequences of the rRNA molecules contain well-defined segments of different evolutionary variability, which in the 16S rRNA molecule are referred to as universal (U), semiconserved (S), and variable (V) regions *(1)*. The universal regions are numbered U1–U8 from the 5'-terminus. A more refined model for the nucleotide substitution rates in bacterial rRNA was recently presented, and it was shown that the nucleotide substitution rates within one of the above regions can vary substantially *(2)*. Ribosomal RNA has the same important function in the cell, irrespective of species, which means that the corresponding genes are under approximately the same evolutionary pressure. These properties together make sequence analysis of rRNA extremely suitable for phylogenetic *(3)* and evolutionary *(4)* studies.

A typical bacterial rRNA operon has the following organization:

5' — 16S rRNA — spacer region — 23S rRNA — 5S rRNA — trailer region — 3' (1)

However, there are many exceptions to this rule, and it has been shown that *Mycoplasma hyopneumoniae* and *Mycoplasma gallisepticum* have an unusual organization of their rRNA genes *(5,6)*. Among mycoplasmas, it is only in the genus *Acholeplasma* where tRNA genes have been found in the spacer region

From: *Methods in Molecular Biology, Vol. 104: Mycoplasma Protocols*
Edited by: R. J. Miles and R. A. J. Nicholas © Humana Press Inc., Totowa, NJ

(7). It is the 1500-nucleotide 16S rRNA molecule particularly that has been used for phylogenetic studies, although the 23S rRNA contains more sequence information with its 3000 nucleotides. Several thousands of complete bacterial 16S rRNA sequences have been deposited in data bases, and many mycoplasma sequences are now available in GenBank or in the European Molecular Biology Laboratory (EMBL) data banks for nucleotide sequences. There are also data banks dedicated for rRNA sequences only ***(8,9)***. The most extensive molecular phylogeny of mycoplasmas was based on 16S rRNA sequence data ***(10)***. It was shown that the mycoplasmas can be arranged in five phylogenetic groups (the hominis, the pneumoniae, the spiroplasma, the anaeroplasma, and the asteroleplasma group). These phylogenetic groups have been further subdivided into 16 clusters, which are named after representative strains, species, or genera belonging to the cluster ***(10–12)***. The taxonomy of the mycoplasmas was recently revised on the basis of 16S rRNA sequence data and also on some other data ***(13)***.

The first 16S rRNA sequence from a mycoplasma to be determined originated from *Mycoplasma capricolum* subsp. *capricolum* ***(14)***. That sequence was determined by cloning the 16S rRNA gene and chemical sequencing of DNA by the Maxam-Gilbert procedure. The second 16S rRNA sequence from a mycoplasma originated from *Mycoplasma* sp. bovine group 7 ***(15)***, and it was determined by cloning the 16S rRNA gene and dideoxynucleotide sequencing of DNA by the Sanger method. Later, many 16S rRNA sequences of mycoplasmas were determined by direct rRNA sequencing with reverse transcriptase ***(10)***, but it is difficult to generate complete and correct sequence data with this method. Solid-phase DNA sequencing has proven to be a very useful method for sequencing both DNA strands ***(16)***, and this method can easily be automated and used for sequencing of PCR products ***(17)***. Automated solid-phase DNA sequencing of PCR products from 16S rRNA genes has been introduced for mycoplasmas, because the procedure is rapid and can be used to generate very accurate sequence data ***(11,12,18–21)***. Another great advantage of direct sequencing of PCR products is that the cloning step is avoided. This means that randomly misincorporated deoxynucleotides, owing to the error frequency of the *Taq* DNA polymerase, will not affect the sequence data.

PCR primers can be designed to be complementary to the universal regions of the 16S rRNA genes of mycoplasmas, and such primers can be used for amplification of the 16S rRNA genes of most species. The PCR products can then be used for direct solid-phase DNA sequencing with sequencing primers complementary to universal regions. The sequence data can be used for similarity searches and for construction of phylogenetic trees. The similarity search will show if the mycoplasma has been sequenced before. If not, a tree can always be constructed and used to determine the phylogenetic cluster to which the new species belongs. Information about its closest relatives will also be obtained. About 100 16S rRNA mycoplasma sequences have so far been deposited in the data banks, which means that the phylogenetic tree can be very informative, if the right species are selected. Sequencing can therefore be used to classify unknown isolates, and the 16S rRNA genes from the majority of the mycoplasmas described so far will, hopefully, be sequenced in the near future. Sequencing of the 16S rRNA genes will then be an extremely powerful tool in the classification of mycoplasmas.

We have developed a set of universal PCR primers that can amplify the 16S rRNA genes of most mycoplasmas ***(11,12,18–21)***. The system is based on seminested PCR with the first primer pair complementary to the universal regions U1 and U8. More than 95% of the gene is amplified with these primers, and a PCR product of about 1500 bp is obtained. The amplicon is then diluted and amplified again in two independent seminested PCR experiments with one primer pair complementary to the universal regions U1 and U5 and another primer pair complementary to the universal regions U2 and U8. These reactions will generate PCR products of about 900 and 1250 bp, respectively, with an overlapping region of about 650 bp. The amplicons can then be sequenced by automated solid-phase DNA sequencing with a set of eight universal sequencing primers ***(11,12,18–21)***. The primers were primarily designed for mycoplasmas, but can also be used to amplify and sequence the 16S rRNA genes from many bacteria related to mycoplasmas. The strategies for design of PCR and sequencing primers for 16S rRNA genes were recently discussed in detail; a combination of primers that can be used for analysis of the 16S rRNA genes of most (eu)bacterial taxa was included ***(20)***.

2. Materials

2.1. Cultivation of Mycoplasmas

Cultivation of mycoplasmas is treated in Chapters 3–5 of this volume. The procedures and materials required for growing mycoplasmas will, therefore, only be discussed briefly under **Subheading 1.** Phosphate-buffered saline (PBS) is used to wash the cells after cultivation.

2.2. In Vitro Amplification of the 16S rRNA Gene by Seminested PCR

Taq DNA polymerase from different commercial companies can be used. In the procedure described below, we have used Amplitaq (Perkin-Elmer, Cetus, Norwalk, CT), and either a thermocycler (model 480 or 9600) from Perkin Elmer or the model Progene with a heated lid from Techne Inc. (Princeton, NJ). It may be necessary to reoptimize the procedure if materials or equipment from other companies are used.

Table 1
Universal PCR Primers for In Vitro Amplification of the 16S rRNA Genes from Mycoplasmas by Seminested PCR *(19)*

Designation	Target region, direction[b]	Nucleotide position[c]	Sequence[d]
593	U1 (F)	10–34	5'-GTTTGATCCT GGCTCAGGAY DAACG-3'
620–B[a]	U8 (R)	1502–1524	5'-RSP[e]-GAAAGGAGGT RWTCCAYCCS CAC-3'
388	U2 (F)	327–348	5'-USP[e]-CCARACTCCT ACGGRAGGCA GC-3'
390-B[a]	U5 (R)	902–924	5'-CTTGTGCGGG YYCCCGTCAA TTC-3'

[a]The reverse primers are biotinylated (B) in the 5'-terminus for solid-phase sequencing with magnetic beads.

[b]Forward (F) or reverse (R).

[c]In the consensus sequence of the 16S rRNA genes from the *rrnB* operons of the members of the *Mycoplasma mycoides* cluster shown in **Fig. 1**. *See* also **ref. *19***.

[d]Degenerated positions are indicated with the corresponding ambiguity code according to the International Union of Biochemistry (IUB).

[e]USP and RSP are universal sequencing handles, forward and reverse, respectively. The corresponding sequencing primers are provided with the AutoRead Sequencing Kit. Usually, we only use the RSP sequencing primer (*see* **Table 2**).

1. DNA template from the mycoplasma to be analyzed. The template can be the DNA in a washed pellet of organisms or DNA prepared by, for instance, phenol extraction according to standard procedures.
2. *Taq* DNA polymerase, 10X PCR buffer, and 25 m*M* $MgCl_2$. The buffer and the $MgCl_2$ are often supplied with the enzyme.
3. Two forward and two reverse PCR primers, which can be combined into three primer pairs (*see* **Table 1**). One of the primers in each pair should be biotinylated.
4. A mixture of the four deoxynucleotides dATP, dCTP, dGTP, and dTTP.
5. Mineral oil, if a thermocycler without a heated lid is used.
6. Microtubes for the PCR reactions.
7. Pipets and tips (0.5–10 µL, 10–100 µL, and 100–1000 µL).
8. Microcentrifuge.
9. Thermocycler.
10. Agarose.
11. 10X TBE-a (electrophoresis) buffer: 0.9*M* Tris, 0.9*M* boric acid, and 26 m*M* EDTA.
12. A stock solution of ethidium bromide (10 mg/mL). Keep the solution protected from light. Ethidium bromide is a strong mutagen and should be handled with great care.
13. Molecular-size marker, for instance *Bgl*I-cleaved pBR 328 DNA and *Hin*fI-cleaved pBR 328 DNA (Boehringer Mannheim, Germany).
14. Gel-loading solution: 30% glycerol and 0.25% bromophenol blue.

Table 2
Universal Sequencing Primers for Analysis of PCR Products of the 16S rRNA Genes from Mycoplasmas *(19)*

Designation	Target region, direction[a]	Nucleotide position[b]	Sequence[c]
583	U1 (F)	12–27	5'-TTGATCCTGG CTCAGG-3'
584	U2 (F)	327–348	5'-CCARACTCCT ACGGRAGGC-3'
631	U3 (R)	512–527	5'-ATTACCGCGG CKGCTG-3'
390	U5 (R)	902–924	5'-CTTGTGCGGG YYCCCGTCAA TTC-3'
538	U4 (F)	792–810	5'-GTAGTCCACG CCGTAAACG-3'
597	U6 (F)	1154–1172	5'-GAGGAAGGYG RGGAYGAYG-3'
585	U7 (R)	1359–1374	5'-ACAAGRCCCG AGAACG-3'
RSP	U8 (R)	—[d]	5'-CACAGGAAAC AGCTATGACC-3'

[a]Forward (F) or reverse (R).

[b]In the consensus sequence of the 16S rRNA genes from the *rrnB* operons of the members of the *M. mycoides* cluster shown in **Fig. 1**. *See* also **ref. *19***.

[c]The sequencing primers are labeled with Cy5 in the 5'-terminus for sequencing with the ALFexpress system. Degenerated positions are indicated with the corresponding ambiguity code according to the IUB.

[d]Sequencing handle, which is used to analyze amplicons generated with the primer 620–B (*see* **Table 1**) as reverse primer.

15. Apparatus for agarose gel electrophoresis (submarine type) of small DNA fragments and a suitable power supply.
16. Equipment for documentation of the gels (a UV table and either a polaroid camera or a solid-state camera).

2.3. Sequence Determination of the PCR Products by the Direct Solid-Phase Method

Different sequencing systems can be used. The automated solid-phase DNA sequencing system described below is based on dideoxynucleotide sequencing according to Sanger ***(21a)***. The ALFexpress™ Automated Laser Fluorescent DNA sequencer from Amersham Pharmacia Biotech (Uppsala, Sweden) was used here to separate the DNA fragments after the sequencing reactions.

1. Biotinylated PCR products.
2. Sequencing primers labeled with the Cy5 indodicarbocyanine phosphoramidite dye (Cy5) (Amersham Pharmacia Biotech) at the 5'-terminus (**Table 2**). Store the stock solutions of Cy5-labeled primers at –20°C, and keep them protected from light. The diluted primer solutions can be stored at +4°C for several months. The Cy5-labeled primers are to be used in concert with the ALFexpress sequencing system.

3. Streptavidin-coated super paramagnetic beads, Dynabeads M280 (Dynal AS, Oslo, Norway).
4. 0.1*M* NaOH (prepare fresh and keep frozen).
5. 0.2*M* HCl.
6. 2X Binding/washing buffer:10 m*M* Tris-HCl buffer, pH 7.5, 1 m*M* EDTA, 2.0 m*M* NaCl.
7. 1.0*M* Tris-HCl, pH 7.4.
8. TE buffer: 10 m*M* Tris-HCl buffer, pH 7.5, 1 m*M* EDTA.
9. A magnetic rack for microcentrifuge tubes.
10. Pipets and tips (as for the PCR experiments).
11. A suitable sequencing kit (Cy5 AutoRead™ Sequencing Kit, Amersham Pharmacia Biotech).
12. The ALFexpress sequencing system with the standard plate kit, a computer, and software (ALFMANAGER for OS/2 or ALFWIN for Windows 95), or another suitable sequencing system.
13. 10X TBE-b (electrophoresis) buffer:1.0*M* Tris, 0.8*M* boric acid, and 10 m*M* EDTA.
14. ReadyMix Gel, ALF-grade (Amersham Pharmacia Biotech).

2.4. Construction of PhylogeneticTrees

Different software are available for computers with various operating systems, and it should be possible to choose a system that is compatible with the overall computer strategy of the department.

1. Computer software for merging sequences into contigs: PC/Gene (Oxford Molecular, Oxford, UK) for MS-DOS may be used for this purpose. Genetics Data Environment (GDE), which is written for Unix, can be used for many different purposes ***(22)***. GDE contains useful tools for handling of sequence data, and can be combined with programes for similarity searches and programs for construction of phylogenetic trees by different algorithms. GDE is a freeware, which is very convenient to use in the different steps in handling of sequence data.
2. Access to Internet with Netscape or another World Wide Web (WWW) reading program is very useful.
3. Access to GenBank (National Center for Biotechnology Information, Bethesda, MD) on line or on CD-ROM and software for handling of the nucleotide sequences from GenBank is necessary. The program package from Genetics Computer Group (GCG) (Madison, WI), which is available for Unix and VMS operating systems, is useful ***(23)***. Access to the data banks for rRNA sequences via Internet is also valuable, because these sequences can be downloaded from the data banks in a prealigned format. The URL address to the home page of the American data bank for rRNA sequences, Ribosomal Database Project (RDP) ***(8)*** from the University of Illinois at Urbana-Champaign is: http://rdp.life.uiuc.edu/. All sequences from RDP can be downloaded to the hard disk of your Unix computer by your system manager and used in concert with GDE. The URL address to the European data bank for rRNA sequences from the University of Antwerp

(9) is: http://rrna.uia.ac.be/. This data bank will be referred to as the rRNA WWW Server below.

4. Software for manual alignment of the new sequences with the prealigned sequences, for instance, GDE.
5. Softwares for discrete character-based and distance matrix-based methods with bootstrap options for computation of phylogenetic trees: The following program packages can be recommended, because they suit most phylogenetic purposes in routine analysis of sequence data. The first is the phylogenetic program package PHYLIP (Phylogenetic Inference Package), which is a freeware program and can be run under Unix, MS-DOS, or on the Macintosh computer *(24)*. PHYLIP can be integrated in GDE. The second is PAUP (Phylogenetic Analysis Using Parsimony), which is another program package that has become popular for Macintosh computers, but versions for MS-DOS and Unix also exist *(25)*. The third is MEGA (Molecular Evolutionary Genetics Analysis), which can be run under MS-DOS or Windows 95 *(26)*. PHYLIP, PAUP, and MEGA have the useful functionalities and contain programs for construction of phylogenetic trees by different algorithms. The three most common methods are maximum parsimony, maximum likelihood, and neighbor-joining. For a review, *see* refs. *(27,28)*. There are also other programs available, and interested readers can visit the home page of Joseph Felsenstein to obtain information about PHYLIP and an almost complete list of the different other programs with short descriptions. The URL address to this home page is: http://evolution.genetics.washington.edu. Some of the freewares are available from ftp servers or by WWW.

3. Methods

3.1. Cultivation of Mycoplasmas

Cultivation of mycoplasmas is treated in chapters 3–5 and therefore, only some comments relevant for PCR analysis are given here. In principle, only a very small volume (100 µL or less) of mycoplasmas is sufficient, but for practical reasons, it is better to prepare larger volumes.

1. Grow the mycoplasmas in, for instance, 10 mL of the appropriate growth medium.
2. Distribute 1 mL of the outgrown suspension culture in microcentrifuge tubes, and centrifuge at about 12,000*g* for 20 min.
3. Wash the pellets three times in 500 µL of PBS by repeated centrifugations and resuspensions. The pellets can be stored frozen for several years at –70°C.
4. Suspend the pellets in 100 µL of H_2O and heat them in a boiling water bath for 10 min. Use the suspension as DNA template in the PCR reactions (*see* **Note 1**).

3.2. In Vitro Amplification of the 16S rRNA Gene with Universal Primers

The 16S rRNA genes are amplified by seminested PCR with four primers complementary to the universal regions U1, U2, U5, and U8, as defined by Gray et al. *(1)*. Two of the primers (U5 and U8) are biotinylated to make the

PCR products suitable for solid-phase DNA sequencing. The sequences and directions of the primers are given in **Table 1** and in **Fig. 1**. The contamination problem often met in PCR is less serious when relatively large amounts of DNA are amplified (*see* **Note 2**).

1. Prepare a suitable volume of the following mastermix, and distribute 45 μL into the number of reaction tubes required:

H_2O	28.85 μL
PCR buffer (10X)	5.0 μL
$MgCl_2$ (2.5 m*M*)	5.0 μL
dNTP mix (2.5 m*M* of each)	4.0 μL
Primer 593 (10 pmol/μL)	1.0 μL
Primer 620B (10 pmol/μL)	1.0 μL
Taq DNA polymerase (5 U/μL)	0.15 μL

   Keep the tubes on ice until the reaction is to be started.
2. Add 5 μL of lysed mycoplasma suspension (diluted 1:10) to the reaction tubes. Overlay with two drops of mineral oil if a thermocycler without heated lid is used.
3. Place the tubes in the thermocycler, and run the reactions for 30 cycles. Each cycle should consist of the following steps: 96°C for 15 s and 70°C for 2 min. The primers have approximately the same annealing temperature as the temperature optimum for the *Taq* DNA polymerase and a combined annealing-elongation step at 70°C can, therefore, be used.
4. Mix 5 μL of the PCR product with 2 μL of the gel-loading solution. Apply the samples on a 1.5% agarose gel prepared in 1X TBE-a buffer, with an appropriate DNA size marker, and perform the separation for about 30 min at a field strength of 5 V/cm. Stain the gel in ethidium bromide (1 μg/mL) for 20 min, or include ethidium bromide in the gel (0.1 μg/mL) during preparation of the gel. Visualize the PCR products by illumination with UV light. The size of the amplicons should be about 1500 bp (*see* **Fig. 2** and **Note 3**).
5. Dilute the PCR products 1:10, and use the diluted amplicon as template in two new PCR reactions based on the same mastermix as above, except for the choice of primers. In one of the reactions, the primers 593 and 390-B should be used, and in the other reaction, the primers 388 and 620-B should be used. The temperature in the combined annealing-elongation step should now be 68°C.
6. Analyze the PCR products as described above. The sizes of the amplicons generated in these reactions are 900 and 1250 bp, respectively (**Fig. 2**).

3.3. Sequence Determination of Biotinylated PCR Products

It is only possible to determine about 96% of the sequence of the 16S rRNA gene by the procedure described here, since the sequences of the target regions and their flanking segments cannot be analyzed. Eight different sequencing primers have to be used to determine the complete sequence of the PCR products from the 16S rRNA gene in both directions (*see* **Note 4**). Four primers, two forward and two reverse, are used for each amplicon. The sequences and

```
10                                       50                                               100
        Seq 583 →
AAATGAGAGTTTGATCCTGGCTCAGGATAAACGCTGGCGGCATGCCTAATACATGCAAGTYGAACGGRGGTGCTTGCACCTYAGTGGCGAACRGGTGAG
              PCR 593 →
110                                      150                                              200
AACACGTATCTAACCTACCTYATAGCGGGGGATAACTTTTGGAAACGAAAGATAATACCGCATGTAGATCTYATTATCRCATGAGAAAAGATCAAAAGA
210                                      250                                              300
CCGTTTGGTTCACTATGAGATGGGGATGCGGCGTATTAGCTAGTAGGTGAGATAATAGCCCACCTAGGCGATGATACGTAGCCGAACTGAGAGGTTGAT
310                                      350                                              400
                              Seq 584 →
GGCCACATTGGGACTGAGATACGGCCCAGACTCCTACGGGAGGCAGCAGTAGGGAATTTTTCACAATGGACGAAAGTCTGATGAAGCAATGCCGCGTGA
                                  PCR 388 →
410                                      450                                              500
TGMTGACGGCCTTCGGGTTGTAAAKCTCTGTTGTAAGGGAAGAAAAAATARAGTAGGAAATGACTTTATCTTGACAGTACCTTACCAGAAAGCCACGGC
510                                      550                                              600
          ← Seq 631
TAACTATGTGCCAGCAGCCGCGGTAATACATAGGTGGCAAGCGTTATCCGGATTTATTGGGCGTATAGGGTGCGTAGGCGGTTTTGCAAGTTTGAGGTTA
610                                      650                                              700
AAGTCYGGAGCTCAACTCCGGTTCGCCTTGAARACTGTWTTACTAGAATGCAAGAGAGGTAAGCGGAATTCYATGTGTAGCGGTGAAATGCGTAGATATA
710                                      750                                              800
                                                                                       Seq 538 →
TGGAAGAAYACCTGTGGCGAAAGCGGCTTACTGGCTTGTTATTGACGCTGAKGCACGAAAGCGTGGGGAGCAAATAGGATTAGATACCCTAGTAGTCCAC
810                                      850                                              900
Seq 538 →
GCCGTAAACGATGAGTACTAAGTGTTGGGGWAAYTCAGCGCTGYAGCTAACGCATTAAGTACTCCGCCTGMGTAGTATGCTCGCAAKAGTGAAACTCAAA
910                                      950                                             1000
   ← Seq 390
GGAATTGACGGGGACCCGCACAAGTGGTGGAGCATGTGGTTTAATTCGAAGCAACACGAAGAACCTTACCAGGGCTTGACATCCAGTGCAAAGCTATAGA
   ← PCR 390-B
1010                                     1050                                            1100
GATATAGTRGAGGTTAACATTGAGACAGGTGGTGCATGGTTGTCGTCAGTTCGTGCCGTRAGGTGTTGGGTTAAGTCCYRCAACGAACGCAACCCTTGTC
1110                                     1150                                            1200
                                                      Seq 597 →
GYTAGTTACTAACATTAAGTTGAGAACTCTAACGAGACTGCTAGTRTAAGYTAGAGGAAGGTGGGGATGACGTCAAATCATCATGCCCCTTATGTCCTGG
1210                                     1250                                            1300
GCTACACACGTGCTACAATGGCTGGTACAAAGAGTTGCAATCCTGTGAAGGGGARCTAATCTCAAAAAaaCCAGTCTCAGTTCGGATTGAAGTCTGYAAC
1310                                     1350                                            1400
                                                         ← Seq 585
TCGACTTCATGAAGCCGGAATCACTAGTAATCGCGAATCAGCTATGTCGCGGTGAATACGTTCTCGGGTCTTGTACACACCGCCCGTCACACCATGAGAG
1410                                     1450                                            1500
TTKGTAATACCAGAAGTAGGTAGMTTAACCRTTTGGAGAGCGCTTYCYAAGGTAGGACTAGCGATTGGGGTGAAGTCGTAACAAGGTATCCGTACGGAA
1510
CGTGCGGATGGATCACCTCCTTTCT
     ← PCR 620-B
                     ← RSP
```

Fig. 1. Positions of the primers relative to the consensus sequence of the 16S rRNA genes from the *rrnB* operon of the members of the *M. mycoides* cluster ***(19)***. The target regions for the PCR primers (PCR) and the sequencing primers (Seq) are underlined, and the directions are indicated with arrows. Positions where nucleotide differences exist in the 16S rRNA genes of the members of the *M. mycoides* cluster are denoted with the ambiguity letter code in boldface according to IUB. The sequence length variation of *Mycoplasma mycoides* subsp. *mycoides* SC is indicated with aa in boldface.

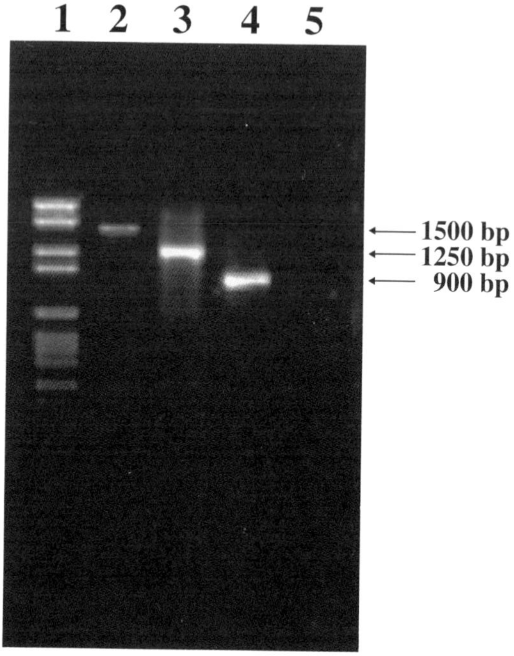

Fig. 2. Analysis of PCR products from *Mycoplasma capricolum* subsp. *capripneumoniae* obtained by amplification with primers complementary to the universal regions U1 and U8 (lane 2), U2 and U8 (lane 3), and U1 and U5 (lane 4). Molecular-size markers (*see* **Subheading 2.2.**) were applied in lane 1 and a negative control in lane 5. The approximate sizes of the amplicons are given in bp.

directions of the primers are given in **Table 2** and in **Fig. 1**. Mycoplasmas have one, two, or three rRNA operons, and nucleotide differences (*see* **Note 5**) as well as length variations (*see* **Note 6**) may occur between the 16S rRNA genes of the different rRNA operons.

1. Wash 20 µL of Streptavidin-coated paramagnetic beads (Dynabeads M280; DYNAL AS, Oslo, Norway) twice in 20 µL PBS and once in 20 µL 1X binding/washing (B/W) buffer. Use a magnetic tube rack to sediment the beads while the supernatant is removed.
2. Resuspend the washed beads in 40 µL of 2X B/W buffer, add 40 µL of the PCR product and leave the suspension for 15 min at room temperature.
3. Discard the supernatant after the magnetic separation, and wash the beads in 40 µL of 1X B/W buffer. The DNA can be stored immobilized on the beads at 4°C for several weeks.
4. Discard the supernatant after the magnetic separation, resuspend the beads in 8 µL of 0.1*M* NaOH, and leave the suspension at room temperature for 10 min.
5. Place the tube in the magnetic tube rack, remove the alkaline supernatant, containing the eluted single strand, and transfer it to a new tube. Neutralize the super-

natant with 4 µL of 0.2*M* HCl, and add 1 µL of 1*M* Tris-HCl buffer, pH 7.4 (*see* **Note 7**).

6. Wash the beads first with 50 µL of 0.1*M* NaOH, then with 40 µL of 1X B/W buffer, and finally with 50 µL of TE buffer. Discard the supernatants after each washing.
7. Dilute,the beads with the immobilized DNA strand in 13 µL of distilled water.
8. Perform the sequencing reactions with both the eluted and the immobilized DNA strand according to the protocol provided with the sequencing kit.
9. Prepare a sequencing gel with ReadyMix Gel in the standard plate kit. The gel must be prepared at least 1 h before the electrophoresis and can be stored at 4°C for up to 1 wk.
10. Introduce the gel cassette into the automated DNA sequencer and fill up the buffer vessels with 1X TBE-b buffer. Apply the samples onto the sequencing gel. The electrophoretic separation, on-line detection, and computerized sequence reading will be performed automatically after initiation of the process.
11. Perform the electrophoresis according to the Standard Operational Procedure suggested by the manufacturer. The sequences will be shown on the computer display (*see* **Fig. 3**) and are obtained as computer files, which can be edited and transformed into a format suitable for evaluation of the data.

3.4. Evaluation of Sequence Data

The 16S rDNA sequence data will play an important role in the future for the classification of mycoplasmas as more species are described ***(12)***, because conventional methods provide less discriminatory information (*see* **Note 8**). The number of deposited 16S rRNA sequences from mycoplasmas is close to 100, and it will, therefore, be possible to determine the phylogenetic cluster and close relatives to a new mycoplasma. There is a constant debate concerning the construction of phylogenetic trees, and it is impossible to cover all aspects in this chapter. However, some general advice is provided that can be used to get started.

1. Merge the sequences of the amplicons to a contig representing the nearly complete 16S rRNA sequence by using a suitable computer program.
2. Use the contig for a BLAST (Basic Local Alignment Search Tool) search ***(29)***. This kind of search can be performed on-line from the home page of the National Center for Biotechnology Information, which has the following URL address: http://www.ncbi.nlm.nih.gov/. It is also possible to use a similarity search program in GCG, for instance, WORDSEARCH, BLAST, or FASTA. Any of these searches will give a list of the species with the most similar 16S rRNA sequences.
3. Select all relevant species from the list, and download those that are available from RDP or the rRNA WWW Server as prealigned sequences into GDE (*see* **Note 9**). Examples of species representing different phylogenetic groups and clusters are given in **Table 3**. Retrieve other relevant sequences from GenBank or EMBL, and align them manually with the prealigned sequences from RDP. Note

Table 3
Examples of Mycoplasmas Representing All Eight Genera *(13)* and All Five Phylogenetic Groups *(10)*

Species[a]	Phylogenetic group[b]	Phylogenetic cluster[c]	Accession no. in GenBank[d]	Reference[e]
Asteroleplasma (As.) anaerobium	Asteroleplasma (1)	*As. anaerobium*	M22351	***(10)***
M. mycoides subsp. *mycoides* SC	Spiroplasma (4)	*M. mycoides* (a)	U26038/U26039	***(19)***
Mycoplasma putrefaciens[f]	Spiroplasma	*M. mycoides* (a)	U26055	***(12,19)***
Mesoplasma (Me.) entomophilum	Spiroplasma	*M. mycoides* (a)	M23931	***(10***
Entomoplasma ellychniae	Spiroplasma	*M. mycoides* (a)	M24292	***(10)***
Spiroplasma apis	Spiroplasma	*S. apis* (b)	M23937	***(10)***
Spiroplasma citri	Spiroplasma	*S. citri* (c)	M23942	***(10)***
Spiroplasma sp. strain Y-32	Spiroplasma	*Spiroplasma* sp. strain Y-32 (d)	M24477	***(10)***
Mycoplasma pneumoniae	Pneumoniae (3)	*M. pneumoniae* (a)	M29061	***(10)***
Mycoplasma muris	Pneumoniae	*M. muris* (b)	M23939	***(10)***
Ureaplasma urealyticum	Pneumoniae	*U. urealyticum* (c)	M23935	***(10)***
Mycoplasma synoviae	Hominis (6)	*M. synoviae* (a)	L07757	***(11,30)***
Mycoplasma lipophilum	Hominis	*M. lipophilum* (b)	M24581	***(10)***
Mycoplasma agalactiae	Hominis	*M. lipophilum* (b)	U44763	***(11)***
Mycoplasma hominis	Hominis	*M. hominis* (c)	M24473	***(10)***
Mycoplasma pulmonis	Hominis	*M. pulmonis* (d)	M23941	***(10)***

"Mycoplasma agassizii"	Hominis	*M. pulmonis* (d)	U09786	***(31)***
Mycoplasma neurolyticum	Hominis	*M. neurolyticum* (e)	M23944	***(10)***
M. hyopneumoniae	Hominis	*M. neurolyticum* (e)	Y00149	***(32)***
Mycoplasma sualvi	Hominis	*M. sualvi* (f)	M23936	***(10)***
Mycoplasma mobile	Hominis	*M. sualvi* (f)	M24480	***(10)***
Anaeroplasma (An.) abactoclasticum	Anaeroplasma (2)	*Anaeroplasma* sp. (a)	M25050	***(10)***
"Phytoplasma" sp., strain STOL	Anaeroplasma	*Phytoplasma* sp (b)	X76427	***(33)***
Acholeplasma laidlawii	Anaeroplasma	*Acholeplasma* sp. (c)	M23932	***(10)***

[a]These species can be selected to construct a phylogenetic tree for the first preliminary characterization of an unknown mycoplasma and are examples of species that may be chosen for the sequence alignments. Nonstandard abbreviations used in **Fig. 4** are given within parentheses. Taxon names that have not been approved are given within quotation marks.

[b]Five phylogenetic groups and 15 phylogenetic clusters were originally defined ***(10)***. One new cluster *(M. synoviae)* was introduced later ***(11)***. The number of phylogenetic clusters so far defined within each phylogenetic group is given within parentheses. The designations (a–f) for the phylogenetic clusters used in **Fig. 4** are given within parentheses after the respective cluster.

[c]If the 16S rRNA sequences of both rRNA operons have been deposited, it is indicated as xxx/yyy for the sequences from the *rrnA* and the *rrnB* operons, respectively. Some of the sequences can be downloaded from RDP or from the rRNA WWW Server.

[d]References to the articles where the sequences were reported and the phylogenetic cluster described.

[e]It has recently been suggested that *M. putrefaciens* should be included in a new cluster ***(12)***.

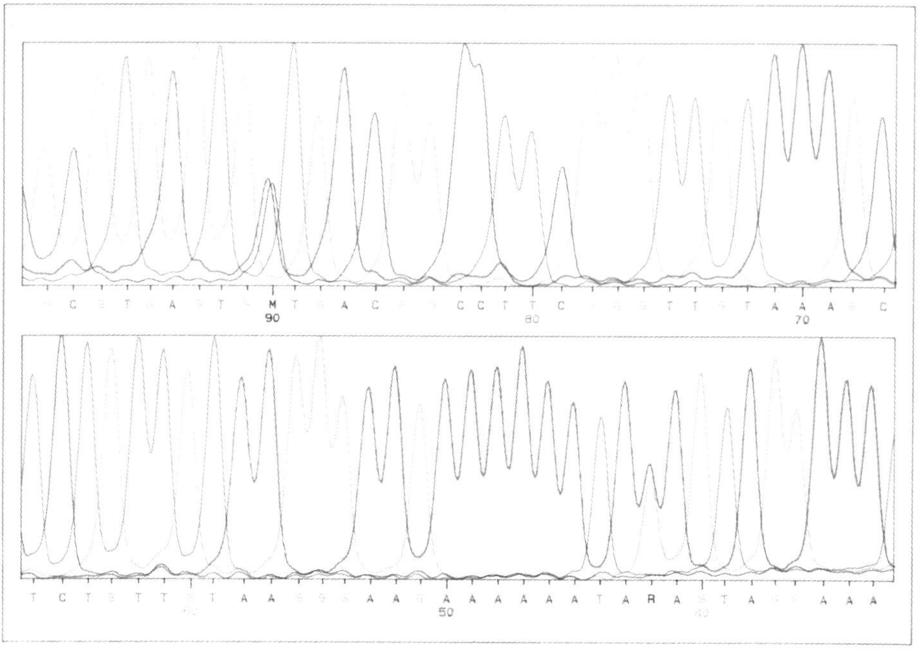

Fig. 3. Sequence analysis of a segment of the V6 region of the 16S rRNA gene of *M. c. capripneumoniae* by direct automated solid-phase sequencing with the ALFexpress system. The PCR product was generated with the primers 593 and 390-B as second primer pair, and the primer 631 was used as a reverse sequencing primer, which explains the reverse numbering order in this figure. Note the polymorphisms in positions 90 (C/A) and 42 (G/A), corresponding to positions 404 and 452, respectively, in the consensus sequence in **Fig. 1** ***(19)***.

that some of the deposited sequences contain segments or positions that have not been determined. Some of the sequences also contain errors. Sequence differences may occur between strains of the same species (*see* **Note 10**). The alignment has to be performed manually, because homologous positions have to be compared, which are not necessarily those that show the best match. The secondary structure models of mycoplasmal 16S rRNA sequences, which also can be retrieved from RDP or from the rRNA WWW Server, are useful for proper identification of nucleotides situated in stems or loops. Models for *M. c. capricolum, M. gallisepticum,* and *M. hyopneumoniae* are available ***(34)***.

4. Remove positions containing gaps and ambiguously aligned positions before the phylogenetic analysis.
5. Select the species to be displayed in the phylogenetic tree, and choose a suitable species as the outgroup. The outgroup should preferably be a species that can be assumed to branch early from the studied group. Information about this can be obtained from other phylogenetic trees ***(10,11)*** or by constructing a preliminary

phylogenetic tree with more species than in the final one. If the outgroup is too distantly related from the species to be studied, the resolution of the tree will be low. If it is too close to the studied group, there is a risk that it is not a true outgroup. Furthermore, the branching order is sometimes dependent on the choice of outgroup.

6. Perform the tree construction according to the manual in the program package for the chosen algorithm.
7. Check the stability and the validity of the tree by bootstrapping the data set, and analyze the original data set for sequence motifs, which support or contradict the obtained branching order. The phylogenetic tree in **Fig. 4** shows the relationship between mycoplasmas of the different phylogenetic groups and clusters. However, it would have been possible to select other species as well.

4. Notes

1. Sample preparation: Mycoplasmas lack a cell wall, and they are, therefore, easy to lyse. It is in most cases possible to use the DNA from lysed mycoplasmas as template for the first PCR without any purification. However, sometimes the results can be improved by phenol extraction of the DNA. Such DNA preparations can also be sent by ordinary mail to another laboratory for analysis. It is also possible to dry different kinds of samples on filter paper, send it to another laboratory, and amplify by PCR for subsequent sequencing.
2. The contamination problem: The contamination problem is less pronounced when the purpose of the PCR is to sequence amplicons, where a relatively large amount of template DNA was used. It is in general not necessary to use dedicated rooms for the different activities. However, all precautions to prevent carryover contamination with amplicons have to be taken, if experiments are performed routinely, where the purpose is to sequence PCR products from a clinical material where the amount of template DNA can be expected to be very small.
3. Visualization of the first PCR product: It is sometimes difficult to visualize the PCR product in the agarose gel after the first PCR with the primer pair complementary to U1 and U8. This could be owing to sequence variations in the target regions for these primers. However, even if PCR products cannot be seen after the first PCR, the amount should in most cases be sufficient for the second PCR. Therefore, perform the second PCR experiments even if the first amplicon cannot be visualized. However, after the second PCR, a strong band should be seen for good results of the sequencing experiments. Negative controls should always be included. The results can sometimes be improved by using phenol-extracted DNA instead of lysed mycoplasmas in the first PCR reaction (compare also **Note 1**). To generate a complete 16S rRNA sequence, it is in general sufficient to produce 3×50 µL of the U1 to U5 amplicon and 2×50 µL of the U2 to U8 amplicon.
4. The importance of sequencing of both strands: The two complementary DNA strands can easily be sequenced by the solid-phase method. For generation of accurate sequence data, it is in fact important that both strands are sequenced. Furthermore, polymorphisms and sequence length variations (*see also* **Notes 5**

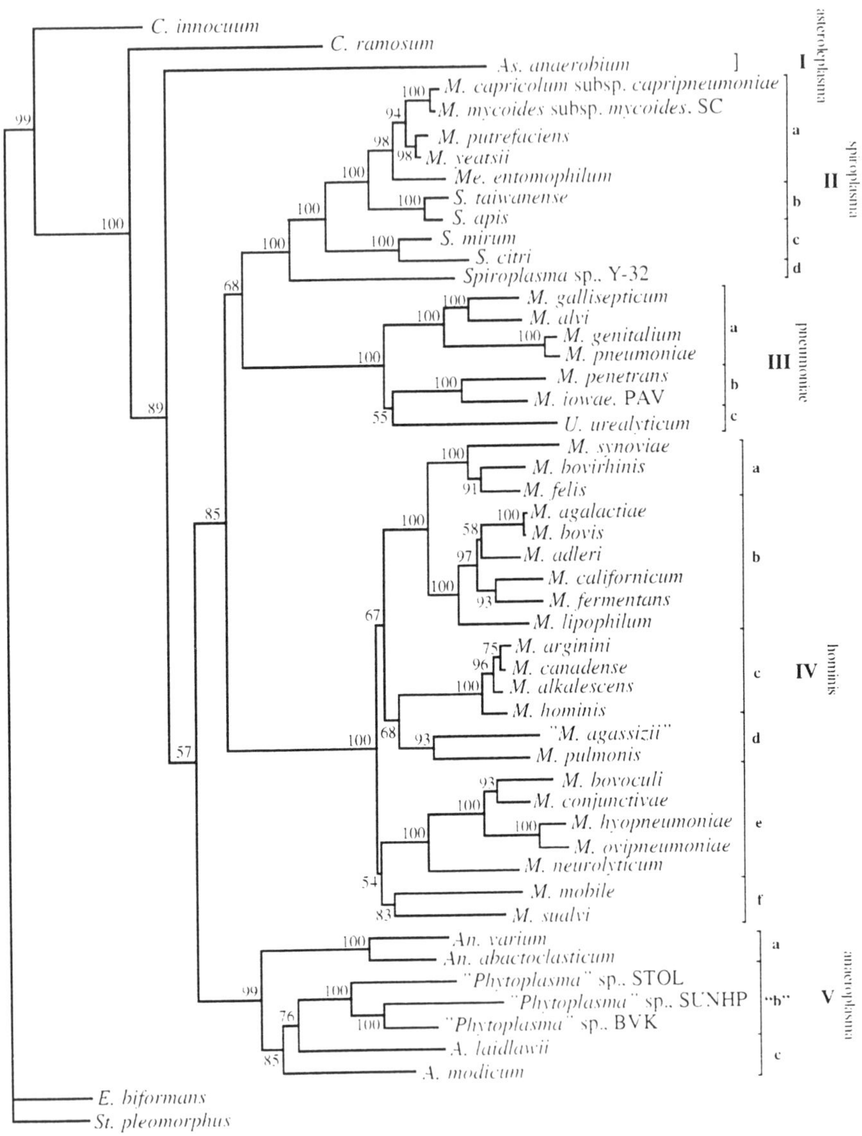

Fig. 4. Phylogenetic tree based on 16S rRNA sequences of mycoplasmas representing all groups and clusters ***(10,11)***. Representatives of the closely related genera *Clostridium* and *Eubacterium* were also included in the tree. *Streptococcus (St.) pleomorphus and Eubacterium biformans* were selected as outgroups. *M. c.* subsp. *capripneumoniae* grouped in the *M. mycoides* cluster. The tree was constructed by neighbor-joining ***(35)*** from a distance matrix corrected by the one parameter nucleotide substitution model ***(36)*** by using the programs NEIGHBOR and DNADIST, respectively.

and **6**) can easily be resolved by sequencing of both strands. Ambiguities caused by polymerase pausing, false stops, and band compression during the electrophoresis step can be resolved by sequencing the other strand. We have compared data generated by conventional sequencing of the 16S rRNA genes of both rRNA operons after cloning of restriction fragments ***(37)*** with solid-phase sequencing of PCR products ***(19)*** and obtained exactly the same sequence data. This shows that solid-phase sequencing of PCR products is the preferred method, since it is much faster.

5. Polymorphisms: The genetic information for rRNA is organized in the genome into so-called rRNA operons. Bacteria can have 1–14 different rRNA operons, and mycoplasmas have been shown to have 1 or 2 and in one case 3 rRNA operons, which are designated *rrnA, rrnB,* and *rrnC.* It has been shown for mycoplasmas that when they have more than one rRNA operon, there could be sequence differences (polymorphisms) between the 16S rRNA genes ***(11,12,18–21)***. In most cases, the PCR primers described in this work will amplify the corresponding segment of the two rRNA genes. The polymorphisms are easy to identify, because they appear as double peaks in the electropherogram when a segment with polymorphisms is sequenced by an automated procedure (*see* **Fig. 3**). The areas under the two peaks in such a double peak should represent about 50% each, if the organism contains two rRNA operons and if both 16S rRNA genes are amplified with the same efficiency. The 16S rRNA sequences of many mycoplasmas with two rRNA operons have been determined, and usually there are only few (0–3) polymorphisms ***(11,12,18–21)***. However, strains of the species *M. capricolum* subsp. *capripneumoniae* have been found to have an unusual large number (11–17) of polymorphisms ***(21)***. These polymorphisms were used as epidemiological markers and to study the evolution of the species. The sequences of the 16S rRNA genes from the individual operons can be determined by using an operon-specific primer set ***(19)***, by cloning and sequencing of the individual operons ***(11,37)***, or by separation of suitable restriction fragments by agarose gel electrophoresis and amplification of the individual bands with general primers ***(38)***.
6. Sequence length variations: It has been shown that sequence length variations can occur in polyA regions between the 16S rRNA genes of the two operons from mycoplasmas. One such polyA region has been identified in *M. mycoides* subsp. *mycoides* SC ***(19)*** and another polyA region was identified in certain strains of *M. c. capripneumoniae* ***(21)***. A sequence length variation will give

---

These programs are included in the phylogenetic program package PHYLIP. Gaps were removed from the final alignment, which comprised 1168 positions. The five phylogenetic groups are shown beside the vertical bars, indicating the phylogenetic groups. The phylogenetic clusters are indicated with letters (a–f), which are explained in **Table 3**. The bootstrap values are given at each node. The scale bar indicates substitutions per 100 nucleotide positions. *See* **Table 3** for nonstandard abbreviations and cluster designations. Taxon names that have not been approved are given within quotation marks.

seemingly confusing sequencing results, but if both DNA strands are analyzed, the sequence can easily be determined, since the data are difficult to interpret only downstream of the sequence length variation ***(19)***.

7. The difficulties in sequencing the eluted DNA strand in the supernatant: It is very important that the alkaline supernatant from the denaturation of DNA immobilized onto the magnetic beads is properly neutralized before the sequencing reactions are performed. Always use the same calibrated pipet when aliquoting both NaOH and HCl, because even very small volume differences may result in a pH that cannot be buffered to pH 7.4. It can also be recommended to prepare large volumes of the two solutions, and use an indicator paper to check that a mixture of the expected volumes is neutral. When such solutions have been prepared, aliquot them in volumes sufficient for 10 sequencing reactions and store them in the freezer until needed. A frozen NaOH solution is quite stable.
8. Why is it important to identify polymorphisms? If 16S rRNA sequence data is to be used for construction of phylogenetic trees for closely related species that have more than one rRNA operon, it is important that sequences of homologous operons be compared. Otherwise small sequence differences can affect the topology of the tree ***(19)***. Polymorphisms can also be utilized for design of detection systems based on, for instance, PCR and restriction enzyme analysis. This has proven particularly useful for closely related species, like *M. c. capripneumoniae* ***(37)*** and *M. mycoides* subsp. *mycoides* SC ***(39)***, both of which are members of the closely related *M. mycoides* cluster, and also for the two closely related species *M. agalactiae* and *Mycoplasma bovis* ***(38)***.
9. Data bank sequences: It is very convenient to be able to download the prealigned 16S rDNA sequences from RDP ***(8)*** or from the rRNA WWW Server ***(9)***. However, it is important to keep in mind that these data banks are not regularly updated, and it is, therefore, necessary to check GenBank or the EMBL data bank for newly deposited relevant mycoplasmal sequences. The sequences in GenBank or EMBL can also be retrieved and manually aligned with the prealigned sequences from RDP or from the rRNA WWW Server by using the GDE program.
10. Sequence variations between strains or species: Ribosomal RNA genes are in general more conserved than protein genes, although the rRNA genes also have regions of high evolutionary variability. The sequence differences between strains or isolates of a species is, therefore, small, and for most mycoplasmas, so far analyzed, on the order of 0–7 nucleotide differences. However, it should be kept in mind that the species concept was invented by taxonomists and is somewhat artificial. It is not possible to give a general rule about how many nucleotide differences there should be in the 16S rRNA genes to justify the organization of two strains into different species. The variation within one species can be greater than the variability within another species, also within the same genus. The two nucleotide differences in the 16S rRNA sequences of *M. gallisepticum* and *Mycoplasma imitans* would normally be regarded as too small to justify the arrangement into different species ***(40)***. Thus, sequence analysis alone is not always sufficient for species designation ***(41)***, and it would be useful to construct a molecular phylogeny based

on other genes for these species. An example of the opposite situation is the seven-nucleotide differences between certain strains of *M. c. capripneumoniae* *(21)*, which would normally place these strains in at least different subspecies.

References

1. Gray, M. W., Sankoff, D., and Cedergren, R. J. (1984) On the evolutionary descent of organisms and organelles: a global phylogeny based on a highly conserved structural core in small subunit ribosomal RNA. *Nucleic Acids Res.* **12,** 5837–5852.
2. Van de Peer, Y., Chapelle, S., and De Wachter, R. (1996) A quantitative map of nucleotide substitution rates in bacterial rRNA. *Nucleic Acids Res.* **24,** 3381–3391.
3. Olsen, G. J. and Woese, C. R. (1993) Ribosomal RNA: a key to phylogeny. *FASEB J.* **7,** 113–123.
4. Woese, C. R. (1987) Bacterial evolution. *Microbiol. Rev.* **51,** 221–271.
5. Taschke, C., Klinkert, M.-Q., Wolters, J., and Herrmann, R. (1986) Organization of ribosomal RNA genes in *Mycoplasma hyopneumoniae*: the 5S rRNA is separated from the 16S and 23S rRNA genes. *Mol. Gen. Genet.* **205,** 428–433.
6. Chen, X. and Finch, L. R. (1989) Novel arrangement of rRNA genes in *Mycoplasma gallisepticum*: separation of the 16S gene of one set from the 23S and 5S genes. *J. Bacteriol.* **171,** 2876–2878.
7. Nakagawa, T., Uemori, T., Asada, K., Kato, I., and Harasawa, R. (1992) *Acholeplasma laidlawii* has tRNA genes in the 16S-23S spacer of the rRNA operon. *J. Bacteriol.* **174,** 8163–8165.
8. Maidak, B. L., Olsen G. J., Larsen N., Overbeek R., McCaughey M. J., and Woese C. R. (1996) The ribosomal database project (RDP). *Nucleic Acids Res.* **24,** 82–85.
9. Van de Peer, Y., Jansen, J., De Rijk, P., De Wachter, R. (1997) Database on the structure of small ribosomal subunit RNA. *Nucleic Acids Res.* **24,** 111–116.
10. Weisburg, W. G., Tully, J. G., Rose, D. L., Petzel, J. P., Oyaizu, H., Yang, D., Mandelco, L., Sechrest, J., Lawrence, T. G., van Etten, J., Maniloff, J., and Woese, C. R. (1989) A phylogenetic analysis of the mycoplasmas: basis for their classification. *J. Bacteriol.* **171,** 6455–6467.
11. Pettersson, B., Uhlén, M., and Johansson, K.-E. (1996) Phylogeny of some mycoplasmas from ruminants based on 16S rRNA sequences and definition of a new cluster within the hominis group. *Int. J. Syst. Bacteriol.* **46,** 1093–1098.
12. Heldtander, M. U. K., Pettersson, B., Tully, J. G., and Johansson, K.-E. (1998) Sequences of the 16S rRNA genes and phylogeny of the goat mycoplasmas; *Mycoplasma adleri, Mycoplasma auris, Mycoplasma cottewii,* and *Mycoplasma yeatsii*. *Int. J. Syst. Bacteriol.* **48,** 263–268.
13. Tully, J. G., Bové, J. M., Laigret, F., and Whitcomb, R. F. (1993) Revised taxonomy of the class *Mollicutes*: proposed elevation of a monophyletic cluster of arthropod-associated mollicutes to ordinal rank (*Entomoplasmatales* ord. nov.), with provision for familial rank to separate species with nonhelical morphology (*Entomoplasmataceae* fam. nov.) from helical species *(Spiroplasmataceae),* and emended descriptions of the order *Mycoplasmatales,* family *Mycoplasmataceae.* *Int. J. Syst. Bacteriol.* **43,** 378–385.

14. Iwami, M., Muto, A., Yamao, F., and Osawa, S. (1984) Nucleotide sequence of the *rrnB* 16S ribosomal RNA gene from *Mycoplasma capricolum. Mol. Gen. Genet.* **196,** 317–322.
15. Frydenberg, J. and Christiansen, C. (1985) The sequence of 16S rRNA from *Mycoplasma* strain PG50. *DNA* **4,** 127–137.
16. Hultman, T., Ståhl, S., Hornes, E., and Uhlén, M. (1989) Direct solid phase sequencing of genomic and plasmid DNA using magnetic beads as solid support. *Nucleic Acids Res.* **17,** 4937–4946.
17. Hultman, T., Bergh, S., Moks, T., and Uhlén, M. (1991) Bidirectional solid-phase sequencing of in vitro-amplified plasmid DNA. *BioTechniques* **10,** 84–93.
18. Pettersson, B., Johansson, K.-E., and Uhlén, M. (1994) Sequence analysis of 16S rRNA from mycoplasmas by direct solid-phase DNA sequencing. *Appl. Environ. Microbiol.* **60,** 2456–2461.
19. Pettersson, B., Leitner, T., Ronaghi, M., Bölske, G., Uhlén, M., and Johansson, K.-E. (1996) Phylogeny of the *Mycoplasma mycoides* cluster as determined by sequence analysis of the 16S rRNA genes from the two rRNA operons. *J. Bacteriol.* **178,** 4131–4142.
20. Pettersson, B. (1997) Direct solid-phase 16S rDNA sequencing: a tool in bacterial phylogeny. PhD thesis. Royal Institute of Technology, Stockholm, Sweden.
21. Pettersson, B., Bölske, G., Thiaucourt, F., Uhlén, M., and Johansson, K.-E. (1998) Molecular evolution of *Mycoplasma capricolum* subsp. *capripneumoniae* strains, based on polymorphisms in the 16S rRNA genes, submitted for publication.
21a. Sanger, F., Nicklen, S., and Couldson, A. R. (1977) DNA sequencing with chain-terminating inhibitiors. *Proc. Natl. Acad. Sci USA* **74,** 5463–5467.
22. Smith, S. (1992) Genetic Data Environment (Version 2.2). Millipore Imaging Systems, Ann Arbor, MI.
23. Program Manual for the Wisconsin Package (Version 8). Genetics Computer Group, Madison, WI.
24. Felsenstein, J. (1993) PHYLIP: Phylogeny inference package (Version 3.52). University of Washington, Seattle, WA.
25. Swofford, D. L. (1991) PAUP: Phylogenetic analysis using parsimony (Version 3.1.1.) Illinois Natural History Survey, Champaign, IL.
26. Kumar, S., Tamura, K., and Nei, M. (1993) MEGA: molecular evolutionary genetic analysis (Version 1.01). The Pennsylvania State University, University Park, PA.
27. Felsenstein, J. (1988) Phylogenies from molecular sequences: inference and reliability. *Annu. Rev. Gen.* **22,** 521–565.
28. Swofford, D. L., Olsen, G. J., Waddell, P. J., and Hillis, D. M. (1996). Phylogenetic inference, in *Molecular Systematics,* 2nd ed. (Hillis, D. M., Moritz, C., and Mable, B. K., eds.), Sinauer Associates, Sunderland, MA, pp. 407–514.
29. Altschul, S. F., Gish, W., Miller, W., Myers, E. W., and Lipman, D. J. (1990) Basic local alignment search tool. *J. Mol. Biol.* **215,** 403–410.
30. Morrow, C. J. (1990) Pathogenicity, immunogenicity and strain identification of Australian isolates of *M. synoviae,* PhD thesis, University of Melbourne, Australia.

31. Brown, D. R., Crenshaw, B. C., McLaughlin, G. S., Schumacher, I. M., McKenna, C. E., Klein, P. A., Jacobsen, E. R., and Brown, M. B. (1994) Taxonomic analysis of the tortoise mycoplasmas *Mycoplasma agassizii* and *Mycoplasma testudinis* by 16S rRNA gene sequence comparison. *Int. J. Syst. Bacteriol.* **45,** 348–350.
32. Taschke, C., Ruland, K., and Herrmann, R. (1987) Nucleotide sequence of the 16S rRNA of *Mycoplasma hyopneumoniae. Nucleic Acids Res.* **15,** 3918.
33. Seemüller, E., Schneider, B., Mäurer, R., Ahrens, U., Daire, X., Kison, H., Lorenz, K.-H., Firrao, G., Avinent, L., Sears, B., and Stackebrandt, E. (1994) Phylogenetic classification of phytopathogenic mollicutes by sequence analysis of 16S ribosomal DNA. *Int. J. Syst. Bacteriol.* **44,** 440–446.
34. Gutell, R. R. (1994) Collection of small subunit (16S- and 16S-like) ribosomal RNA structures: 1994. *Nucleic Acids Res.* **22,** 3502–3507.
35. Saitou, N. and Nei, M. (1987) The neighbor-joining method: a new method for reconstructing phylogenetic trees. *Mol. Biol. Evol.* **4,** 406–425.
36. Jukes, T. H. and Cantor, C. R. (1969) Evolution of protein molecules, in *Mammalian Protein Metabolism,* vol 3. (Munro, H. N., ed.), Academic Press, New York, pp. 21–132.
37. Ros Bascuñana, C., Mattsson, J. G., Bölske, G., and Johansson, K.-E. (1994) Characterization of the 16S rRNA genes from *Mycoplasma* sp. strain F38 and development of an identification system based on PCR. *J. Bacteriol.* **176,** 2577–2586.
38. Johansson, K.-E., Berg, L.-O., Bölske, G., Deniz, S., Mattsson, J., Persson, M., and Pettersson, B. (1996) Specific PCR systems based on the 16S rRNA genes of *Mycoplasma agalactiae* and *Mycoplasma bovis,* in *COST 826. Agriculture and Biotechnology. Mycoplasmas of Ruminants: Pathogenicity, Diagnostics, Epidemiology and Molecular Genetics* (Frey, J. and Sarris, K., eds.), European Commission, Brussels, Belgium, pp. 88–90.
39. Persson, A., Pettersson, B., Johansson, K.-E. (1996) Identification of *Mycoplasma mycoides* subsp. *mycoides* SC type by PCR and restriction enzyme analysis with *Alu*I. *IOM Lett.* **4,** 79–80.
40. Boyle, J. S., Bradbury, J. M., and Morrow, C. J. (1993) Further evidence that *M. imitans* is closely related to *M. gallisepticum,* unpublished.
41. Bradbury, J. M., Abdul-Wahab, O. M. S., Yavari, C. A., Dupiellet, J.-P., and Bové, J. M. (1993) *Mycoplasma imitans* sp. nov. is related to *Mycoplasma gallisepticum* and found in birds. *Int. J. Syst. Bacteriol.* **43,** 721–728.

19

PCR and RFLP Methods for the Specific Detection and Identification of *Mycoplasma mycoides* subsp. *mycoides* SC

John B. Bashiruddin

1. Introduction

The polymerase chain reaction (PCR), DNA hybridization, and sequence analysis have been valuable in the study of the phylogenic relationships between members of the *Mycoplasma mycoides* "cluster" ***(1)***. They have confirmed the very close relationships between these organisms suggested by others ***(2,3)***, and in the case of *M. mycoides* subsp. *mycoides* SC (*Mmm*SC), provided a rapid diagnostic test for the detection and identification of this organism based on DNA amplification by PCR and restriction fragment length polymorphism (RFLP) between *Mmm*SC and (*Mmm*LC) ***(4)***.

Contagious bovine pleuropneumonia (CBPP), which is caused by *Mmm*SC, is a serious disease of cattle worldwide, particularly in sub-Saharan Africa. Recently, it has reappeared in parts of Mediterranean Europe and has persisted stubbornly in the Iberian peninsula. The biological characteristics of the causative agent, its place within the *M. mycoides* "cluster," and some of the diagnostic tests currently available have been reviewed recently ***(5,6)***.

Several authors have reported the development and use of PCR for the detection of *Mmm*SC from clinical samples of lung tissue, pleural fluid, tracheal scrapings, and nasal swabs ***(4–10)***. *Mmm*SC may be detected with these tests at very low levels, of <100 organisms, which may be made more sensitive by the addition of further steps to probe immobilized DNA, for the detection of the PCR product. These procedures report an increase in sensitivity to 1 CFU of mycoplasma. Other approaches to the detection of PCR products have focused on the hybridization of immobilized single-stranded DNA bound to microtiter plates to denatured amplified DNA labeled with biotin followed by immuno-

From: *Methods in Molecular Biology, Vol. 104: Mycoplasma Protocols*
Edited by: R. J. Miles and R. A. J. Nicholas © Humana Press Inc., Totowa, NJ

logical detection of the hybrid. The numerous variations of oligonucleotide-based detection systems have been described, and offer more rapid and cost-effective product detection than the traditional gel-based systems (*see* **Note 1**).

PCR is useful for the rapid clarification of suspicious clinical situations where persistent positive serological tests have made the search for the causative agent in the lung necessary ***(11)***. In these cases, postmortem clinical material, such as necrotic lung tissue, lymph nodes, tracheal srcapings, and pleural fluid, may be taken for PCR and bacteriological culture. DNA extracted from tissues as well as mycoplasma isolates may be used for PCR to verify the presence or absence of *Mmm*SC. Confirmation of suspicious lung pathology in the course of abattoir surveillance may also be done by this method. The testing of nasal swabs by PCR has also been described, which may be applicable to the screening of large numbers of samples important in the control and surveillance of CBPP ***(4,12)***.

The PCR technique with modified primers coupled to a detection method that specifically captures amplified double-stranded DNA in microtiter plates is described here ***(13)***. Since it is based on 96-well microtiter plate technology, it offers a major advantage in research involving large sample numbers. The procedure described covers specimen collection, sample preparation, DNA amplification by PCR, colorimetric detection of PCR product, and specific identification of *Mmm*SC by restriction enzyme digestion of PCR product.

A routine diagnostic procedure with PCR as its core technique, by its very nature, is susceptible to contamination with amplified DNA. To be sure that a positive result is possible only from specimen-derived template DNA, several precautions are necessary. The first is the structural location of the diagnostic facility, which must separate "clean" pre-PCR areas from "dirty" post-PCR areas. The second is the flow of reagents and personnel through the facility. The third is the flow of air through the facility, with respect to the fact that aerosols are a common source of contamination. Generally, three dedicated areas are required in which PCR reaction mixes, sample preparation, and product analysis are handled exclusively (*see* **Note 2**.) Therefore preplanning to accommodate these ideas is essential. The description that follows assumes that adequate consideration has been given to the establishment of a PCR laboratory.

2. Materials

2.1. Specimen Collection

2.1.1. Lung, Lymph Node Tissue and Tracheal Scrapings

1. Access to postmortem tissue.
2. Protective clothing, e.g., overalls including footwear.
3. Gloves.

4. Sterile disposable scalpel blades.
5. Receptacles for tissue samples.
6. Specimen tubes with 3 mL of mycoplasma transport media.

2.1.2. Nasal Swabs

1. Access to infected restrained animals preferably in a crush.
2. Protective clothing, e.g., overalls including footwear.
3. Gloves.
4. Sterile rayon swabs on plastic sticks.
5. Specimen tubes with 3 mL of mycoplasma transport media.

2.2. Sample Preparation

2.2.1. DNA Extraction from Tissues

1. Lung or lymph node tissue: Trachael scrapings may also be used.
2. Eppendorf microtubes and tissue homogenizer.
3. TNE buffer: 0.01*M* Tris-HCl, pH 8.0, 0.01*M* NaCl, 0.01*M* EDTA.
4. 10% SDS: (w/v) in water.
5. Sarcosine: 10% (w/v) in water.
6. Proteinase K: 20 mg/mL in water.
7. Phenol saturated with 0.2*M* Tris-HCl, pH 7.2.
8. Phenol:chloroform:isoamyl alcohol, 25:24:1.
9. Sodium acetate: 3.0*M* in water.
10. Heating block at 56°C.
11. 100% ethanol at –20°C.
12. Bench centrifugation capable of 14,000*g* for Eppendorf tubes.
13. 80% Ethanol at –20°C.
14. PCR-grade water

2.2.2. Specimen Preparation from Cultures

1. Inoculated culture medium or well-separated colonies on agar.
2. Bench centrifugation capable of 14,000*g* for Eppendorf microtubes.
3. PBS (Life Technologies, Renfrewshire, Scotland).
4. PCR-grade water.
5. Heating block at 100°C.

2.3. DNA Amplification by PCR

1. Thermal cycler (Perkin-Elmer, Norwalk, CT, 480, or 9600 for 96-well format compatibility).
2. Oligonucleotides (*see* **Note 3**):
   450B 5'-Biotin-GTATTTTCCTTTCTAATTTG
   451+ 5'-GGATGACTCATTTAATAAATCAAATTAATAAGTGTG
   Adjust each primer concentration to 50 pmol/μL.
3. dNTPs (Perkin-Elmer).

4. *Taq* polymerase, 10X reaction buffer, 25 m*M* $MgCl_2$ (AmpliTaq DNA polymerase 10X PCR buffer II, Perkin-Elmer).

2.4. Colorimetric Detection

1. *pleuro*TRAP kit (AMRAD, Australia) (*see* **Note 4**).
2. Prepare all reagents supplied as stocks according to the manufacturer's instructions.
3. Stop solution (0.5*M* sulfuric acid).
4. An 8- or 12-channel multipipet (not essential).
5. Suitable 96-well microtiter plate or strip washer (not essential).
6. Suitable 96-well microtiter plate reader capable of a 450-nm wavelength.

2.5. Agarose Gel Electrophoresis

1. DNA-grade agarose for 1% gels (Boehringer Mannheim, Germany) and 2% gels SDF agarose (Amersham, Buckinghamshire, UK).
2. TAE, DNA typing grade, 50X stock (Life Technologies).
3. Ethidium bromide, 10 mg/mL, (Life Technologies).
4. Gel loading dye (Sigma, St. Louis, MO).
5. DNA size markers, Amplisize (Bio-Rad, Milan, Italy).
6. Submarine DNA electrophoresis tank, gel tray, and comb.
7. Power supply.
8. UV transilluminator.
9. Gel documentation equipment.

2.6. Restriction Enzyme Digestion

1. DNA; PCR product.
2. Restriction enzyme *Asn*1 (Boehringer Mannheim, Germany) or *Vsp*1 (Promega, Madison, WI) and matching buffers.
3. Heating block at 37°C.

3. Method

3.1. Specimen Collection

3.1.1. Lung and Lymph Node Tissue

1. Wear protective clothing and gloves during sample collection, and change gloves between specimens. Be sure to discard protective wear before leaving the site.
2. With a fresh sterile scalpel, excise about 1 g of lung or lymph node tissue selected from an area at the border of a typical lesion or a necrotic area of the lung, and/or the interior of a lymph node (*see* **Note 5**)
3. Place the tissue into a sterile container, and transport to the laboratory at 4°C within 24 h. Tissue may also be transported in mycoplasma transport medium.

3.1.2. Tracheal Scrapings

1. Wear protective clothing and gloves during sample collection, and change gloves between specimens.

2. Make a longitudinal slit along the trachea and expose the lumen.
3. Scrape the mucosal layer with a fresh scalpel, and transfer the scrapings into a tube with 3 mL of mycoplasma transport media.
4. Transport to the laboratory may be at ambient temperature, but the time should not exceed 24 h.

3.1.2. Nasal Swabs

1. Wear protective clothing and gloves during sample collection, and change gloves between specimens if necessary.
2. Make sure that the animal is held firmly in the crush.
3. Take a fresh rayon swab, and swab the nasal passage of the animal with a quick, firm motion. Be short and precise in this action, since this causes the animal some distress. Dislodge the collected mucus into a tube with 3 mL of mycoplasma transport media with a twirling motion of the swab, and squeeze the swab against the side of the tube.
4. Discard the swab.
5. Transport to the laboratory may be at ambient temperature, but the time should not exceed 24 h.

3.2. Sample Preparation

3.2.1. DNA Extraction from Tissues

1. Select 1 g of tissue from an area at the border of a typical lesion or a necrotic area of the lung, and/or the interior of a lymph node.
2. Homogenize the tissue in 1 mL of transport medium.
3. To 200 µL of the suspension, add 100 µL of TNE, 10 µL of 10% *N*-lauroylsarcosine, and 10 µL of 10 mg/mL proteinase K.
4. Incubate the mixture at 56°C for 15 min and then 37°C for 45 min.
5. Extract the DNA from the lysate with equal volumes of phenol, followed by phenol:chloroform:isoamyl alcohol.
6. Precipitate the DNA from the aqueous phase with 3*M* sodium acetate and ethanol.
7. Pellet the DNA by centrifugation at 14,000*g* for 10 min, and wash the pellet in 80% ethanol.
8. Resuspend the DNA thoroughly in 20 µL of PCR-grade water, and prepare a 1:50 dilution with PCR-grade water.
9. Use 1 µL of each neat and 1:50 dilution in PCR-grade water as template for PCR.

3.2.2. Specimen Preparation from Cultures

1. Centrifuge 200 µL of culture medium inoculated at 14,000g for 10 min. Medium inoculated with tracheal scrapings as well as nasal swabs may be used (*see* **Note 6**).
2. Wash the pellet in 200 µL of PBS.
3. If there are colonies on agar plates, pick one well-separated colony from the agar surface avoiding the agar, and resuspend the organisms in 200 µL of PBS.

Table 1
Preparation of PCR Master Mix

No. of tests	1	3	5	10	15	20	30	50
dH_2O	30.5	91.5	152.5	305	457.5	610	915	1525
RB x 10	5	15	25	50	75	100	150	250
25m*M* $MgCl_2$	3	9	15	30	45	60	90	150
dNTP 1.25 m*M*	8	24	40	80	120	160	240	400
Primer 450B	1	3	5	10	15	20	30	50
Primer 451+	1	3	5	10	15	20	30	50
Taq polymerase	0.5	1.5	2.5	5	7.5	10	15	25

4. Collect the organisms in 200 µL PBS, and resuspend the pellet in 50 µL of PCR-grade water.
5. Heat the suspension to 100°C for 7 min, and immediately chill on ice just before use.
6. Use 1 µL as template for PCR.

3.3. DNA Amplification by PCR

1. Prepare the "master mix" for the number of tests required (in the clean room) in the order that they appear in **Table 1**. If more tubes are required, make convenient multiples of any tube.
2. Aliquot 49 µL of master mix into each reaction tube. These may be stored at 4°C for up to 1 wk and at –20°C for up to 1 mo. With the addition of 1 µL of template, this results in 50 µL of reaction mix of 10 m*M* Tris-HCl (pH 8.3), 50 m*M* KCl, 1.5 m*M* $MgCl_2$, 0.001% (w/v) gelatin, 200 µ*M* each dATP, dGTP, dTTP, dCTP, 50 pmol of each primer, and 1.5 U of *Taq* polymerase.
3. Take the number of tubes needed for that particular run into the sample preparation room.
4. Add 1 µL of the desired sample into each tube. Include batch positives and batch negatives with each plate as indicated in **Fig. 1** (for Perkin-Elmer 480 machines, use a drop of mineral oil in each reaction tube) (*see* **Note 7**).
5. Program the thermal cycler to perform the following steps: 94°C for 5 min followed by 30 cycles each of 94°C for 30 s, 50°C for 30 s, 72°C for 30 s (or by 30 cycles each of 94°C for 30 s, 50°C for 30 s, 72°C for 30 s) with a final extension step of 72°C for 5 min (*see* **Note 8**).

3.4. Colorimetric Detection

The *pleuro*TRAP assay is designed to distinguish between amplified target DNA and unamplified DNA. It is therefore critical that the user of this assay ensures that a standard, optimal PCR regime is established prior to commencing a large-scale program. Optimization of the PCR format should be evaluated using agarose gel electrophoresis. The assay, once started, should be performed to completion without interruption.

	1	2	3	4	5	6	7	8	9	10	11	12
A												
B												
C												
D												
E												
F												
G	PN	PP										
H	BN	BP	BN	BP			BN	BP			BN	BP

Fig. 1. Layout of *pleuro*TRAP assay.

1. Bring all of the reagents except streptavidin-peroxidase conjugate to room temperature.
2. Prepare 1X wash buffer, 1X buffer, and conjugate according to procedures detailed in the section "Preparation of Reagents" (*see* **Note 4**).
3. Assemble required number of strips in the plate holder.
4. Add 100 µL of 1X buffer solution to each well, and incubate for 15 min at room temperature. Aspirate solution from coated plate, and blot the inverted plate by tapping firmly, on absorbent paper towel (*see* **Note 9**).
5. Add sample diluent followed by DNA sample to wells to a final volume of 50 µL followed immediately by 50 µL 2X buffer (as supplied). Avoid the introduction of mineral oil into the wells by wiping the outside of the tube with a piece of tissue. Add the batch-positive control and batch-negative control to the wells as indicated in **Fig. 1**, followed immediately by 50 µL 2X buffer (as supplied) (*see* **Note 10**).

   To quality control the assay, use 50 µL of positive plate control and negative plate control, as indicated in **Fig. 1**. Then add 50 µL 2X buffer. Incubate for 30 min at room temperature (*see* **Note 11**).
6. Aspirate sample from plate and wash four to six times using 1X wash buffer (approx 250 µL/well/wash). After the last wash, thoroughly blot the inverted plate by tapping firmly on absorbent paper towels (*see* **Note 12**).
7. Add 100 µL of diluted conjugate to each well. Incubate for 30 min at room temperature (*see* **Note 13**).
8. Prepare substrate prior to the completion of this incubation as detailed in the section "Preparation of Reagents" (*see* **Notes 4** and **14**).
9. Aspirate conjugate from plate, and wash four to six times using 1X wash buffer (approx 250 µL/well/wash). After the last wash, thoroughly blot the inverted plate by firmly tapping on absorbent paper towel (*see* **Note 12**).
10. Add 100 µL of diluted substrate to each well. Incubate in the dark for 15 min at room temperature.

11. Add 100 μL of 0.5*M* H_2SO_4 to terminate the enzymatic reaction.
12. Read absorbance on a microtiter plate reader within 30 min after adding the stop solution using the 450-nm filter with 620-nm filter as the reference, or visually interpret results within 30 min of adding the Stop Solution.
13. Determine results either by use of a spectrophotometric plate reader or visually, following the optimization of the PCR regimen and selection of batch control samples.
14. Express results obtained with a plate reader either as a positive to negative ratio, or as absolute absorbance values. Under normal conditions, the *pleuro*TRAP assay gives positive to negative ratio values of 10:1–50:1. When results are expressed as absolute values, the values obtained for samples should be compared with the absolute values of the batch control samples.
15. Interpret results visually by judging samples that give reaction colors close to the batch-positive control color as positive and judging samples that give a color close to the batch-negative control as negative. It is recommended that any reaction colors that are not clearly defined as positive or negative are either retested using increased sample DNA volumes or tested using another confirmation method (*see* **Note 15**).

3.5. Agarose Gel Electrophoresis

PCR products may be detected by electrophoresis in 1% agarose, but 3% agarose is necessary to resolve the DNA fragments produced by restriction enzyme digestion of PCR products (*see* **Note 16**).

1. Assemble a clean gel tray and the appropriate size comb to contain the agarose gel on a flat, even surface.
2. To prepare a 1% minigel of (7 × 7 cm) measure 40 mL of 1X TAE into a 100-mL beaker, and add 0.4 g of agarose. For a 3% gel, add 1.2 g of agarose to 40 mL of 1X TAE (*see* **Note 17**).
3. Heat the mixture to boiling point, and melt the agarose to a homogenous solution. A microwave oven is the most convenient device for this purpose. Typically, gels may be prepared in 2 min. However, a hot plate at 100°C or a boiling water bath may also be used.
4. Hold the mixture at room temperature for about 1 min until it stops to steam, and add 1 μL of ethidium bromide into the liquid.
5. Avoiding bubbles, swirl to mix the ethidium bromide into the molten gel, and cool the mixture at room temperature until it can be held.
6. Pour the gel into the tray avoiding bubbles. Remove any accidental bubbles to the side of the apparatus with a disposable pipet tip, and allow 20 min for the solution to gel.
7. During this time, prepare the PCR products to be analyzed by adding 5 μL of each reaction to 1 μL of gel loading dye into sterile microtubes. Avoid the introduction of mineral oil into the tubes by wiping the outside of the tube with a piece of tissue. At the same time, prepare the DNA size markers by adding 5 μL of the thawed marker solution to 1 μL of gel loading dye. If necessary, collect the liquid in the bottom of the tube by centrifugation.

8. Remove the comb and any material used to seal the sides, and place the gel into the tank with 1X TAE. Additional 1X TAE may be added as required to ensure that the gel is just submerged.
9. Load all of the 6 µL of the samples into the wells avoiding well-to-well spillover.
10. Replace the tank cover, attach the plugs to the power supply, and apply voltage at 100 V for 1 min to drive the DNA into the gel. Reduce the voltage to 80 V, and continue electrophoresis for about 30 min or until the dye reaches 2 cm from the bottom of the tray.
11. Turn off the power, and disconnect the gel apparatus from the power supply, remove the gel tray to the transilluminator, and view the fluorescent DNA bands with UV radiation. A permanent record of the result may be documented by Poloroid photographic of electronic imaging systems.
12. Dispose of the gel containing ethidium bromide appropriately for eventual decontamination by incineration.

3.6. Restriction Enzyme Digestion

Restriction enzymes *Asn*1 or *Vsp*1 are used for the differentiation of the product of *Mmm*SC from that of *Mmm*LC and *M. mycoides* subsp. *capri (Mmc)*.

1. To clean microtubes, add 3 µL of PCR-grade water, 5 µL of positive PCR product, 1 µL of restriction enzyme buffer, and 1 µL of restriction enzyme.
2. Centrifuge the reactants to the bottom of the tube, and incubate the mixture for 1 h at 37°C.
3. Add 1 µL of gel loading dye to the tubes just before analysis of the products by electrophoresis in a 3% agarose gel as described in **Subheading 3.5.**
4. *Mmm*SC produces two restriction products of about 380 and 180 bp, whereas *Mmm*LC and *Mmc* produce three bands of about 230, 180, and 150 bp.

4. Notes

1. Post-PCR DNA analysis with anion-exchange, high-performance liquid chromatographic (HPLC) has been shown to be useful and suitable for automation ***(14)***. Immunological methods for the detection of PCR products with biotin–streptavidin conjugates and probe capture of PCR product have also been described ***(15)***.
2. False-positive results from the contamination of the reaction mix with DNA products from previous positive reactions is a most serious risk connected with the use of PCR for the analysis of large numbers of samples. Methods have been described that destroy the PCR product with UV or enzymatically prior to PCR, but precautions, which include the structure of the laboratory, correct sample, and product handling procedures, must be installed and maintained to produce consistent and useful results ***(16)***.
3. The oligonucleotide primers required for this test are supplied in the *pleuro*TRAP kit.

4. Refer to *pleuro*TRAP kit instructions (Cat. No. 9001240) available from AMRAD Operations Pty. Ltd., 34 Wadhurst Drive, Boronia Vic 3155, Australia. E-mail: amradpb2@ozemail.com.au.
5. The sensitivity of PCR is such that consecutive positive results are likely if the same blade is used to incise the positive tissue, and then for subsequent tissues and animals. Since it is the common practice of veterinary personnel and meat inspectors to use the same knife for all animals, care must be taken to sample from fresh areas of tissue or to sample the animals first for PCR. Since the same personnel are often responsible for sample taking, time must be taken for the education and in the training of these people to provide adequate samples for PCR. In the field, if there is any doubt of crosscontamination, take the appropriate precautions, e.g., change gloves, tubes, and so forth. In the laboratory, follow the same rule.
6. Samples incubated for 24 h are adequate for PCR. Samples incubated for a longer period without prior filtration may be overgrown with other mycoplasmas or bacteria inhibiting the growth of the relevant mycoplasma. Filtered cultures in which mycoplasmas are suspected make excellent samples for PCR.
7. The batch-positive and batch-negative controls should be known positive and known negative samples, which when tested under the same conditions and at the same time as the batch of test samples give positive and negative results, respectively. The *pleuro*TRAP kit has double-stranded DNA from *Mmm*SC to use as a positive PCR control.
8. Amplitaq Gold is recommended; the thermal cycling conditions should be altered to 94°C for 7 min followed by 30 cycles each of 94°C for 30 s and 72°C for 30 s with a final extension step of 72°C for 5 min.
9. At no time should 2X buffer be added to the coated plate prior to addition of the PCR sample.
10. One to 50 µL of the PCR sample can be used per well. The sample must be added to sample diluent to give a final volume of 50 µL. Dilution of samples can be performed in the coated plate. It is important that the sample diluent is added to the wells first, followed by the PCR sample and then the 2X buffer. The 2X buffe is added to the diluted PCR sample prior to transfer to the coated plate. Alternatively, sample preparation can be conducted in a separate blank microtiter plate. The *pleuro*TRAP assay normally gives excellent discrimination between positive and negative results with 5 µL of the PCR sample.
11. The positive plate control and a negative plate control included are utilized to determine the kit's integrity. Under normal operating conditions, the positive plate control and the negative plate control should have absorbance values of ≥ 0.500 and <0.100, respectively. An absorbance value for the negative plate controls of >0.200 or a positive plate control value of <0.400 may indicate that one or more of the reagents in the kit have deteriorated or that the procedure for the assay was not followed correctly. These values are a guide to the integrity of the kit and are not intended for use as a cutoff value for results.
12. Automated plate or strip washers may be used with this system.

13. The HRPO conjugate should be made fresh for each plate and used within 90 min.
14. The substrate solution should be made fresh for each plate in a clean tube protected from light and used immediately. The substrate should be colorless to very pale yellow and should not be used if it has a distinct yellow-orange color, which indicates contamination or deterioration of the reagent. The working substrate reagent should be colorless. A distinct blue color indicates that the reagent is contaminated and should be discarded.
15. In the event that the batch-negative control results in strong color, check for the formation of primer-dimers by gel electrophoresis, and reoptimize the PCR. The amplification conditions have been designed to circumvent the formation of primer dimers and false priming, but consideration should be given to the following factors.
    a. Unbalanced primer concentrations may influence primer-dimer formation. Therefore, the primers should be equimolar and between 10 and 50 pol/50 µL (0.2–0.8 µ*M*) reaction; excess primers will enable primer-dimers to form more readily.
    b. Excessive cycling may also promote primer-dimers and therefore not >30 cycles are recommended.
    c. "Hot-start" PCR inhibits false priming of oligonucleotides and therefore helps prevent primer-dimers. "Hot-start" is performed by omitting one or more of the essential components of the PCR (such as *Taq* polymerase or dNTP), while the remaining components are heated to 95°C for 5 min. After the 5 min, the tubes are lowered to 80°C and not allowed to go below this temperature, whereas the final component is added and thermal cycling is begun immediately. Alternatively, wax can be used to separate the reagents while the temperature is being raised to the denaturing temperature, such as detailed for AmpliWax™ (Perkin Elmer). *Taq* GOLD™ may also be used.
    d. Concentration of each nucleotide should be between 20 and 200 µ*M,* since this influences the free magnesium concentration, which in turn influences *Taq* polymerase activity and fidelity.
    e. Mg^{2+} concentration should be approx 1.5–3.5 m*M* and should be optimized for each primer pair.
    f. *Taq* polymerase concentration should be as low as possible to minimize nonspecific amplification e.g., 1.0–1.5 U/50 µL reaction. Note: The PCR is covered by US Patents 4,683,195 and 4,683,202 and corresponding patent rights in other territories, owned by Hoffmann-La Roche Inc. and F Hoffmann-La Roche Ltd.
16. Wear gloves during the preparation and performance of agarose gel electrophoresis. Usually the equipment is contaminated with ethidium bromide.
17. The buffer used for the preparation of the gel must be from the same batch as the buffer used in the tank.

References

1. Taylor, T. K., Bashiruddin, J. B., and Gould, A. R. (1992) Relationships between the members of the *Mycoplasma mycoides* cluster as shown by DNA probes and sequence analysis. *Int. J. System. Bacteriol.* **42,** 593–601.

2. Askaa, G., ErnØ, H., and Ojo, M. O. (1978) Bovine mycoplasmas: classification of groups related to *Mycoplasma mycoides. Acta. Vet. Sand.* **19,** 166–178.
3. Costas, M., Leach, R. H., and Mitchelmore, D. L. (1987) Numerical analysis of PAGE protein patterns and the taxonomic relationships within the "*Mycoplasma mycoides* cluster". *J. Gen. Microbiol.* **133,** 3319–3329.
4. Bashiruddin, J. B., Taylor, T. K., and Gould, A. R. (1994) A PCR based test for the specific identification of *Mycoplasma mycoides* subspecies *mycoides* SC. *J. Vet. Diagn. Invest.* **6,** 428–434.
5. Egwu, M. O., Nicholas, R. A. J., Ameh, J. A., and Bashiruddin, J. B. (1996) Contagious bovine pleuropneumonia: an update. *Vet. Bull.* **66,** 875–888.
6. Nicholas, R. A. J. and Bashiruddin, J. B. (1996) *Mycoplasma mycoides* subspecies *mycoides* (small colony variant): the agent of contagious bovine pleuropneumonia and a member of the "*Mycoplasma mycoides* cluster". *J. Comp. Pathol.* **113,** 1–27.
7. Bashiruddin, J. B., Nicholas, R. A. J., Santini, F. G., Woodward, M. J., and Taylor, T. K. (1994). Use of Polymerase chain reaction to detect mycoplasma DNA in cattle with contagious bovine pleuropneumonia. *Vet. Rec.* **134,** 240–241.
8. Brandão, E. (1995) Isolation and identification of *Mycoplasma mycoides* subsp. *mycoides* SC strains in sheep and goats. *Vet. Rec.* **136,** 98–99.
9. Brandão, E., Botelho, A., Bashiruddin, J. B., and Nicholas, R. A. J. (1994) Use of polymerase chain reaction for identification of *Mycoplasma mycoides* subsp. *mycoides* SC isolated from sheep and goats. *IOM Letters* **3,** 24–25.
10. Dedieu, L., Mady, V., and Lefevre, P. C. (1994) Development of a selective polymerase chain reaction assay for the detection of *Mycoplasma mycoides* subsp. *mycoides* SC (contagious bovine pleuropneumonia agent). *Vet. Microbiol.* **42,** 327–339.
11. De Santis, P. and Bashiruddin, J. B. (1996) Application of *Mycoplasma mycoides* subsp. *mycoides* PCR to Italian field samples, in *Mycoplasmas of Ruminants: Pathogenicity, Diagnostics, Epidemiology and Molecular Genetics* (Frey, J. and Sarris, K., eds.), EUR 16394 EC, Brussels, pp. 141–143.
12. Kaura, H. T. and Hübschle, O. J. B. (1996) Use of PCR to detect Mycoplasma mycoides subspecies *mycoides* from nasal filter strips. *Vet. Rec.* **138,** 444–445.
13. Kemp, D. J., Smith, D. B., Foote, S. J., Samaras, N., and Peterson, M. G. (1989) Colorimetric detection of specific DNA segments amplified by polymerase chain reactions. *Proc. Natl. Acad. Sci. USA* **86,** 2423–2427.
14. Katz, E. D. (1993) Quantitation and purification of polymerase chain reaction products by high-performance liquid chromatography, in *PCR Protocols: Current Methods and Applications,* vol. 15 (White, B., ed.), Humana Press, Totowa, NJ, pp. 63–74.
15. Lazar, J. G. (1996) Immunological detection of PCR products, in *PCR Primer: A Laboratory Manual* (Dieffenbach, C. W. and Dveksler, G. S., eds.), pp. 177–192.
16. Dieffenbach, C. W., Dragon, E. A., and Dveksler, G. S. (1996) Setting up a PCR Laboratory, in *PCR Primer: A Laboratory Manual* (Dieffenbach, C. W. and Dveksler, G. S., eds.), Cold Spring Harbor Laboratory Press, NY, pp. 7–16.

20

Characterization of Mycoplasmas by RAPD Fingerprinting

Georges A. Rawadi

1. Introduction

The diagnosis and typing of microorganisms of human and veterinary significance are vitally important in the rapid and effective treatment of infectious diseases. Traditionally, the growth of bacteria in specific media and/or the immunological detection of surface antigens is used to identify and compare the infectious agents. Although these procedures are considered routine techniques, many microorganisms, including mycoplasmas, are difficult to grow or share common antigens responsible for serological crossreactivity. Diagnostic microbiology laboratories are continuously updating methods in an effort to provide highly accurate, inexpensive, and rapid results for clinicians and research scientists. Genomic methods, such as DNA fingerprinting, DNA and RNA probes, hybridization, and PCR, meet most of these criteria.

After the spectacular expansion of PCR techniques during the last decade, and its widespread application in both basic research and clinical diagnosis, some derivative techniques, such as random amplification polymorphic DNA (RAPD) or arbitrarily primed-PCR (AP-PCR) have been developed. These techniques allow the typing of a given DNA (eukaryotic or prokaryotic DNA) *(1)*, and when applied to the diagnosis and typing of mycoplasmas, overcome many of the problems encountered with these fastidious bacteria significantly improving results obtained with traditional microbiological methods.

Classical PCR uses two well-defined primers (forward and reverse) to amplify a known DNA segment with the specificity of the amplification verified by the size of the amplicon (base pair length). RAPD, on the other hand, involves arbitrarily chosen primers without prior knowledge of the nucleotide sequence of the target DNA. The generated RAPD fingerprint results from the

From: *Methods in Molecular Biology, Vol. 104: Mycoplasma Protocols*
Edited by: R. J. Miles and R. A. J. Nicholas © Humana Press Inc., Totowa, NJ

chance annealing of the arbitrarily chosen primers to selective sequences and the amplification of the flanked regions. Depending on the frequency of these selective sequences within the genomic DNA, the fingerprints obtained may vary from one species to another, and in some particular cases, from one strain to another within the same species. To allow adequate matching of these primers, the DNA synthesis is initiated at low stringency for the initial PCR cycles (typically 5–10 cycles). High complementarity of the 3'-end bases of the primers is crucial for the effective priming and amplification during these cycles. Additional PCR cycles (25–30 cycles) are performed at higher annealing temperatures (up to 50°C), since enough DNA template has typically been generated. The reproducibility of the resulting pattern, or fingerprint, depends on numerous factors, such as the amount of $MgCl_2$, template concentration, and thermostable polymerase ***(2,3)***.

RAPD fingerprinting can be performed on genomic DNA as well as on RNA. RNA fingerprinting, also referred to as "differential display," is mainly used for identification of new and/or differentially expressed genes. For typing and identification of cells (prokaryotic or eukaryotic cells), the DNA template is a more suitable choice. Once the RAPD procedure is optimized to generate a reproducible genomic fingerprint, it can be used as a "personal signature" of the particular species. Differences in DNA fingerprints between two cells is a sign of polymorphism during the evolutionary process or mutations that may have occurred throughout the generation. Thus, the application of RAPD could be highly useful in biological, ecological, evolutionary, and epidemiological studies.

The application of RAPD for bacteria typing and identification requires a reproducible and well-defined amplification protocol. In order to set up such a protocol, an appropriate arbitrary primer set must be chosen that generates an amplification pattern from a genomic DNA of interest (in this particular case, mycoplasma DNA). Following this selection, the different parameters that influence the fingerprint and the analytical method applied should then be carefully examined and optimized in order to establish an RAPD amplification protocol. As mentioned above, there are no unique or standard protocols for RAPD fingerprinting, and a protocol must be established for each specific application. This chapter describe the general principles and guidelines for the researcher to establish such a protocol.

2. Materials

1. Thermal cycler (Perkin Elmer 460, Branchburg, NJ).
2. Thin-wall PCR tubes (Perkin-Elmer).
3. Micropipets (20 and 200 µL).
4. Sterile mineral oil.

5. Mycoplasma lysis solution: 100 m*M* EDTA, 150 m*M* NaCl, pH: 8.0, 2.5% SDS, 100 mg/mL Proteinase K, and 50 μg/mL RNaseA.
6. 10X standard PCR buffer for AmpliTaq: 500 m*M* KCl, 100 m*M* Tris-HCl, pH 8.5, 15 m*M* $MgCl_2$, and 0.1% gelatin.
7. 10X PCR buffer without $MgCl_2$ at pH value ranging from 8.0–10.0.
8. 25 m*M* $MgCl_2$ solution.
9. 25 m*M* dNTP solutions: add to the AP-PCR mixture at final concentration of 200 μ*M* each.
10. *Taq* polymerase (AmpliTaq, Perkin Elmer): add to the AP-PCR mixture at final concentration of 0.05 U/μL.
11. Arbitrary primers at 40 μ*M*: add to the AP-PCR mixture to obtain the expected concentration as described below.
12. TAE: 40 m*M* Tris-HCl, 10 m*M* sodium acetate, and 2 m*M* EDTA.
13. TE: 10 m*M* Tris-HCl and 1 m*M* EDTA.
14. PBS, pH 7.4.
15. Phenol/chloroform/isoamyl alcohol (25/24/1).
16. Absolute and 70% cold ethanol.
17. Agarose and metaphor agarose (FMC, Rockland, ME).
18. Arbitrary primers for *Mycoplasma mycoides* AP-PCR fingerprinting:
    Mlip1: 5'-GGTCATTCAATGGGTGGA-3'
    Mlip4: 5'-TCCTCCCATTGAGTGACC-3'

3. Methods

3.1. Mycoplasma DNA Purification

Unlike classic PCR, RAPD amplification cannot be performed directly on a cell lysate or in the presence of foreign DNA, since these can introduce artifacts in the genomic fingerprint. Therefore, generation of a reproducible pattern requires purified DNA.

1. Cultivate mycoplasma species in a suitable media and at density $>10^6$ CCU/mL ***(4)***.
2. Centrifuge at 10,000*g*, at 4°C for 30 min. Wash the pellet twice with cold PBS, and resuspend the bacteria in the lysis solution.
3. Incubate the suspension at 60°C for 15 min, and allow the mixture to cool to room temperature.
4. Extract the lysate with 1 vol of phenol/chloroform/isoamyl solution, and precipitate the DNA from the aqueous phase by adding 2.5 vol of absolute ethanol.
5. Centrifuge the precipitate DNA at 10,000*g* for 30 min at 4°C prior to washing with 70% cold ethanol.
6. Dry the DNA pellet (e.g., use a Speed Vac), and resuspend in 100 μL of TE buffer. Determine the DNA concentration at OD_{260}.

3.2. Assessment of Arbitrarily Chosen Primers Sets

1. Perform PCR using a standard buffer and the mycoplasma DNA of interest. The following cycling conditions are considered applicable in most RAPD protocols: (a)

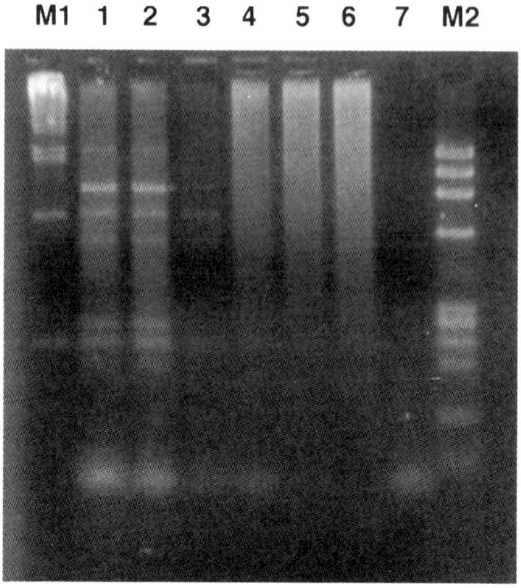

Fig. 1. Optimization of AP-PCR amplification conditions: effect of DNA concentration on fingerprint. AP-PCR amplification was performed using the standard PCR buffer and primers at 1.6 μ*M*. Different amounts of genomic DNA prepared from *M. capricolum* subsp. *capricolum* were added to the reactions: (1) 100 ng, (2) 500 ng, (3) 10 ng, (4) 1 μg, (5) 1.5 μg, (6) 2 μg, and (7) no DNA. M1 and M2 molecular weights were, respectively, λ-*Bst*EII and ϕX174-*Hae*III.

5 cycles as follow: 94°C for 30 s, 37°C for 45 s, and 72°C for 1.5 min; (b) 30 cycles as follow: 94°C for 30 s, 50°C for 45 s, and 72°C for 1.5 min.

2. Run 10–20 μL of the amplification products through 1% agarose gel to reveal the presence of a DNA pattern by ethidium bromide staining. If one (or more than one) of the tested primer sets (*see* **Notes 1** and **2**) demonstrates amplification during the test, continue as in **Subheading 3.3.** Where primers do not initiate amplification, test other primers.

3.3. Optimization of Amplification Parameters

1. To optimize the template and the primer concentrations (*see* **Note 3**), perform an AP-PCR amplification using a standard PCR buffer and the same cycling conditions as above. Vary the DNA concentration from 1 ng to 1 μg, and the primer concentration from 1–3.5 m*M*.
2. Analyze an aliquot of 10–20 μL of the amplification products by gel electrophoresis and ethidium bromide staining.
3. Compare the amplification patterns obtained in these conditions, which will determine the optimum concentration. **Figure 1** illustrates an example of the effect of DNA amount on the genomic fingerprint obtained by AP-PCR.
4. Establish the reproducibility in at least three independent experiments once optimum concentrations are defined (*see* **Note 4**).

5. Determine the most appropriate PCR buffer by varying $MgCl_2$ concentration, and the pH range at the optimum concentrations of DNA and primers.
6. Establish the reproducibility of the final buffer selected (*see* **Note 5**).

3.4. DNA Fingerprint Analysis

1. See **Note 6** and 7.

3.5. AP-PCR Protocol for M. mycoides Characterization

Based on genomic homologies and biochemical assays, the *M. mycoides* cluster comprises six mycoplasma species that are closely related. AP-PCR fingerprinting was developed and applied to differentiate each species within this cluster (*see* **Note 8**). AP-PCR is performed in 50 µL of reaction.

1. Prepare genomic DNA from each of the mycoplasma species included in the *M. mycoides* cluster.
2. Use arbitrary primers for *M. mycoides* AP-PCR fingerprinting.
3. Prepare the following AP-PCR mixture in thin-wall tubes (optimized for this particular application as described above): 50 m*M* KCl, 10 m*M* Tris-HCl, pH 8.5, 2.5 m*M* $MgCl_2$, 200 µ*M* of each of dNTP, 2 µ*M* of each of the primers, and 1.75 U of AmpliTaq DNA polymerase (for 50 µL reaction).
4. Add 300 ng of mycoplasma DNA to each reaction, and overlay the mix with 50 µL of sterile mineral oil.
5. Transfer the tube to the thermal cycler, and run the following program:
   a. 5 cycles as follow: 94°C for 30 s, 37°C for 45 s, and 72°C for 1.5 min;
   b. 30 cycles as follow: 94°C for 30 s, 50°C for 45 s, and 72°C for 1.5 min;
   c. 1 extension cycle at 72°C for 10 min.
6. Run 20 µL of the reaction on the 2.5% metaphor agarose gel in TAE (100 V, for 1.30 h). Stain with ethidium bromide, and wash with distilled water. Examine bands on the UV transluminator (*see* **Note 9**).

4. Notes

1. Primers for RAPD fingerprinting are available from commercial vendors of molecular biology reagents. However, oligonucleotides can be randomly designed or arbitrarily chosen from primers available in the laboratory. It is worth noting that not all primers are able to generate DNA patterns from a genomic template. Therefore, different primers should be assessed for their ability to initiate RAPD amplification. In addition, the length of the primers may also affect the number of amplified fragments in the fingerprint. Long primers (more than 10 mers) are generally used for more complex fingerprints.
2. The *Taq* polymerase selected with a given primer set also affects the complexity of the DNA fingerprint. Thus, these two parameters (i.e., length of the primers and thermostable DNA polymerase) should be considered carefully when designing an RAPD protocol. When using *Taq* DNA polymerase, such as AmpliTaq (Perkin Elmer), 20-mer primers are recommended, although with the Stoffel frag-

ment, a 10-mer primer, is preferred. The high A/T percentage of mycoplasmas genome (up to 80% A/T) is not important when designing primers for their characterization by RAPD, since primers that have been described for mycoplasmas DNA fingerprinting so far are surprisingly rich in G/C.

3. The reproducibility of RAPD fingerprinting is highly dependent on the amplification conditions. Therefore, all parameters that may affect the DNA pattern should be optimized, including template and primer concentrations, PCR buffer mixture ($MgCl_2$ concentration and pH), and thermostable DNA polymerase.
4. In cases where the genomic fingerprint lack reproducibility under optimized AP-PCR conditions, the purity of the stock DNA or the newly prepared one should be suspected. To overcome this problem, the DNA should be subjected to a midiprep Quiagen column as follows: wash the column twice with 5 mL of Quiagen QC buffer, elute the DNA with 5 mL of Quiagen QF buffer, and precipitate by adding 0.7 vol of isopropanol; centrifuge, wash with 70% cold ethanol, air-dry the DNA pellet, and resuspend in 100 µL of TE.
5. In order gain a clear understanding of the identity of the DNA fragments generated by AP-PCR, these fragments can be easily cloned by mean of TA-cloning technique. Sequencing and sequence analysis will provide the expected information. As a result, new coding regions can be identified and used to probe and isolate the corresponding complete genes. Moreover, these cloned fragments can also be used to establish specific probes for DNA hybridization purposes.
6. For any new batch of DNA polymerase, all conditions should be reoptimized. Even under optimized conditions, it is worth noting that DNA fingerprints vary as a result of the thermal cycler used. In fact, thermal cyclers based on a distinct heating and cooling process (e.g., Peltier-effect) are available commercially. In order to generate a reproducible result using other protocols, it is highly recommended to use the same type of thermal cycler.
7. DNA fingerprints obtained by RAPD or AP-PCR amplification represent the specific signature for genomic DNAs of a group of individuals and/or of each individual within the same group. Fingerprints have two types of DNA bands: those present in all individuals of the same group and known as monomorphic, and those present only in some individuals or showing a distinct mobility, which are known as polymorphic. As a result, DNA fingerprint analysis is crucial for distinguishing between these bands, particularly the polymorphic ones. Gel electrophoresis and ethidium bromide staining are widely used to analyze PCR products. However, classic agarose gel electrophoresis is usually not capable of distinguishing tiny variations in band sizes (<50 bp). Specific agarose types are often used for this task. In the case of DNA fingerprint analysis, we recommend the metaphor agarose gel at 2.5% (in TAE buffer), which allows an accurate resolution for bands ranging from 100 bp to 1.3 kb. The percentage of metaphor can be modified to accommodate differences in the expected size range of the fingerprints. In some particular cases, it is very useful to look for variations in bands that are apparently monomorphic using distinct electrophoresis techniques, such as single-strand polymorphism analysis ***(5)*** or denaturing gradient gel electrophoresis (DGGE) ***(6)***. In both methods, the distinction is based on

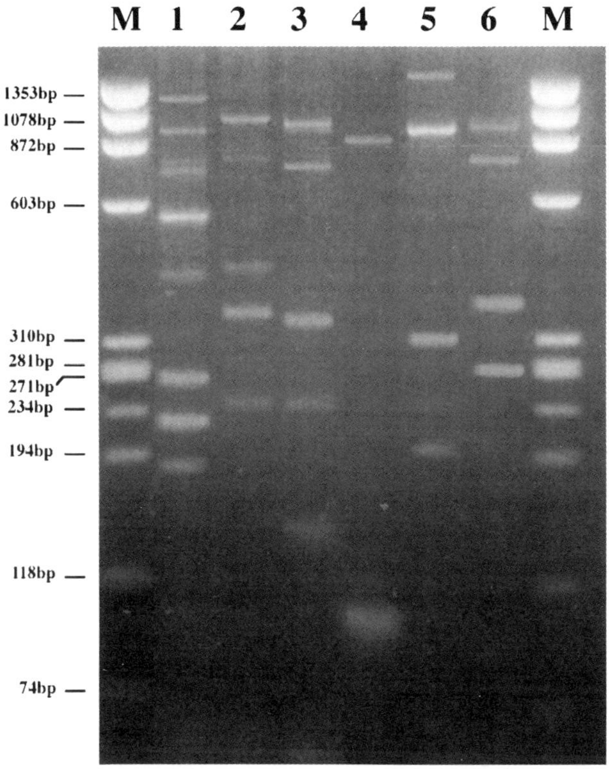

Fig. 2. RAPD fingerprints obtained from type strains of the six mycoplasma groups included in *Mycoides* cluster: (1) *M. capricolum* subsp. capricolum (california kid), (2) *M. mycoides* subsp. *capri* (PG3), (3) *M. mycoides* subsp. *mycoides* LC (Y-goat), (4) *M. mycoides* subsp. *mycoides* SC (PG1), (5) *M. capricolum* subsp. *capripneumoniae* (F38) and (6) *M.* sp. group 7 of Leach (PG50). M, molecular weight was ϕX174-*Hae*III; 20 μL of the AP-PCR were analyzed through 2.5% metaphor agarose gel and stained with ethidium bromide.

the fact that DNA molecules of equal length, but differing by a single base change will migrate differently on a polyacrymalide gel under denaturing or nondenaturing conditions, respectively. Finally, analysis of RAPD fingerprints may be improved via computer-assisted band scanning technology ***(7,8)***. The computer-generated dendrograms can identify separate groups and ultimately distinguish between individuals that were originally considered identical. This approach is particularly useful in bacterial taxonomy and epidemiological studies.

8. Different RAPD fingerprinting protocols have been reported for characterization of different mycoplasma species, including *M. pneumoniae* ***(9)***, *M. hyopneumoniae* ***(10)***, *M. gallisepticum* ***(11)***, and *M. mycoides* cluster ***(12)***.
9. The RAPD fingerprinting technique was performed on the type strains of each of the six species included in the *M. mycoides* cluster. As shown in **Fig. 2**, it is

possible to unambiguously distinguish among all of the six groups by comparing the DNA patterns. Although *M. capricolum* subsp. *capricolum* (**Fig. 2**, lane 1) and *M. c. capripneumoniae* (**Fig. 2**, lane 5) are considered highly related by DNA–DNA hybridization, protein fingerprints, and serological methods, they were clearly differentiated by their respective genomic fingerprints obtained by AP-PCR. This demonstrates the sensitivity of RAPD fingerprinting to distinguish between highly related species. The application of this technique allows a rapid one-step reaction to characterize isolates belonging to the *M. mycoides* cluster through the comparison of their DNA fingerprints to those of the standard or type strains of each species of the cluster. Also note that even when the same primer set is used, the complexity of the RAPD patterns markedly varies from one species to another. Finally, when this AP-PCR technique was used on different strains of the same species or subspecies, the polymorphic bands distinguished between each of the individuals, whereas monomorphic ones displayed their common identity ***(12)***.

References

1. Caetono-Anollés, G. (1993) Amplifying DNA with arbitrary oligonucleotide primers. *PCR Methods Appl.* **3,** 85–94.
2. Bassam, B. J., Caetano-Anollés, G., and Gresshoff, P. M. (1992) DNA amplification fingerprinting of bacteria. *Appl. Microbiol. Biotech.* **38,** 70–76.
3. Meunier, J. R. and Grimont, P. A. D. (1993) Factors affecting reproducibility of random amplified polymorphic DNA fingerprinting. *Res. Microbiol.* **144,** 373–379.
4. Taylor-Robinson, D. and Furr, P. M. (1981) Recovery and identification of human genital tract mycoplasmas. *Isr. J. Med. Sci.* **17,** 648–653.
5. Orita, M., Suzuki, M., Sekiya, T., and Hayashi, K. (1989) Rapid and sensitive detection of point mutations and DNA polymorphisms using the polymerase chain reaction. *Genomics* **5,** 874–879.
6. Sheffield, V. C., Cox, D. R., and Myers, R. M. (1990) Identifying DNA Polymorphisms by Denaturing Gradient Gel Electrophoresis, in *PCR Protocols: a Guide to Methods and Applications* (Innis, M. A., Gelfand, D. H., Sninsky, J. J., and White, T. J., eds.), Academic, New York, pp. 206–218.
7. Sneath, P. H. A. and Sokal, R. R. (1973) The principles and practice of numerical classification, in *Numerical Taxonomy: The Principles and Practice of Classification*, W. H. Freeman, San Francisco.
8. Swofford, D. (1993) PAUP: Phylogenetic Analysis Using Parsimony, version 3.1. Illinois Natural History Survey, Champaign.
9. Ursi, D., Ieven, M., van Bever, H., Quint, W., Niesters, H. G. M., and Goossens, H. (1994) Typing of *Mycoplasma pneumoniae* by PCR-mediated DNA fingerprinting. *J. Clin. Microbiol.* **32,** 2873–2875.
10. Artiushin, S. and Minion, F. C. (1996) Arbitrarily primed PCR analysis of Mycoplasma hyopneumoniae field isolates demonstrates genetic heterogeneity. *Int. J. Syst. Bacteriol.* **46,** 324–328.

11. Geary, S. J., Forsyth, M. H., Aboul Saoud, S., Wang, G., Berg, D. E., and Berg, C. M. (1994) *Mycoplasma gallisepticum* strain differentiation by arbitrary primer PCR (RAPD) fingerprinting. *Mol. Cell. Probes* **8,** 311–316.
12. Rawadi, G., Lemercier, B., and Roulland-Dussoix, D. (1995) Application of arbitrarily-primed polymerase chain reaction to mycoplasma identification and typing within the Mycoplasma mycoides cluster. *J. Appl. Bacteriol.* **78,** 586–592.

21

Classification of Isolates by DNA–DNA Hybridization

Konrad Sachse and Helmut Hotzel

1. Introduction

In routine diagnosis, bacterial field isolates are identified by serological and/or biochemical tests. Occasionally, however, isolates deviate from the generally accepted patterns, i.e., they may react with diagnostic antisera of more than one type strain or fail to react with any of them, or their abilities to ferment certain substrates may be altered. In such a situation, or in the case where no diagnostic antiserum is available for the respective species, the diagnostician can turn to alternative methods of classification based on genotypic rather than phenotypic characteristics, such as DNA–DNA hybridization. Genotypic classification may become even more important in the light of recent findings concerning antigenic variation of mycoplasmas, i.e., the ability of several species to vary size, structure, and composition of major surface antigens with high frequency. Moreover, the investigation of genetic relatedness between bacterial strains in the absence of genomic sequence data, as well as the introduction and definition of new species require information on DNA homology.

DNA–DNA hybridization in its classical membrane filter mode is based on the ability of single-stranded nucleic acid fragments to form duplexes (hybrids) via basepairing with filter-immobilized counterparts of complementary nucleotide sequence. Usually, chromosomal DNA of strains to be investigated is loaded onto nitrocellulose or nylon membranes, and probed with a solution of labeled single-stranded genomic DNA of a reference strain. Under well-defined conditions, the extent of duplex formation can serve as a universal measure of DNA sequence homology between different samples.

A number of factors affecting hybridization rates have to be considered when selecting experimental conditions. First of all, the formation of duplexes is

From: *Methods in Molecular Biology, Vol. 104: Mycoplasma Protocols*
Edited by: R. J. Miles and R. A. J. Nicholas © Humana Press Inc., Totowa, NJ

temperature-dependent with the optimum at 20–25°C below the theoretical melting temperature T_m. The latter is defined as the temperature at which half the strands are dissociated or denatured, and represents an important parameter of DNA hybrid stability. It can be calculated from the following relationship *(1)*:

$$T_m = 81.5 + 16.6 (\log M) + 0.41 (\%G + C) - 0.72 (\% \text{ formamide}) \qquad (1)$$

where M is the molarity of the monovalent cation in the hybridization buffer, and (% G + C) is the molar percentage of guanine and cytosine residues in the DNA.

In practice, most hybridizations in aqueous solution are carried out at 68°C or, alternatively, in solutions containing 50% formamide at 42°C. Since the addition of formamide leads to a decrease of Tm (*see* **Eq. 1**), this is a practical way to lower the hybridization temperature, which may be convenient in many cases, because probes are more stable and nucleic acid noncovalently bound to filter membranes is better retained at lower temperature. Formamide can also be used to alter the stringency of reaction conditions. If DNA homology among strains of different species is to be investigated in the same experiment, it is useful to hybridize at temperatures below the above-mentioned optimum in order to include partially mismatched duplexes and have a wider range of differentiation.

High concentration of the single-stranded probe DNA in the reaction mix can seriously impair the accuracy of hybridization results owing to sizeable reassociation of probe strands and nonspecific binding to the filter membrane. The effects of other parameters, such as ionic strength and G + C content of the DNA, are usually negligible at conditions used in standard protocols.

In the present chapter, two methods are described that have proven useful for qualitative screening or quantitative evaluation of DNA homology, respectively. Dot-blot hybridization using a 96-well manifold represents a rapid method for screening large numbers of bacterial isolates. The protocol presented here is partly based on instructions in the DIG DNA Labeling and Detection Kit (Boehringer Mannheim, Germany). Formation of duplexes between target DNA and digoxigenin-labeled probe is visualized by immunostaining using an alkaline phosphatase-conjugated antidigoxigenin antibody and a suitable substrate. Since the procedure does not require radioactive labeling, it can be carried out in any laboratory equipped for molecular biological work. In order to obtain semiquantitative data, the colored dots can be scanned using a densitometer. However, this requires considerable expertise and extensive preliminary studies on calibration. The filter disk hybridization procedure should be used when quantitative data are needed *(2)*, but is more laborious in comparison to dot-blot hybridization, because each filter has to be loaded and processed individually. Moreover, radioactively labeled probes are used, which requires special equipment.

2. Materials

2.1. Dot-Blot Hybridization

All buffers and water must be sterile. Water should be deionized.

1. Mycoplasma chromosomal DNA: must be purified by phenol-chloroform extraction or chromatography. Check purity spectrophotometrically. The criterion is A_{260nm}/A_{280nm} = 1.7–2.0.
2. Filter membranes: Hybond N (Amersham, Braumschweig, Germany).
3. Dot-blot apparatus: Minifold I -SRC96 D (Schleicher & Schuell, Dassel, Germany).
4. TE buffer: 10 m*M* Tris-HCl, pH 7.5, and 1 m*M* EDTA.
5. Denaturation buffer: 1.5*M* NaCl and 0.5*M* NaOH.
6. Neutralization buffer: 0.5*M* Tris-HCl, pH 8.0, and 3*M* NaCl.
7. Restriction enzymes: *Eco*RI, *Hind*III, or other suitable nonrare-cutting enzymes.
8. Reagent for phenol-chloroform extraction: prepare saturated solution of phenol in TE buffer and chloroform/isoamyl alcohol = 24:1 (v/v). Mix equal volumes of both solutions prior to extraction.
9. Microcentrifuge, e.g., desktop centrifuge # XX42C CFO 50 (Millipore, Eschborn, Germany).
10. Vortex shaker, e.g., Microshaker VF-2, Janke & Kunkel (Staufen, Germany)
11. Hexanucleotide mixture: mixture of hexamer nucleotides of all possible sequences for random-primed DNA labeling (#1277 081, Boehringer Mannheim).
12. dNTP labeling mixture: DIG DNA Labeling Mix (#1277 065, Boehringer Mannheim).
13. Klenow enzyme.
14. 0.2*M* EDTA solution.
15. 4*M* LiCl solution.
16. Hybridization flasks, e.g., # 052–003 (Biometra, Göttingen, Germany) .
17. Hybridization oven, e.g., OV2 (Biometra).
18. 20X SSC buffer: 3*M* NaCl and 0.3*M* Na-citrate, pH 7.0.
19. Blocking reagent: Stock solution is prepared as a 10% (w/v) solution of #1096 176 (Boehringer Mannheim) in buffer A, heated at 70°C for 1 h, autoclaved, and stored at –20°C.
20. Hybridization buffer: 5X SSC, 0.1% (w/v) *N*-lauroylsarcosine, 0.02% (w/v) SDS, and 1% (w/v) blocking reagent. Store at –20°C.
21. Washing buffer I: 2X SSC and 0.1% (w/v) SDS.
22. Washing buffer II: 0.1X SSC and 0.1% (w/v) SDS.
23. Horizontal shaker, e.g., IKA-VIBRAX-VXR (Janke & Kunkel).
24. Buffer A: 0.1*M* maleic acid and 0.15*M* NaCl. Adjust with NaOH to pH 7.5, and autoclave.
25. Buffer B: 0.1*M* Tris-HCl, 0.1*M* NaCl, and 0.05*M* $MgCl_2$. Adjust to pH 9.5.
26. Buffer C: 10 m*M* Tris-HCl, pH 8.0, and 1 m*M* EDTA.
27. Antidigoxigenin antibody (Fab fragments) conjugated to alkaline phosphatase (Boehringer Mannheim). Once opened, store at 4°C.
28. Color substrate solution: Mix 66 µL of a 5% (w/v) stock solution of nitroblue tetrazolium in 70% (v/v) dimethyl formamide (DMF), 33 µL of 6% (w/v)

stock solution of bromo-chloro-indolyl phosphate in DMF, and 10 mL of buffer B. Store stock solutions at –20°C.

2.2. Filter Disk Hybridization

1. Buffers: SSC and 0.2*M* EDTA as above.
2. Labeling reagents and Klenow enzyme as above.
3. Filter holder, diameter 47 mm, e.g., Millipore # XX1004700.
4. Filters, e.g., Millipore #GSWP 04700.
5. α-^{32}P-deoxyadenosine triphosphate (Amersham), SA 220 TBq/mmol.
6. QIAquick spin columns (QIAGEN).
7. Denhardt's solution: 0.02% (w/v) Ficoll (Pharmacia, Freiburg, Germany), 0.02% (w/v) polyvinyl pyrrolidone, and 0.02% (w/v) bovine serum albumin.
8. 20-mL vials for liquid scintillation counting (LSC): Polyvials #307 0001 (Zinsser, Frankfurt, Germany).
9. LSC cocktail, e.g., Rotiszint eco plus (Roth, Karlsruhe, Germany).

3. Methods

3.1. Dot-Blot Hybridization

3.1.1. Immobilization of DNA on the Membrane

1. Place membrane sheet of 120 × 80 mm into 96-well dot-blot apparatus, and load wells with 3 µg of purified chromosomal DNA (contained in approx 10 µL TE buffer) of each strain, including homologous and heterologous controls. Apply gentle vacuum to ensure slow passage of the solution (*see* **Notes 1** and **2**).
2. Put the loaded and air-dried membrane on filter paper soaked with denaturation buffer for 5 min.
3. Put membrane on filter paper soaked with neutralization buffer for 5 min.
4. Air-dry the membrane by maintaining it on dry filter paper for 30 min.
5. Immobilize DNA by heating ("baking") the membrane at 80°C for 1 h.

3.1.2. Labeling of the Reference DNA (Random-Primed Labeling with Digoxigenin)

1. Cleave chromosomal DNA of the reference strain by digestion of 1 µg DNA with 10 U of restriction enzyme at 37°C for 4–16 h. The standard volume of the digestion mix is 20 µL.
2. Remove enzyme by phenol-chloroform extraction: Add sterile deionized water to make up to 100 µL, add 100 µL of extraction reagent, vortex, microcentrifuge for 30 s, transfer aqueous phase into another tube, and precipitate DNA with 2 vol of cold ethanol.
3. Collect precipitate by centrifugation at 12,000*g* for 20 min, discard supernatant, and put the tube upside down for 10 min to allow the liquid to drain. Redissolve in 15 µL of TE buffer. Prior to labeling, denature DNA by heating in a boiling water bath for 5 min and subsequent cooling in ice.

4. For labeling, add the following reagents to the denatured DNA: 2 µL hexanucleotide mixture, 2 µL dNTP labeling mixture, and 1 µL Klenow enzyme. Vortex, microcentrifuge a few seconds and incubate at 37°C for at least 60 min.
5. Stop the reaction by adding 2 µL of 0.2*M* EDTA.
   Steps 6–8 are optional (*see* **Note 6**).
6. To precipitate DNA, add 2.5 µL of 4*M* LiCl and 75 µL of cold ethanol, vortex, and maintain the tube at –70°C for 30 min or at –20°C for 2 h.
7. Centrifuge at 12,000*g* for 20 min, wash pellet with cold 70% (v/v) ethanol, and recentrifuge for 5 min.
8. Dry DNA pellet under vacuum, and redissolve in 50 µL TE buffer.

*3.1.3. Hybridization (see **Notes 3–5**)*

1. Prehybridization: Place loaded and dried membrane sheet into a hybridization flask, add 20 mL of hybridization buffer/100 cm^2, and rotate the flask in an hybridization oven at 68°C for 1 h.
2. Remove liquid, add 2.5 mL of hybridization buffer/100 cm^2 and 100 ng of fresh denatured, digoxigenin-labeled reference DNA. Carry out hybridization by rotating the flask at 68°C for at least 6 h.
3. Wash membrane twice with 50 mL of washing buffer I/100 cm^2 at room temperature for 5 min.
4. Wash membrane twice with the same volume of washing buffer II at 68°C for 15 min. The membrane is now ready for immunological staining, or it may be air-dried and stored in a desiccator at room temperature (*see* **Notes 6** and **7**).

3.1.4. Immunological Staining of Hybridization Membranes

The following incubations and reactions are carried out at room temperatures with shaking. Volumes are calculated/100 cm^2 of membrane area.

1. Incubate membrane in 100 mL of buffer A for 1 min.
2. Incubate membrane in 100 mL of blocking reagent diluted 1:10 in buffer A for 30 min.
3. Incubate membrane in 20 mL of a 1:5000 dilution of antidigoxigenin antibody in the same solution as in **step 2** for 30 min.
4. Wash twice in 100 mL of buffer A for 15 min.
5. Wash twice in 100 mL 0.5X SSC at 65°C for 10 min.
6. Equilibrate membrane in 20 mL of buffer B for 2 min.
7. Place the soaked membrane on the bottom of a suitable box, and add 10 mL of color substrate solution. Close the box, and leave in the dark without disturbing.
8. When the spots have developed to the desired intensity, stop the reaction by washing in 50 mL of buffer C for 5 min. The membrane may be air-dried and stored.
9. If semiquantitative evaluation by densitometric scanning is intended, the wet membrane should be photocopied onto overhead transparencies or photographed on reversal film.

3.2. Filter Disk Hybridization

1. Prepare solutions containing 300 µg of purified chromosomal DNA of each strain in 6 mL 6X SSC, and subject them to one heat-denaturation cycle.
2. Place nitrocellulose filter disks of 47-mm diameter previously soaked in 6X SSC for 20 min into a filter holder. Apply gentle suction to draw the DNA solution slowly through the filter (*see* **Note 8**).
3. Dry filters at room temperature, and then at 80°C for 4 h. Store them until use in a desiccator at 4°C.
4. For the hybridization assay, punch 20 disks of 6-mm diameter out of each filter.
5. Carry out steps 1–5 as in **Subheading 3.1.2.**, except that the dNTP labeling mixture (step 4) has to be replaced with 1 µL32P-adATP and 1 µL 2 mM dCTP, 1 µL 2 mM dGTP, 1 µL 2 m*M* dTTP (*see* **Note 9**).
6. Remove nonincorporated radioactive label using a QIAquick spin column according to the manufacturer's instructions (*see* **Note 10**).
7. Place each filter disk into a 20-mL LSC vial, and add 500 µL of Denhardt's solution (*see* **Note 11**). Carry out prehybridization by gently shaking the vials at the temperature of hybridization for 3–4 h.
8. Remove liquid from each vessel. Add 200 µL 3X SSC and formamide, if applicable (*see* **Note 12**), and 10 ng of 32P-labeled fresh denatured probe DNA.
9. For hybridization, incubate the closed vials at 68°C (or at a lower temperature, if formamide is included) on gentle shaking overnight (16–24 h).
10. Collect hybridization mixture in a special container for radioactive waste. Wash filter disks three times with 500 µL of cold 3X SSC.
11. Add 5 mL of LSC cocktail to each vial, and allow filters to dissolve.
12. Measure count rates.
13. Calculate hybridization rates (relative binding rates) for each strain. The value of the respective homologous control is set at 100% (*see* **Note 13**).

4. Notes

1. In dot-blot hybridization, characteristic homologous and heterologous controls have to be included on each membrane. This is the most straightforward method of checking whether stringency conditions are correct. Usually the type strains of all species in question are used as controls.
2. It is very important to know the concentration of each DNA preparation in order to make sure that defined (and equivalent) quantities are loaded onto the filter membranes. UV absorption measurement is an easy and reliable method. Spectrophotometers of several manufacturers are equipped with DNA quantitation software. If no such model is available, a calibration curve can be set up at 260 nm using purchased DNA (e.g., calf thymus DNA) and quantified on this basis. In any case, concentration data are only reliable for sufficiently pure DNA containing no or very little protein. Therefore, each preparation should be checked for purity (A260nm/A280nm = 1.7–2.0) prior to quantitation.
3. In the hybridization mix, the concentration of single-stranded probe DNA should be <100 ng/mL in order to rule out significant reassociation of probe

strands and ensure an optimal quantitative ratio between probe and (immobilized) target DNA. The latter should be on thc order of 1:1000.

4. Fragmentation of probe DNA, i.e., chromosomal DNA of reference strains, is necessary to attain optimal kinetics of duplex formation. Hybridization rates would be very low owing to steric hindrance if chromosomal DNA strands in their native length were allowed to interact with target DNA. The conditions of complete cleavage by restriction endonuclease digestion, particularly time of incubation, have to be established empirically, since enzyme qualities differ considerably between manufacturers.
5. The times indicated for digoxigenin labeling, prehybridization, hybridization and washing steps represent the minimum and can be extended without adverse consequences. For instance, the labeling reaction (**step 4** in **Subheading 3.1.2.**) can be allowed to proceed for up to 20 h, which can bring about some increase in specific labeling rates. Also, it is often convenient to carry out the hybridization step overnight.
6. After completion of the digoxigenin labeling reaction, the mixture can be subjected to one denaturation cycle (5 min of boiling and quick cooling in ice) and used for hybridization immediately. If labelled DNA is to be stored, **steps 6–8** in **Subheading 3.1.2.** are necessary.
7. The hybridization mix containing digoxigenin-labeled DNA can be reused several times if it is stored at –20°C and heat-denatured prior to each hybridization. The probe is stable for up to 1 yr.
8. On average, 200–250 μg of DNA can be retained by a filter. The amount of bound DNA may be checked by measuring UV absorption at 260 nm before and after passage through the filter. Slow passage of the solution is important to attain high loads. Each 6-mm disk punched out of the large filter should carry approx 10 μg DNA.
9. Radioactive labeling can also be done by nick translation. The final specific activity of the probe DNA should be ≥5 × 106 dpm/μg.
10. Labeled DNA can also be separated from nonincorporated radioactive dATP by passing the reaction mix spiked with some Dextran Blue (Pharmacia) as indicator over a small chromatographic column filled with 1 mL Sephadex G-50 (Pharmacia). The high mol-wt DNA fraction is eluted immediately after the blue color zone (elution buffer: TE).
11. Blocking reagent prepared as mentioned under **item 19** in **Subheading 2.1.** can be used instead of Denhardt's solution.
12. Using formamide in the hybridization buffer can be a convenient practical alternative. Calculate the necessary volume proportion from the formula given in **Subheading 1.**
13. To obtain reproducible results, it is recommended that two series of each trial should be conducted with reference strains tested in quadruplicate and all other strains in triplicate.

References

1. McConaughy, B. L., Lairds, C. D., and McCarthy, B. J. (1969) Nucleic acid reassociation in formamide. *Biochemistry* **8,** 3289–3295.
2. Schimmel, D. and Sachse, K. (1993) Classification of *Pasteurella* field strains isolated from farms in Germany using traditional methods and DNA–DNA hybridization. *Zbl. Bakt.* **279,** 125–130.

22

Insertion Sequence Analysis

Joachim Frey

1. Introduction

Prokaryotic insertion sequences (IS) are transposable genetic elements with a length ranging from 800–2500 bp. They are found at a remarkable variety in the genomes of many different bacteria and mycoplasmas at a multiplicity of between a few and several hundred per genome. In bacteria, they are very frequently found as part of natural plasmids. Genetic phenomena associated with transposition of IS elements are: spontaneous, often highly polar mutations, sometimes associated with deletion or inversion of adjacent DNA segments, as well as activation of the transcription of flanking genes ***(1)***. Insertion sequences therefore contribute basically to the variability of the prokaryotic genome and phenotype, and are thought to play an important role in the adaptability of prokaryotes to the environment. Insertion sequences provide a valuable source of experimental material for the study of gene expression, recombination and repair events, population dynamics, and horizontal transmission of genes. They are also frequently used in molecular genetics as markers to facilitate gene mapping, as mutagens, and for gene cloning. Several insertion sequences have been used as markers for molecular DNA fingerprints for identification and subtyping of bacterial and mycoplasmal strains, such as *Mycobacterium tuberculosis* ***(2,3)***, *Salmonella typhimurium* ***(4)***, *Vibrio cholerae* ***(5)***, and *Mycoplasma mycoides* subsp. *mycoides* small-colony type (SC) ***(6)***.

Identification of individual strains or of lineages of strains of mycoplasmas with conventional methods is often impossible because of lack of appropriate phenotypic markers. In addition, ribotyping, which is generally the method of choice for genetic subtyping of strains, is not applicable to mycoplasmas owing to the low copy number of ribosomal RNA genes *(rrn)*. IS fingerprinting, however, provides a particularly useful method for strain identification and

From: *Methods in Molecular Biology, Vol. 104: Mycoplasma Protocols*
Edited by: R. J. Miles and R. A. J. Nicholas © Humana Press Inc., Totowa, NJ

Table 1
Known Insertion Sequences in *Mycoplasmas*

Name	Size, bp	Species	Copy number average	Sequence accession number	Ref.
IS*1138*	1288	*Mycoplasma pulmonis*	8	Z16416	*(8)*
IS*1221*	1511	*Mycoplasma hyopneumoniae*	3	L33924	*(9)*
IS*1221*	1519	*Mycoplasma hyorhinis*	3	L33925	*(9)*
IS*1296*	1485	*Mycoplasma m. mycoides* SC	19	X84021	*(7)*
IS*1296*$_{vLC}$	1485	*Mycoplasma m. mycoides* LC	5		*(7)*
IS*1296*$_{vGr7}$	1485	*Mycoplasma* sp. bovine gr 7 (PG50)	2		*(7)*
IS	2210	*Mycoplasma incognitus*	10	M58403	*(10)*

subtyping of mycoplasmas that contain known insertion sequences or that are closely related to known IS. The method that is basically a Southern blot DNA:DNA hybridization using a given IS as probe can be used for the identification of a particular strain, such as a vaccine strain, or for whole lineages of strains in epidemiological investigations ***(6)*** (*see* **Note 1**). Thus far, only a small number of mycoplasmal IS elements are known from some mycoplasma species (**Table 1**). However, IS elements occur as structurally closely related elements in families, of which the IS*3* family is by far the largest one known ***(1)***. Among mycoplasmas, IS elements closely related to IS*1296* were found in *M. m. mycoides* LC and in *Mycoplasma* sp. bovine group 7 ***(7)***. This indicates that IS elements might be spread among many more mycoplasmal species from which no IS have yet been reported. It is therefore possible for strain typing purposes to use an IS element from another mycoplasma species or subspecies as a probe, provided that it shares enough homologous sequences with IS elements in the strains to be analyzed. The following method includes a short paragraph on a general strategy for the identification of a suitable IS element for insertion sequence analysis, followed by the description of the method of the analysis of *M. mycoides* subsp. *mycoides* SC strains using IS*1296* as an example.

1.1. General Strategy for the Identification of a Suitable IS Element

1. Search the GenBank/EMBL data library for known insertion sequences in *Mycoplasma*.
2. If necessary, keep your query broad enough (include as key words in the query related micro-organisms like *Acholeplasma, Asteroplasma, Spiroplamsa, Ureaplasma,* or if necessary *Firmicutes*).

3. If no IS are known for the mycoplasma species to be analyzed, choose about three DNA sequences of IS elements from species that are most closely related to the species of your investigation, and proceed for each as follows.
4. For each IS, select sequences for two oligonucleotides as primers for PCR amplification of a central, 500–1500 bp long DNA fragment. Avoid including sequences from the inverted repeats in the oligonucleotides. If possible, find oligonucleotides with relatively high melting temperatures.
5. Prepare genomic DNA from the mycoplasma (or related microorganism) that is reported to contain the IS element and from a few strains to be investigated (*see* **Subheading 3.2.**).
6. Prepare a digoxygenin-labeled probe of the IS element using the selected oligonucleotide primers and genomic DNA from the species containing the IS (**Subheading 3.4.**, preparation of a digoxygenin-labeled probe).
7. Perform a Southern blot analysis of restriction enzyme-digested genomic DNA of the strains to be analyzed and include the DNA of the species containing the IS as control (*see* **Subheading 3.3., steps 1–3**).
8. If the strains to be analyzed give clear multiple signals, optimize the restriction enzyme to be used for the digestion of chromosomal DNA (*see* **Subheading 3.4.**). If no signal is visible, try another IS element (it is of course possible that no signal will be obtained owing to the fact that the mycoplasma species to be analyzed do not contain IS elements resembling the ones that are known).

2. Materials

The following method applies to the insertion sequence analysis of *M. m. mycoides* SC or *M. m. mycoides* LC using IS*1296* as a probe.

1. Culture 10 mL of *M. m. mycoides* SC type strain PG1 (or other mycoplasmas) at end of exponential growth phase.
2. Culture 10 mL, of each of the strains *M. m. mycoides* SC (or other mycoplasmas) to be typed at end exponential growth phase.
3. TES buffer: 10 m*M* Tris-HCl, pH 7.9, 0.1 m*M* EDTA, 140 m*M* NaCl
4. GES buffer: 5*M* guanidium thiocyanate, 100 m*M* EDTA, 0.5% *N*-lauroylsarcosin sodium salt (Sarkosyl).
5. TE buffer: 10 m*M* Tris-HCl, pH 7.9, 0.1 m*M* EDTA.
6. TSM buffer: 100 m*M* Tris-HCl, pH 9.5, 100 m*M* NaCl, 50 m*M* $MgCl_2$.
7. TBE buffer: 0.045 *M* Tris, 0.045 boric acid, 1 m*M* EDTA.
8. TS buffer: 100 m*M* Tris-HCl pH 7.9, 140 m*M* NaCl.
9. 7.5*M* Ammonium acetate, pH 7.7.
10. 500 m*M* EDTA.
11. PCIA:phenol:chloroform:isoamyl alcohol (49.5:49.5:1).
12. Isopropanol.
13. Oligonucleotide primer IS1296P1–L 25 μM, 5'-AAGCGTTTAGAATAGAA-GGGCTA-3'.

14. Oligonucleotide primer IS1296P1–R 25 μM, 5'-CTGAATTGTACAGGAGA-CAATCC-3'.
15. 10 m*M* dATP.
16. 10 m*M* dCTP.
17. 10 m*M* dGTP.
18. 10 m*M* dTTP.
19. 1 m*M* Digoxygenin-11-dUTP (Boehringer Mannheim).
20. *Taq* DNA polymerase.
21. *Taq* buffer (10X): 100 m*M* Tris-HCl, pH 8.3, 15 m*M* $MgCl_2$, 500 m*M* KCl, 0.05% Tween 20 (10X *Taq* buffer provided by the supplier of *Taq* DNA polymerase can be used instead).
22. Antidigoxygenin antibodies (Boehringer Mannheim).
23. Agarose.
24. Ethanol 80%.
25. Nylon membrane positively charged (Boehringer Mannheim # 1209 299).
26. 10X SSC: 1.5 *M* NaCl, 150 m*M* sodium citrate, pH 7.7 (stock solution).
27. 1% *N*-lauroylsarcosin sodium salt (Sarkosyl) (stock solution).
28. 1% Sodium dodecyl sulfate (SDS) (stock solution).
29. Antidigoxygenin antibodies alkaline phosphatase-labeled Fab fragments (Boehringer Mannheim #1093 274).
30. Nitroblue-tetrazolium chloride (NBT) (Boehringer Mannheim #1585-029): 3% in 70% dimethylformamide (stock solution).
31. 5-Bromo-4chloro-3 indolyl phosphate (BCIP) (Boehringer Mannheim #1585-002): 1.5% in dimethylformamide (stock solution).
32. Centrifuge for conical reaction tubes (type Eppendorf).
33. Thermocycler.
34. Horizontal gel-electrophoresis equipment inclusive power supply.
35. Equipment for blotting of gels inclusive power supply.
36. Hybridization oven or water bath at 68°C to incubate the membrane sealed in a plastic bag.

3. Methods

3.1. Purification of Genomic DNA (see Note 2)

1. Centrifuge for 10 mL mycoplasma culture 15 min at 13,000*g*.
2. Resuspend the mycoplasmas in 5 mL TES buffer for washing the cells.
3. Centrifuge for 15 min at 13,000*g*.
4. Repeat the washing twice.
5. Resuspend the cells in 100 μL TE buffer.
6. Add 2 μL 500 m*M* EDTA.
7. Add 500 mL GES buffer, vortex, and incubate at room temperature 5 min.
8. Cool on ice.
9. Add 250 μL ammonium acetate 7.5 *M* (4°C) and vortex.
10. Incubate for 10 min on ice cooling.
11. Add 500 mL PCIA, and vortex thoroughly.

12. Centrifuge for 15 min at 13,000*g* at 4°C.
13. Take the aqueous phase carefully, and repeat the PCIA extraction twice.
14. Estimate the volume of the aqueous phase, and add 0.7 vol of isopropanol (4°C).
15. Mix by inverting the tube, and incubate for 30 min at –20°C.
16. Centrifuge for 15 min at 13,000*g* at 4°C.
17. Remove liquid, and carefully add 500 µL 80% ethanol (–20°C) to wash the precipitate.
18. Centrifuge for 5 min at 13,000*g* at 4°C.
19. Repeat the washing of the precipitate.
20. Remove all liquid, and dry the precipitate.
21. Suspend the DNA in 100 µL TE buffer and estimate the DNA concentration.

3.2. Preparation of a Digoxygenin-Labeled Probe

1. Mix the following items in a PCR reaction tube:
   10 µL Taq buffer 10X.
   1.7 µL dATP 10 m*M*.
   1.7 µL dCTP 10 m*M*.
   1.7 µL dTTP 10 m*M*.
   1.7 µL dGTP 10 m*M*.
   5 µL Digoxygenin-11-dUTP 1 m*M*.
   0.5 µL oligonucleotide primer IS1296P1-L.
   0.5 µL oligonucleotide primer IS1296P1-R.
   0.5 U *Taq* polymerase.
   1 ng Template DNA (genomic DNA from *M. m. mycoides* SC strain PG1).
   H_2O to 100 µL.
2. Submit to 35 cycles amplification in the thermocycler with the following parameters: 94°C 1 min; 56°C 1 min; 72°C 1 min.
3. Analyze 1 µL of the amplicon on a 0.7% agarose TBE gel by electrophoresis, stain with ethidium bromide, and inspect under UV light to estimate DNA concentration.

3.3. Southern Blot Analysis

1. Select suitable restriction enzymes to digest the genomic mycoplasma DNA (*see* **Note 3**). Generally use a restriction enzyme with a recognition sequence of 6 bp containing 4 A/T and 2 G/C. In the example, *Hind*III (A'AGCTT) and *Bgl*II (A'GATCT) was used.
2. Digest 200 ng genomic DNA of the strains to be analyzed (e.g., *M. m. mycoides* SC type^T strain PG1 and a European strain L2 ***[6]***).
3. Separate the fragments of the digested DNA by electrophoresis on a 0.7% agarose gel (approx 10 × 10 cm).
4. Use digoxygenin-labeled *Hind*III-digested bacteriophage λ DNA as molecular-mass standards (Boehringer Mannheim).
5. Transfer the DNA from the gel to nylon membranes by Southern blot and denature DNA ***(11)***.

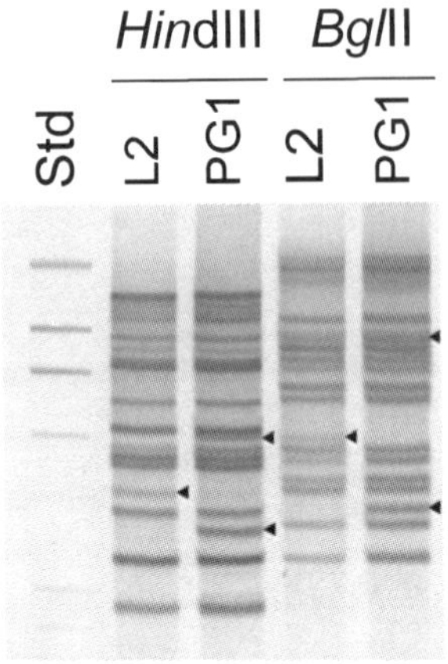

Fig. 1. Southern blot of genomic DNA from *Mycoplasma mycoides* subsp. *mycoides* SC strains PG1 (type strain) and L2 (European field strain) digested with *Hind*III or *Bgl*II, separated on a 0.7% agarose gel, and hybridized with digoxygenin-labeled IS*1296* probe. Std indicates standard: digoxygenin-labeled bacteriophage λ DNA (23.1, 9.4, 6.6, 4.4, 2.3, 2.0 kb). The filled triangles indicated the positions of the differences.

6. Prepare hybridization buffer (5 x SSC, 0.1% N-lauroylsarcosine, 0.02% SDS, and 1% blocking reagent [Boehringer Mannheim]).
7. Preincubate the membrane in 20 mL hybridization buffer at 68°C for 2 h.
8. Hybridize the membrane with 2.5 mL hybridization buffer containing 0.5 μg digoxygenin-labeled IS*1296* probe (corresponding to approx 5 μL PCR-amplified labeled probe *see* **Subheading 3.4.**) at 68°C for 18 h.
9. Wash the membrane with 2X SSC containing 0.1% SDS at room temperature for 5 min.
10. Wash the membrane with 2X SSC containing 0.1% SDS twice at 68°C for 15 min.
11. Wash the membrane with 20 mL TS buffer.
12. Incubate the membrane with 20 mL 1% blocking reagent (Boehringer Mannheim) in TS buffer at room temperature for 5 min.
13. Incubate the membrane in 2.5 mL antidigoxygenin antibodies phosphatase-labeled diluted 1:5000 in TS buffer at room temperature for 30 min.
14. Wash filter in 20 mL TS buffer at room temperature twice for 15 min.
15. Wash filter in 20 mL TSM buffer at room temperature for 2 min.
16. Prepare color reaction solution:
    20 mL H_2O.
    90 μL NBT 3%.
    70 μL BCIP 1.5%.
17. Incubate membrane in color reaction solution until the bands are clearly visible.
18. Stop the color reaction by washing the filter in H_2O.
19. Dry the membrane, and analyze the bands (**Fig. 1**) (*see* **Note 4**).

3.4. Interpretation of the Results

1. Before strains can be subtyped by insertion sequence analysis, a convenient restriction enzyme that resolves all copies of the insertion sequence in the Southern blot must be found. In the case of typing of *M. m. mycoides* SC strains with IS*1296, Hind*III was found to be the most suitable enzyme (**Fig. 1**). Several enzymes should be used initially in order to determine if the differences found in the profiles are owing to transposition events (in this case, the same number of differences in profiles between given strains are found with several restriction enzymes) or if some differences are owing to variance of flanking restriction enzymes recognition sites (certain enzymes give more differences in patterns than others). In order to obtain a good interpretation of the results, all strains to be analyzed should be treated on the same Southern blot membrane. If a very high number of samples is to be used, it is advisable to include on each membrane one or several strains that were chosen to give standard profiles (e.g., use the type strain).
2. For the clustering of the strains, it is recommended to establish an "evolutionary network" of the different genotypes *(7)* (*see* **Note 5**). In order to establish this network, one has to assume that every additional IS element discovered between the different strains arose from a transposition event (supposing that deletion formation of IS elements would play a minor role). Find a pattern of a strain that represents a subset of bands common to all other patterns as the ancestral strain in your set analyzed. Often, however, a hypothetical pattern of an ancestral strain having the most possible bands in common with all patterns observed must be constructed from which the evolutionary network can be drawn. In the example of the two strains shown in **Fig. 1**, such an ancestral strain "X" having one copy of IS*1296* less than strain PG1 and two copies less than strain L2 (marked with a black triangle) would fulfill this requirement. From this hypothetical ancestral train "X," PG1 would have arisen by two additional transposition events and strain L2 by a single transposition. The method therefore is not only appropriate to cluster strains, but also to give a short-term evolutionary picture of the different mycoplasma strains analyzed *(7)*. This is especially interesting when type strains of unknown origin are included in such a study, which allows, in some cases, the tracing back of the putative origin of such strains.

4. Notes

1. The reproducibility of the Southern blots is essential for subtyping of microorganisms. In order to obtain satisfactory results with insertion sequence analysis, high-resolution Southern blots are imperative. This requires:
   a. A high degree of purity of genomic DNA used for the analysis.
   b. Restriction enzyme digestion, which results in clearly separated fragments containing the hybridizing IS elements.
   c. A detection system for the hybridizing DNA that gives focused pictures.
2. High purity of DNA was constantly achieved with the described method, which was derived from the extraction method of Pitcher et al. *(12)*. It is, however, important to note that thorough washing of the mycoplasma cells (**steps 4** and **6**)

prior to DNA extraction is imperative. The number of three PCIA extractions (**steps 13–15**) must not be reduced. For some strains, additional PCIA extractions might be necessary to obtain high-purity DNA.

3. The choice of a suitable restriction enzyme for the analysis must be made carefully prior to the analysis of large series of strains. Restriction enzymes cutting inside the IS element are generally not used. However, if they are chosen, this fact must be considered in the interpretation of copy number and differences of profiles.
4. The clearest pictures of the Southern blots are achieved using digoxygenin-labeled DNA probes and phosphatase-labeled antidigoxygenin antibodies combined with NBT-BCIP substrate for alkaline phosphatase as the detection system. The use of chemiluminescent substrate for alkaline phosphatase combined with autoradiography, although being more sensitive, is not recommended, since bands located closely to each other are not easily resolved owing to diffusion effects. The same problem is encountered with radioactive labeling of the probes using ^{32}P-labeled dNTP.
5. Owing to the relatively small number of differences in the patterns of the IS Southern blots, the use of the UPGMA algorithm ***(13)*** for clustering of the strains is generally not recommended. The method described above gives a more direct picture of the interrelationships of the different strains and has the advantage of giving an evolutionary approach to the analysis.

References

1. Galas, D. J. and Chandler, M. (1989) Bacterial insertion sequences. *Mobile DNA* (Berg, D. E. and Howe, M. M., eds.), American Society for Microbiology, Washington, DC, pp. 109–162.
2. van Embden, J. D., Cave, M. D., Crawford, J. T., Dale, J. W., Eisenach, K. D., Gicquel, et al. (1993) Strain identification of *Mycobacterium tuberculosis* by DNA fingerprinting: recommendations for a standardized methodology. *J. Clin. Microbiol.* **31,** 406–409.
3. Small, P. M., Hopewell, P. C., Singh, S. P.; Paz, A., Parsonnet, J.; Ruston, D. C., et al. (1994) The epidemiology of tuberculosis in San Francisco. A population-based study using conventional and molecular methods. *N. Engl. J. Med.* **330,** 1703–1709.
4. Stanley, J., Baquar, N., and Threlfall, E. J. (1993) Genotypes and phylogenetic relationships of *Salmonella typhimurium* are defined by molecular fingerprinting of IS*200* and 16S rrn loci. *J. Gen. Microbiol.* **139,** 1133–1140.
5. Bik, E. M., Gouw, R. D., and Mooi, F. R. (1996) DNA fingerprinting of *Vibrio cholerae* strains with a novel insertion sequence element -a tool to identify epidemic strains. *J. Clin. Microbiol.* **34,** 1453–1461.
6. Cheng, X. X., Nicolet, J., Poumarat, F.; Regalla, J., Thiaucourt, F., and Frey, J. (1995) Insertion element IS1296 in *Mycoplasma mycoides* subsp. *mycoides* small colony identifies a European clonal line distinct from African and Australian strains. *Microbiology* **141,** 3221–3228.

7. Frey, J., Cheng, X. X., Kuhnert, P., and Nicolet, J. (1995) Identification and characterization of IS*1296* in *Mycoplasma mycoides* subsp. *mycoides* SC and presence in related mycoplasmas. *Gene* **160,** 95–100.
8. Bhugra, B. and Dybvig, K. (1993) Identification and characterization of IS*1138*, a transposable element from *Mycoplasma pulmonis* that belongs to the IS*3* family. *Mol. Microbiol.* **7,** 577–584.
9. Ferrell, R. V., Heidari, M. B., Wise, K. S., and McIntosh, M. A. (1989) A Mycoplasma genetic element resembling prokaryotic insertion sequences. *Mol. Microbiol.* **3,** 957–967.
10. Hu, W. S., Wang, R. Y., Liou, R. S., Shih, J. W., and Lo, S. C. (1990) Identification of an insertion-sequence-like genetic element in the newly recognized human pathogen *Mycoplasma incognitus. Gene* **93,** 67–72.
11. Ausubel, F. M., Brent, R., Kingston, R. E., Moore, D. D., Seidman, J. G., Smith, J. A., et al. (1990) *Current Protocols in Molecular Biology.* Wiley Interscience, New York.
12. Pitcher, D. G., Saunders, N. A., and Owen, R. J. (1989) Rapid extraction of bacterial genomic DNA with guanidium thiocyanate. *Lett. Appl. Microbiol.* **8,** 151–156.
13. Sneath, P. H. A. and Sokal, R. R. (1973) *Numerical Taxonomy. The Principles and Practice of Numerical Classification.* W. H. Freeman, San Francisco.

23

Detection of Mycoplasmas in Cell Cultures by Cultural Methods

Gerald K. Masover and Frances A. Becker

1. Introduction

Workers using cell cultures for research, diagnostics, or production of biopharmaceuticals have in common the need to maintain their cultures in a state of control. The living cell-culture system that is expected to produce reliably scientific and diagnostic data or a product must not be encumbered with additional unknown living systems. Therefore, cell culturists are traditionally very diligent in selection and preparation of raw materials, equipment, and processes that will minimize the opportunity for infection of their cultures with adventitious biological agents, and they employ a program of regular testing for the presence of contaminants. Cell cultures without antibiotics are growth-permissive systems in which bacterial and fungal contaminations can become apparent, because they produce turbidity and effects on the cell cultures that are easily seen. However, mycoplasma contaminations of cell cultures are not as readily observed and are, therefore, the subject of a separate set of diagnostic tests. The classical cultural methodology that remains the primary diagnostic test for mycoplasmas in cell culture is the subject of this chapter. However, *Mycoplasma hyorhinis* DBS 1050, a strain that commonly infects cell cultures, is not readily grown on the commonly used agar preparations ***(1)***. An agar medium formulation that will permit growth of this strain has been reported recently ***(2)***, but until it has been more widely tested, it will be necessary to complement the cultural method with another established method for detection of this contaminant. All *M. hyorhinis* strains cytoadsorb strongly to cells in culture and are readily viewed by the use of DNA binding fluorochrome dyes (DNAFs). Therefore, a method taking advantage of these properties should be included in any mycoplasma detection program (*see* Chapter 24).

From: *Methods in Molecular Biology, Vol. 104: Mycoplasma Protocols*
Edited by: R. J. Miles and R. A. J. Nicholas © Humana Press Inc., Totowa, NJ

1.1. Epidemiology of Mycoplasmas in Cell Cultures

The mycoplasmas that have been cultivated and identified are parasites of humans, animals, arthropods, and plants. The primary habitats of human and animal mycoplasmas are the mucous surfaces of the respiratory and genital tracts. In some animals, mycoplasmas are found in the eyes, alimentary canal, mammary glands, and joints. The widespread use of cell cultures for biomedical research and other applications stimulated increased interest in the occurrence of mycoplasmas in cell cultures as well. The epidemiology of cell-culture contaminant mycoplasmas was recently reviewed ***(3)***. More thorough discussions on the biology of mycoplasmas can be found in comprehensive microbiological handbooks ***(4,5)*** or in special publications on the subject ***(6,7–10)***.

1.2. Basis for the Cultural Method

Amplification of viable mycoplasmas in broth and subsequent demonstration of typical colonies on agar are a powerful classical microbiological method. In a broth culture, individual cells may replicate to a high concentration that, alone, often provides diagnostic information by production of a pH change, seen as a medium color change or slight turbidity. More importantly, the increased concentration also increases the probability of including viable cells within a given sample volume. When an aliquot of a broth culture is subcultured onto agar and incubated, the viable mycoplasmas replicate to form visible colonies from single cells or small clumps of cells (colony-forming units, CFUs). This technique allows a much larger volume to be tested by inoculation into broth and later subculture to agar than would be feasible by direct inoculation onto agar alone. The subcultures must be performed at various times to allow amplification and recovery of viable cells of both rapid- and slow-growing organisms in the broth.

2. Materials

2.1. Preparation of Media

The majority of species classified as *Mycoplasma* or *Acholeplasma* will grow well on the type of media originally devised by Edward ***(11)*** and modified by Hayflick ***(12)***. The essential and major nutrient constituents are beef heart infusion, peptone, fresh yeast extract, and unheated horse serum. The heart infusion and peptone, together with a small amount of sodium chloride, are generally supplied in commercial pleuropneumonia-like organism (PPLO) broth or agar preparations. Various minor constituents, such as specific sugars, selected amino acids and pH indicators, are also generally included in media formulations. In addition, because mycoplasmas do not synthesize nucleic acid precursors, many formulations contain either native DNA or its precursors. It is now known that the capacity of mycoplasmas to synthesize lipids and amino acids is also limited ***(13)*** and that these nutrients must be supplied. This has traditionally been done by the inclusion in media of nutrient-rich undefined components, such as serum and yeast extract.

Formulations for unsupplemented and supplemented modified Edward broth and agar media are given below.

2.1.1. Modified Edward Mycoplasma Broth Medium, Unsupplemented

The following components are mixed, sterilized, and stored prior to addition of supplements: PPLO broth without crystal violet (Difco, Detroit, MI, 0554-01 or equivalent), 21 g; phenol red 0.5% w/v aqueous stock solution, 10 mL; calf thymus DNA (Sigma, St. Louis, MO, D-1501 or equivalent), 0.02 g; distilled water, sufficient quantity to make 692 mL. Adjust the pH to 7.4 ± 0.1 with 1*N* NaOH or 1*N* HCl. Sterilize by autoclaving at 121°C for 15 min. Store for up to 6 mo at 2–30°C. The formulation given is for a volume of 692 mL, which is sufficient to make 1 L of supplemented medium.

2.1.2. Modified Edward Mycoplasma Broth Medium, Supplemented (1 L)

The following sterile components are combined aseptically: unsupplemented mycoplasma broth, 692 mL; glucose-arginine stock solution (*see* **Note 1**), 8 mL; fresh yeast extract (*see* **Note 2**), 100 mL; unheated horse serum, 200 mL. Aseptically adjust the pH of the complete broth to 7.4 ± 0.1, if necessary, by dropwise addition of sterile 1*N* NaOH or 1*N* HCl. Store at 2–8°C for up to 90 d (*see* **Note 3**).

2.1.3. Modified Edward Mycoplasma Agar Medium

The formulation for unsupplemented agar medium is identical to that for the unsupplemented broth, except that 14 g of purified agar (Difco 0560–01 or equivalent, *see* **Note 4**) are added for each 692-mL portion of unsupplemented agar medium. The supplements are the same for the agar medium as for the broth. Approximately 8 mL of the agar medium are generally poured into a 60-mm Petri dish in a manner that minimizes surface and subsurface bubbles (*see* **Note 5**). The agar medium should be tested to confirm that it is not itself microbially contaminated (*see* **Note 6**) and is capable of detecting a low number of CFUs of potential cell-culture contaminant mycoplasmas (*see* **Note 7**).

3. Method

3.1. Sampling, Sample Size and Sample Storage

1. Testing is done on an antibiotic-free suspension of cells in the culture fluid, and the sensitivity of the test is a function of the volume of sample tested (*see* **Note 8**). Generally, a 1.0-mL sample will suffice for routine mycoplasma testing of cell

cultures for laboratory use. A larger test sample (10 mL) may be inoculated into 100 mL of broth to increase the test sensitivity if desired. Programs that must strive for high assurance of absence of mycoplasmas, such as testing during production of human biopharmaceuticals, require the larger samples and must, in addition, be very rigorously controlled and thoroughly documented ***(14)***.

2. If an adherent cell culture is to be tested, the analyst must assure that an appropriate number of cells is resuspended in the culture medium by scraping them off their growth surface. The cells should not be trypsinized in preparation for mycoplasma testing.
3. On receipt by a testing laboratory, samples are stored at 2–8°C until frozen at –60°C or below. Samples should be frozen if they are not tested within 24 h of sampling. The sample size should be sufficient to freeze a number of aliquots in the event it is necessary to perform retests.

3.2. Inoculation and Incubation

1. The test is initiated by inoculating a minimum of 2 Petri dishes of supplemented (complete) agar medium with 0.2 mL of sample (*see* **Note 9**). An additional 1.0 mL of sample is inoculated into 10 mL or more of complete mycoplasma broth. It is good practice to incubate an equivalent amount of uninoculated broth as a negative control for broth color changes.
2. For laboratories that perform large numbers of tests, positive controls for each test session should also be included. The positive controls (*see* **Note 10**) generally consist of a low number (target 100 CFU) of a glucose-fermenting species, such as *Mycoplasma hyorhinis* or *Acholesplasma laidlawii*, and of a nonfermenting species such as *Mycoplasma orale, Mycoplasma arginini,* or *Mycoplasma hominis.* The positive controls are inoculated onto agar and into broth, and subcultured in exactly the same manner as test samples.
3. The broth containers and half of the inoculated Petri dishes are incubated aerobically at 36°C ± 1°C (*see* **Notes 11** and **12**). Half of the agar plates are incubated at the same temperature, but anaerobically (*see* **Note 13**).
4. On days 3, 7, and 14, 0.2 mL of the broth culture is subcultured to a minimum of two agar plates, and half of the plates from each time-point are incubated aerobically and the other half are incubated anaerobically. The plates are allowed to incubate for at least 7 d or for up to 14 d before they are observed (*see* **Note 14**).

3.3. Observation of Mycoplasma Broth Cultures

A few of the cell culture contaminant mycoplasmas produce modest turbidity in broth culture, but most do not. Most cell culture contaminant mycoplasmas do, however, produce color changes in suitable broth media. Species such as *Mycoplasma pneumoniae*, *M. hyorhinis*, and *A. laidlawii*, which hydrolyze glucose, are generally referred to as "fermenters." They cause an acid color change from cherry red to orange to yellow in a medium containing phenol red pH indicator and starting at a pH between 7.0 and 7.4. Any unequivocal progressive color change of an inoculated container compared to an uninoculated control container can be taken as an indication of growth. Cultures that have progressed to the yellow color are in stationary or death phase, and are not optimal for subculture. Species such as *M. orale*, *M. hominis,* and *M. arginini,* which hydrolyze arginine, are generally referred to as "nonfermenters" and cause an alkaline color change, i.e., from cherry red to deep red in a medium containing phenol red pH indicator and starting at a pH of 7.0–7.4. As with the fermenters, any unequivocal progressive color change of an inoculated container compared to an uninoculated control container is indicative of growth. Cultures having a slight but discernible color change are optimal for subculture.

3.4. Observation of Mycoplasma Agar Cultures

1. Observe mycoplasma agar plates microscopically after at least 7 d of incubation.
2. Observe a minimum of 10 fields within the inoculated area of each 60-mm plate at a total magnification of 40× (*see* **Note 15**). The appearance of mycoplasma colonies (*see* **Note 16**) may be confirmed at 100x total magnification or higher.
3. Observations are recorded as "positive," meaning that mycoplasma colonies were observed, or "negative," meaning that no mycoplasma colonies were observed.
4. An ambiguous result should be checked by a second observer. If an ambiguous result cannot be resolved as either positive or negative, the questionable colony(s) should be subcultured to a new agar plate and a retest should be performed.
5. Agar subculture is accomplished by aseptically excising a block of agar containing the suspected colony, inverting it on a fresh plate, and pushing it across the fresh agar surface to transfer organisms from the old agar surface to the new one.

3.5. Interpretation of Results

1. Valid test: For a test to be considered valid, all negative control plates must be negative for mycoplasma, and at least one of the positive control plates for each positive control mycoplasma species must be positive. Microbially contaminated plates are discarded if the contamination interferes with the observation of mycoplasma colonies. At least one plate from each group must be readable, and all readable plates must be negative for a negative result. A "group" in this context is a set of agar plates that is inoculated at the same time with the same inoculum and incubated under the same conditions.
2. Negative test: For a test sample to be considered negative, all agar cultures must be negative for mycoplasma after at least 7 d of incubation in a valid test.
3. Positive test: One or more mycoplasma colonies on any of the test plates renders the test positive for mycoplasma. Weakly positive test results should be confirmed, and ambiguous results should be resolved.
4. Retest criteria: A retest should be performed if:
   a. A test is invalid;
   b. Results are ambiguous; or
   c. A determinate error is identified in the assay.

4. Notes

1. This is a filter-sterilized aqueous solution containing 25% w/v of glucose and 25% w/v of arginine HCl. It gives a final concentration of 0.2% w/v in the supplemented broth.
2. Fresh yeast extract refers to a recently prepared hot water extract of live yeast cells. It is prepared by slowly adding 250 g of active dry baker's yeast to 1 L of boiling water and then allowing the mixture to simmer (boil slowly) for about 15 min. The boiled preparation is clarified by centrifugation (8000*g*) or filtration. The yield from the clarification step is about 50% of the starting volume. The extract may be sterilized by autoclaving or filtration. We prefer step-down filtration with a 0.2-μm filter as the final step. The final extract is brought to pH 7.4 ± 0.1 and stored frozen (≤–20°C) for not more than 6 mo. It may be reclarified after thawing, if necessary, by a second filtration through a 0.2-μm filter.
3. The 90-d maximum storage time for supplemented medium was conservatively assigned based on rigorous quantitative growth promotion testing of numerous lots of media against a number of different mycoplasma strains in our laboratory.
4. Use of the highest grades of agar is frequently recommended for mycoplasma agar media, because lesser grades are occasionally toxic to some potential isolates ***(3,15)***. The better-quality agars also result in very clear agar media with fewer artifacts that might be confused with or obscure mycoplasma colonies.
5. When preparing supplemented agar medium, components must be mixed at appropriate temperatures. Since agar solidifies at approx 44°C, and the serum protein is denatured and will precipitate at approx 60°C, the medium components must be combined at an intermediate temperature. Sterile unsupplemented agar may be held at 50–55°C in a water bath after it has been liquefied in a microwave oven, an autoclave, or boiling water. The other sterile components are best warmed to avoid local solidification of the agar when adding components. Best mixing occurs when the supplements are heated briefly to near the agar hold temperature (50–55°C). The complete agar medium is mixed well without excessive agitation and should be dispensed (approx 8.0 mL/60-mm Petri dish) immediately after it is prepared. Bubbles, if any, that form on the agar surface may be removed by "brushing" lightly with a Bunsen burner flame. It is important to avoid subsurface bubbles by carefully dispensing, because large numbers of subsurface bubbles will render the agar unfit for use. The final product should be clear. The finished plates should be packaged and stored in a manner that maintains their sterility and cleanliness, and prevents excessive water loss.
6. The entire batch of supplemented mycoplasma broth or agar is incubated at 35–37°C for 2 d as a sterility check. All units that are observed to be microbially contaminated are discarded. The entire batch is discarded if more than 10% of the units are contaminated.
7. Growth-promotion testing is accomplished by inoculating media with about 100 CFU (*see* **Notes 8** and **10**) of different mycoplasma species, e.g., *M. hyorhinis*, a glucose-fermenting, acid producing species; *M. orale*, an arginine-hydrolyzing mycoplasma; and *M. pneumoniae*, a slow-grower. A batch of

supplemented medium meets the growth promotion requirements if each of the positive control organisms is positive for growth. For agar, each of the organisms must form at least one colony on at least one plate. For broth, each of the organisms must form at least one colony on at least one subculture plate. Medium batches that fail to meet these criteria may be retested once. The growth-promotion test may be performed concurrently with the use of the medium in mycoplasma detection tests. However, the mycoplasma detection test results must be considered invalid if the medium used fails to meet the requirements of the growth promotion test.

8. A CFU is defined as the viable unit (single cell or a small clump of cells) that grows to form one colony. Identification of a colony as a mycoplasma colony is the basis for diagnosis in this test. Therefore, the sensitivity of the test is 1 CFU in whatever volume is tested, and for that reason, it is desirable to test as large a sample as possible for greatest sensitivity.
9. Agar inoculations are performed in a laminar flow biosafety cabinet. The 0.2 mL of broth used for agar inoculation is placed in the center of the Petri dish and is not spread out. The inoculum is allowed to absorb into the agar in a minimal area. This allows the observer to identify the "edge of the inoculum," which is valuable for comparing inoculated with uninoculated areas during microscopic examination. It is also useful to "stab" the agar in the center of the inoculum with the pipet. This small hole in the agar will serve as a reference to locate the area inoculated, as well as the agar surface, which is otherwise not as easily found during microscopic examination of clean and clear agar media.
10. Mycoplasma stocks for positive controls and growth promotion testing of agar are 1.0-mL aliquots of log-phase cultures that are stored frozen at ≤–60°C. The titer on thawing is determined for each stock by performing plate counting, preferably in duplicate on three different cryo-vials. The grand mean titer is the basis for calculating dilutions aimed at delivering 100 CFU. A detailed description of mycoplasma frozen stock preparation has been published recently ***(14)***.
11. Although most of the cell-culture contaminant mycoplasmas will grow in anaerobic environments, aerobic incubation should also be performed because *M. hyorhinis,* one of the more common cell culture contaminant mycoplasmas, is markedly inhibited by the GasPak environment, which is generally used to achieve the suggested anaerobic conditions ***(16)***.
12. It is best to incubate the inoculated mycoplasma agar in closed water-tight containers that have been disinfected with 70% v/v isopropanol or ethanol before use. The disinfection is necessary to assure that the agar dishes do not become contaminated with fungi or bacteria during the prolonged incubation period. The closed container is necessary to prevent loss of moisture in the incubation environment.
13. The anaerobic environment may be generated by use of the GasPak Plus system (BBL, Cockeysville, MD) or equivalent.
14. The known cell-culture contaminant mycoplasmas are expected to grow on agar within 7 d. Very thorough programs that strive for high assurance of mycoplasma

absence, such as testing during production of human biopharmaceuticals, require a 14-d incubation *(14)*.

15. Systematic observation of mycoplasma colonies using a microscope is facilitated by use of gridded Petri plates or a fabricated plastic grid on which the Petri plate being observed may rest.
16. Typical mycoplasma colonies are circular and 50–500 μm in diameter. Viewed from the top or bottom, they will have a lighter, thinner peripheral zone surrounding a darker, deeper central zone. The deep character of the central zone is best appreciated by continually moving the focal plane from just above to just below the agar surface. Most often, colonies appear textured, fragile, and similar in appearance to moist sand. Occasionally, smaller colonies may appear glossy. A degree of confidence in identifying mycoplasma colonies is readily achieved with practice.

References

1. Hopps, H. E., Meyer, B. C., Barile, M. F., and Del Giudice, R. A. (1973) Problems concerning "noncultivable" mycoplasma contaminants in tissue cultures. *Ann. NY Acad. Sci.* **225,** 265–276.
2. Gardella, R. S. and Del Giudice, R. A. (1995) Growth of *Mycoplasma hyorhinis* cultivar α on semisynthetic medium. *Appl. Environ. Microbiol.* **61,** 1976–1979.
3. Barile, M. F. and Rottem, S. (1993) Mycoplasmas in cell culture, in *Rapid Diagnosis of Mycoplasmas*" (Kahane, I. and Adoni, A., eds.), Plenum, New York, pp. 155–193.
4. Masover, G. and Hayflick, L. (1981) The genera *Mycoplasma, Ureaplasma*, and *Acholeplasma*, and associated organisms (Thermoplasmas and Anaeroplasmas), in *The Procaryotes* (Starr, M. P., Stolp, H., Truper, H. G., Balows, A., and Schlegel, H. G., eds.), Springer-Verlag, Berlin, pp. 2247–2270.
5. Razin, S. (1991) The genera *Mycoplasma, Ureaplasma, Acholeplasma, Anaeroplasma*, and *Asteroleplasma,* in *The Procaryotes* (Starr, M. P., Stolp, H., Truper, H. G., Balows, A., and Schlegel, H. G., eds.), Springer-Verlag, Berlin, pp. 1937–1959.
6. Maniloff, J., McElhaney, R. N., Finch, L. R., and Baseman, J. B. (eds.) (1992) *Mycoplasmas: Molecular Biology and Pathogenesis.* American Society for Microbiology, Washington, DC.
7. Razin, S. and Tully, J. G. (eds.) (1983) *Methods in Mycoplasmology*, vol. I, *Mycoplasma Characterization.* Academic, New York.
8. Tully, J. G. and Razin, S. (eds.) (1983) *Methods in Mycoplasmology*, vol. II, *Diagnostic Mycoplasmology.* Academic, New York.
9. Tully, J. G. and Razin, S. (eds.) (1995) *Molecular and Diagnostic Procedures in Mycoplasmology*, vols. I and II. Academic, Orlando.
10. Dybig, K. and Voleker, L. L. (1996) Molecular biology of mycoplasmas. *Annu. Rev. Microbiol.* **50,** 25–57.
11. Edward, D. G. (1947) A selective medium for pleuropneumonia-like organisms. *J. Gen. Microbiol.* **1,** 238–243.
12. Hayflick, L. (1965) Tissue cultures and mycoplasmas. *Tex. Rep. Biol. Med.* **23(Suppl. 1),** 285–303.

13. Fraser, C. M., Gocayne, J. D., White, O., Adams, M. D., Clayton, R. A., Fleischmann, R. D., et al. (1995) The minimal gene complement of *Mycoplasma genitalium. Science* **270,** 397–403.
14. Masover, G. K. and Becker, F. A. (1996) Methodology for detection of mycoplasmas in cell cultures used to produce human pharmaceuticals in *Automated Microbial Identification and Quantitation* (Olson, W. P., ed.), Interpharm, Buffalo Grove, IL, pp. 149–177.
15. Olson, L. D. and Barile, M. F. (1988) Mycoplasma infection of cell cultures: Isolation and detection. *J. Tissue Culture Methods* **11,** 175–179.
16. Polak-Vogelzang, A. A., DeHaan, H. H., and Borst, J. (1983) Comparison of various atmospheric conditions for isolation and subcultivation of *Mycoplasma hyorhinis* from cell cultures. *Antonie van Leeuwenhoek* **49,** 31–40.

24

Detection of Mycoplasmas in Cell Cultures by Fluorescence Methods

Gerald K. Masover and Frances A. Becker

1. Introduction

The fluorescence stains used most often for mycoplasma detection are DNA binding fluorochromes (DNAFs) and fluoresceinated antibodies. Both permit direct visualization of individual mycoplasma organisms. DNAFs will bind to any appropriately conformed DNA that is present in a sample preparation and are, therefore, not specific for mycoplasmas. Conversely, fluoresceinated antibodies are highly specific for mycoplasmas, discriminating to species level. The two stains can be used individually or in combination, and they can be used either directly on the cells of interest or they may be used in a system involving mycoplasma-free indicator cells, which further enhances the diagnostic reliability of tests utilizing these fluorescent reagents.

Historically, the use of DNAFs for direct viewing of individual cell-associated mycoplasmas was reported by Russell et al. *(1)* and by Chen *(2)*. Prior to this work, Fogh *(3)* had advocated the use of a cytochemical method that employed uninfected indicator cell cultures. Del Giudice and Hopps *(4)* combined the use of DNAF with the use of uninfected indicator cells to create the method that is most often used today. The indicator cell-culture method has the advantage of being conducive to standardization by the inclusion of positive and negative controls, and by the use of a uniform cell substrate for the test. It also has the advantage of being a biological system in which low-level contaminant mycoplasmas can be enriched in number to make them more readily discernible and more confidently diagnosed.

Because there is a strain of mycoplasma (*Mycoplasma hyorhinis*, strain DBS 1050) that commonly infects cell cultures but is not readily grown on the commonly used mycoplasma media *(5,6)*, it is necessary to complement

From: *Methods in Molecular Biology, Vol. 104: Mycoplasma Protocols*
Edited by: R. J. Miles and R. A. J. Nicholas © Humana Press Inc., Totowa, NJ

the cultural method (*see* Chapter 23) with a method that will detect this particular contaminant. All *M. hyorhinis* species cytoadsorb strongly to cells in culture and are readily viewed by the use of DNAFs. Therefore, the DNAF method is often included in cell-culture mycoplasma detection programs.

In 1978, Del Giudice and Hopps ***(4)*** reported use of fluoresceinated antibody staining of cover slip cell cultures and colonies on agar as well as separate tests of the cover slip indicator cultures using a DNAF. This was done in the interest of demonstrating a strain of *M. hyorhinis* that did not grow on agar. Prior to that time, the principal use of fluoresceinated antibodies in diagnosis of mycoplasmas as cell-culture contaminants was for identification of colonies on agar ***(7)***. In 1990, Freiberg and Masover ***(8)*** showed that the DNAF and fluoresceinated antibody stains could be applied to the same cover slip cell culture and viewed concomitantly. In 1994, Masover and Pleibel ***(9)*** reported that a "cocktail" of at least five fluoresceinated antibodies to different mycoplasma species could be used in this double-stain procedure without interference among specific antibodies.

The use of a DNAF alone and in combination with a fluoresceinated antibody in an indicator cell system is described in this chapter.

2. Materials

1. Fixatives: For DNAF, use 1 part glacial acetic acid plus 2 parts absolute methanol prepared fresh for each use. Prepare at least 10 mL for each cover slip to be stained. (*See* **Note 1** for other fixative options.)
2. DNA binding stain: The DNAF method is described using *bis*-benzimide as the stain. *Bis*-benzimide concentrated stock solution (1000X) contains 5.0 mg *bis*-benzimide (2'-[4-Hydroxyphenyl]-5-[4-methyl-1-piperazinyl]-2,5'-bi-1H-benzimidazole)/100 mL deionized water. Each 100 mL are preserved with 0.01 g thimerosal. Store this solution at 2–8°C in the dark (foil wrapped). *Bis*-benzimide stain working solution (0.05 μg/mL) contains 0.1 mL of the concentrated stock solution plus 99.9 mL citrate/phosphate buffer, pH 5.5. It is prepared fresh for each use.
3. Fluoresceinated antibodies: The methods described for preparation and fluoresceination of polyclonal antimycoplasma antibodies in the 1980s ***(10)*** are still widely employed and are essentially unchanged ***(11)***. Live washed mycoplasma cells can be used as antigen for the production of antimycoplasma antibodies. A typical small-scale immunization, bleeding, serum fractionation, and fluoresceination procedure for rabbit immunoglobulins are described by Freiberg and Masover ***(8)***. The resultant fluoresceinated antibodies are typically species-specific. The florescence titer is determined as the most dilute of serial twofold dilutions of the conjugate that is able to give specific (green) florescence on agar colonies of the homologous mycoplasma species or on cultured cells that are infected with the homologous species. A working dilution, which is generally one or two twofold dilutions more concentrated than the fluorescence titer, is chosen for immunoflorescence staining of mycoplasma-infected cells ***(10)***.

4. Evans blue counterstain: 1% stock solution (20X) containing 1.0 g Evans blue powder (Chroma-Gesellschaft, Koengen, West Germany) in sufficient PBS, pH 7.2 to make 100 mL. Evans blue working solution is 1 part of the 1% stock solution plus 19 parts PBS, pH 7.2.
5. Citrate/phosphate buffer, pH 5.5: Prepare by mixing 44 mL of 0.1*M* citric acid with 56 mL 0.2*M* sodium phosphate (dibasic) for each 100 mL citrate/phosphate buffer.
6. Phosphate-buffered saline (PBS), pH 7.2: 8.5 g sodium chloride, 1.1 g sodium phosphate (dibasic, anhydrous), and 0.32 g sodium phosphate (monobasic, monohydrate) in sufficient deionized water to make 1 L.
7. Mounting solution: The mounting solution for DNAF consists of equal portions of glycerol and citrate/phosphate buffer, pH 5.5, to make the desired final volume (2 or 3 drops/slide are required). Add *n*-propyl gallate to make a final concentration of 0.2% w/v (*see* **Note 2**). Final mounting solution for the double stain (both DNAF and immunofluorescence on one preparation) is equal parts glycerol and PBS, pH 7.2, with *n*-propyl gallate 0.2% w/v added. The mounting solutions are stored at –10°C or below.

3. Method

3.1. Test Samples

Testing is done on a suspension of about 10^4 or 10^5 cells from a culture that has been passaged at least three times in culture fluid without antibiotics or selective agents, and is at least 3 d beyond the last medium change. If an adherent cell culture is to be tested, resuspend a few cells in the culture medium by scraping them off their growth surface, but do not trypsinize them in preparation for mycoplasma testing. Test samples promptly on receipt or store frozen at ≤60°C (*see* **Note 3**).

3.2. Preparation of Indicator Cells

1. Prepare multiwell cell-culture plates or Petri dishes containing 1 presterilized 22-mm^2 number 1 cover slip/well (or plate).
2. Trypsinize an antibiotic-free stock culture of Vero cells (*see* **Note 4**), and prepare a suspension containing 5×10^4 viable cells/mL. For six-well plates, dispense 4.0 mL of cell suspension/well (ca. 3.8-cm diameter). Adjust the volume of the cell suspension proportionately for different-sized culture vessels.
3. Incubate the cover slip cell cultures at 35–37°C in a humidified 4–6% CO_2 in air atmosphere.

3.3. Inoculation of Indicator Cells

1. Inoculate indicator cell cultures 1 d after the cells are plated by adding the following to two wells each:
   a. Test sample, 0.5 mL. Multiple samples may be tested concurrently.
   b. 100 CFU or less (*see* **Note 5**) of positive control *Mycoplasma hominis* (derived from ATCC #23114) in 0.5 mL or less of complete mycoplasma broth (*see* **Note 6**).

c. 100 CFU or less (*see* **Note 5**) of positive control *M. hyorhinis*, strain BTS-7 (derived from ATCC #17981) in 0.5 mL or less of complete mycoplasma broth (supplemented broth; *see* Chapter 23, **Subheading 2.**, item 2.1.2.). (*See* **Note 7**).
d. Negative control, uninoculated wells.

2. Incubate at 35–37°C in a humidified incubator containing a 5% CO_2 in air gaseous environment.

3.4. DNAF Staining

1. Three to 5 d after sample addition, remove medium from the well. Do not allow the cover slip to dry.
2. Add fixative in sufficient volume to completely immerse the cover slip (2–5 mL), and let stand for 2 min.
3. Aspirate the fixative, and then apply fresh fixative. Let stand for 5 min.
4. Aspirate the fixative, and either air-dry (30–60 min) or rinse twice with citrate/phosphate buffer, aspirating after each rinse.
5. Apply working solution of *bis*-benzimide stain (sufficient quantity to immerse the cover slip), and let stand for 5 min in the dark. Aspirate, apply fresh stain, and let stand for 10–30 min in the dark.
6. Aspirate the stain, and wash two times with distilled water.
7. Aspirate the distilled water, and either mount or store if desired. Stained cover slips may be stored refrigerated for up to 7 d either dry or immersed in citrate/phosphate buffer.
8. Mount the cover slips on a slide, cell side down, on a drop of mounting solution. A more permanent mount may be made by sealing the mounted cover slip with petroleum jelly, nail polish, or an equivalent liquid barrier. Cover slips mounted in this manner may be stored refrigerated in the dark for several weeks.

3.5. Double Staining with Fluoresceinated Antibody and DNAF

If fluorescent antibody is available and is to be applied in addition to the DNAF, the following modified staining procedure is used:

1. Three to 5 d after addition of the test samples, remove the supernatant cell-culture medium, and rinse the cell layer twice with PBS.
2. Fix by rinsing the cover slip cell culture twice with 100% ethanol and then adding fresh 100% ethanol for an additional 20 min.
3. Remove the ethanol by aspiration, and rinse twice with PBS, pH 7.2.
4. Add about 0.5 mL of the working dilution of fluoresceinated antimycoplasma antibody preparation, and incubate at room temperature for 30 min. One or more antibody preparations against different mycoplasma species may be combined in this procedure. We have used up to five ***(9)***.
5. Remove the antibodies by aspiration, and rinse twice with PBS, pH 7.2.
6. Add *bis*-benzimide working solution in sufficient quantity to immerse the cover

slip, and let stand in the dark for 5 min. Remove the *bis*-benzimide by aspiration, and add fresh *bis*-benzimide working solution. Allow to stand in the dark for an additional 15 min.

7. Remove the *bis*-benzimide stain by aspiration, and rinse twice with PBS, pH 7.2.
8. Add Evans blue counterstain, and allow to stand at room temperature for 30 min. Remove the Evans blue by aspiration, and rinse twice with PBS.
9. Mount the cover slips cell-side down as described for the DNAF method, but using the PBS/glycerol mounting solution.
10. Observe microscopically.

3.6. Microscopy

Use a microscope equipped with a fluorescence light source, an epi-illumination system, and fluorite objectives. We have used high-pressure mercury lamps (HBO 50) with good results. We prefer magnifications of 630 or 1000× with oil immersion for viewing.

Fluorescence filters similar to those described below are suitable for *bis*-benzimide:

Excitation	BP365/11 (band pass 365 nm; 11 nm band width)
Barrier	LP397 (long-wave pass 397 nm)
Beam splitter	FT395 (dichromatic beam filter 395 nm)

Fluorescence filters similar to those described below are suitable for fluorescein:

Excitation	BP485/20 (band pass 485 nm; 20 nm band width)
Barrier	LP 520 (passes light above 520 nm wavelength)
Beam splitter	FT 510 (dichromatic beam filter 510 nm)

For the double-stain method, the microscope must also be equipped with a carriage that is capable of holding both of the fluorescence filter sets and allowing visualization of a given field by alternating with either set. A field of interest is selected using one of the filter sets, and then the same field is observed using the other filter set. This is done by sliding the filter housing to the second position without making any changes in the field of observation. It is useful to make minor adjustments of focus continuously and to switch the filter sets back and forth several times for each field being observed.

3.7. Interpretation of Results

The *bis*-benzimide–DNA complex emits 475-nm wavelength light, which is a light blue color. The indicator cell nuclei are intensely stained with entire margins and are generally ovoid. Mitochondrial DNA is not visualized with this stain, and therefore, cell cytoplasmic areas are not observed. The background on a good preparation is black. Occasionally, micronuclei formed by anomalous mitotic events are seen, but they are usually much larger than mycoplasmas. Occasional mitotic figures will also be observed. The nuclei are approx 10–20 µm in diameter (or long axis), and the mycoplasmas are seen as small, round bodies approx 0.3 µm in diameter. The mycoplasmas are generally not intracellular, but are adherent to the cell membrane and are seen in the areas where the cytoplasm is expected to be. When the indicator cells are well separated, heavily contaminated cells have a blue fluorescent nucleus with the cytoplasmic region virtually covered with discrete, small, blue fluorescent bodies and the space between the cells will generally be black. Mycoplasmas also cover the membrane above the nucleus, but these are not distinguishable from the very brightly fluorescing nucleus. The extent of cytoadsorption is variable with different mycoplasma species (*see* **Note 8**), and in some preparations, small blue fluorescent bodies are seen adhering to the glass slide between cells. These cannot be confidently diagnosed as mycoplasmas by use of *bis*-benzimide alone (refer to the double stain procedure in this chapter). DNAF staining is illustrated in **Fig. 1A**.

Although, in theory, observation of a single mycoplasma stained by DNAF could be considered a positive result, in practice, confidence in the diagnosis is enhanced by observing numbers of cell-associated fluorescent bodies that are the size and shape of mycoplasmas on an appropriate number of cells. For a test to be considered positive, the positive controls should be positive, the negative controls should be negative, and sufficient numbers (*see* **Note 9**) of blue fluorescent bodies the size of mycoplasmas and cell associated should be seen in the test culture.

In the double stain, both DNAF and fluorescein are observed in the same microscopic field by using different filter sets. The DNAF appears as described above. Specific fluorescein fluorescence is apple green, whereas nonspecific fluorescence is generally yellow. The Evans blue counterstain gives the entire cell a deep red color, which provides a striking background for the mycoplasmas and eliminates cell-associated nonspecific fluorescence. The mycoplasmas are seen as small (ca. 0.3 mm diameter), green, cell-associated, and generally round bodies. They are seen singly and in groups in both the cytoplasmic and nuclear areas of the cell (*see* **Note 10**). Observation of specific fluorescence with fluoresceinated antibody preparations used singly in this method in addition to size, shape, and cell association is sufficient to identify mycoplasmas to the species level.

Use of the double-stain method enhances confidence in diagnosing mycoplasma infection of cell cultures, because it combines the information that both DNA and mycoplasma antigen(s) are associated with the body that has been visualized by use of fluorescence. These factors along with the morphology, size, and context (cell association) are sufficiently powerful that a confident diagnosis can be made by looking at a smaller number of doubly fluorescent

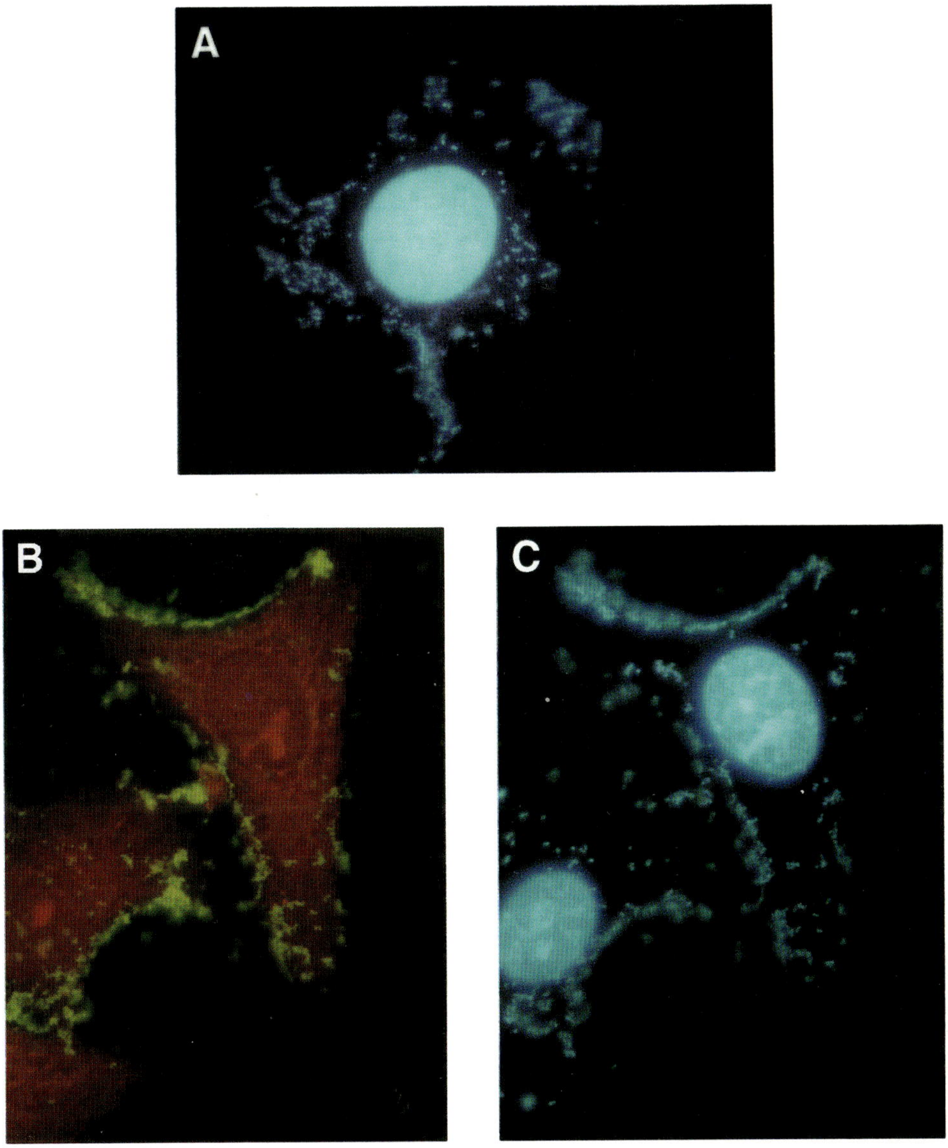

Fig. 1. Vero cells infected with *M. hyorhinis* and stained by the double-stain method with DNAF (*bis*-benzamide) and fluoresceinated anti-*M. hyorhinis* antibodies. **(A)** A single infected cell viewed using the filter set for DNAF. Photographs **(B)** and **(C)** are of the same field. Photograph (B) used the filter set for fluorescein and (C) used the filter set for DNAF.

bodies on a smaller number of host cells, or even appropriately sized doubly fluorescent bodies in the absence of host cells.

The double stain is illustrated in **Figs. 1B** and **1C**.

4. Notes

1. Acetone is commonly used as a fixative for immunofluorescence techniques. If used for the flourescent antibody part of the double-stain procedure, indicator cell cover slips must be removed from their plastic dishes to a glass staining jar for fixation and then returned to the plastic dishes for the rest of the procedure. An alternative is to use 100% ethanol as the fixative. If both a DNAF and a fluoresceinated antibody are being used as stains, fix with ethanol (100%), because it provides acceptable fixation without compromising fluorescence of either of the stains ***(8)***.
2. The *n*-propyl gallate is used to reduce photobleaching ***(12)***. This allows a particular area of a slide to be viewed for a longer period or at several different times. It also enhances maintenance of fluorescence on storage and facilitates photography.
3. Laboratories that routinely test cell cultures for mycoplasmas may want to freeze samples at ≤–60°C as soon as they are received. The advantage of freezing samples is that groups of samples may be tested on a predetermined schedule, and aliquots may be stored for retest. There will be some reduction in the number of viable mycoplasma in the test sample on thawing, but that is expected to have little impact because the numbers of contaminant mycoplasmas are almost always large enough to withstand this reduction without affecting test results.
4. Cells chosen as indicator cells should grow attached to a solid surface and have a large ratio of cytoplasmic to nuclear area. Vero and 3T6 cells are frequently used. They must also be free of mycoplasmas and must, therefore, be tested regularly for mycoplasmas by both cultural and DNAF methods. We test the indicator cells monthly in our laboratory. Cells are best seeded the day prior to inoculation of a test sample and in a density that will be subconfluent at the end of the 3- to 5-d culture period.
5. Log-phase cultures of the positive control mycoplasmas are frozen in 1-mL aliquots at <–60°C. The titer of these frozen stocks is carefully determined by quick-thawing several aliquots in a 37°C water bath, preparing serial 10-fold dilutions, and performing replicate plate counts. The titer, determined in this manner, is the basis for diluting the stocks in a manner that will reliably give 100 or less CFU in the 0.5 mL of inoculum for positive controls. Detailed procedures for preparation of such positive control stocks of mycoplasmas have been published recently ***(13)***.
6. The intent of this positive control is to test a low number of poorly cytodsorbing mycoplasmas. Therefore, *M. orale* strains or other poorly cytodsorbing mycoplasma species may also be used, but they may not give reliably positive results when inoculated in low numbers.
7. The intent of this positive control is to test a low number of strongly cytoadsorbing mycoplasmas. Therefore, other strains of *M. hyorhinis* or other cytoadsorbing mycoplasmas may be used.

8. A weakness of this method is that mycoplasmas that are poorly cytabsorbed, such as *M. orale*, are much more difficult to diagnose.
9. We have established in our laboratory that for a test to be considered positive for mycoplasma, at least 5% of ≥200 cells must have multiple (more than 5) blue fluorescent bodies the size of mycoplasmas associated with the indicator cells. Some samples, notably hybridoma cultures, contain particulate debris that reacts with the stain, rendering the test result ambiguous. In such cases, a retest is done in which 0.5 mL of supernatant medium and indicator cells resuspended by scraping from the original test culture is subcultured into fresh indicator cultures after 3–5 d of incubation. The subculture is stained and observed after an additional 3–5 d of incubation.
10. Some minimum number of green fluorescent bodies in some minimum number of cells may be defined as a diagnostic minimum requirement as described for DNAF above (**Note 9**). In the case of mycoplasmas that cytadsorb well, there is rarely question about whether or not there is an infection. However, those mycoplasmas that cytadsorb poorly do not provide easily diagnosed infections and, to detect them, it is best to observe systematically a minimum of about 200 cells. The major features of immunofluorescence staining are illustrated in **Fig. 1b**. The apple green color of specific fluorescence is easily compromised in the photographic process, but one should expect to *see* the characteristic green color with the microscope.

References

1. Russell, W. C., Newman, C., and Williamson, D. H. (1975) A simple cytochemical technique for demonstration of DNA in cells infected with mycoplasmas and viruses. *Nature* **253,** 461–462.
2. Chen, T. R. (1977) In situ detection of mycoplasma contamination in cell cultures by fluorescent Hoechst 33258 stain. *Exp. Cell Res.* **104,** 255–262.
3. Fogh, J. (1973) Contaminants demonstrated by microscopy of living tissue cultures or of fixed and stained tissue culture preparations, in *Contamination in Tissue Culture* (Fogh, J., ed.), Academic, New York, pp. 65–106.
4. Del Giudice, R. A. and Hopps, H. E. (1978) Microbiological methods and fluorescent microscopy for the direct demonstration of mycoplasma infection of cell cultures, in *Mycoplasma Infection in Cell Cultures* (McGarrity, G. J., Murphy, D. G., and Nichols, W. W., eds.), Plenum, New York, pp. 57–69.
5. Hopps, H. E., Meyer, B. C., Barile, M. F., and Del Giudice, R. A. (1973) Problems concerning "noncultivable" mycoplasma contaminants in tissue cultures. *Ann. NY Acad. Sci.* **225,** 265–276.
6. Gardella, R. S. and Del Giudice, R. A. (1995) Growth of *Mycoplasma hyorhinis* cultivar α on semisynthetic medium. *Appl. Environ. Microbiol.* **61,** 1976–1979.
7. Del Giudice, R. A., Robillard, N. F., and Carski, T. R. (1967) Immunofluorescence identification of *Mycoplasma* on agar by use of incident illumination. *J. Bacteriol.* **93,** 1205–1209.

8. Freiberg, E. F. and Masover, G. K. (1990) Mycoplasma detection in cell culture by concomitant use of bisbenzimide and fluoresceinated antibody. *In Vitro Cell Dev. Biol.* **26,** 585–588.
9. Masover, G. K. and Pleibel, N. (1994) Rapid definitive diagnosis of cell culture contaminant mycoplasmas by concomitant use of two fluorescent stains. *IOM Lett.* **3,** 74.
10. Gardella, R. S., Del Giudice, R. A., and Tully, J. G. (1983) Immunofluorescence in *Methods in Mycoplasmology* (Razin, S. and Tully, J. G., eds.), Academic, New York, pp. 431–439.
11. Tully, J. G. (1996) Introductory remarks, in *Molecular and Diagnostic Procedures in Mycoplasmology*, vol. 2 (Tully, J. G. and Razin, S., eds.), Academic, San Diego, pp. 89–91.
12. Giloh, H. and Sedat, J. W. (1982). Fluorescence microscopy: reduced photobleaching of rhodamine and fluorescein protein conjugate by *n*-propyl gallate. *Science* **217,** 1252–1255.
13. Masover, G. K. and Becker, F. A. (1996) Methodology for detection of mycoplasmas in cell cultures used to produce human pharmaceuticals, in *Automated Microbial Identification and Quantitation* (Olson, W. P., ed.), Interpharm, Buffalo Grove, IL, pp. 149–177.

25

Transformation of Mycoplasmas

F. Chris Minion and Paul A. Kapke

1. Introduction

Transformation of mycoplasmas was not clearly demonstrated until 1987 when Dybvig and Cassell were able to show the introduction of the Gram-positive transposon Tn*916* into *Acholeplasma laidlawii* and *Mycoplasma pulmonis* ***(1)***. This was the first direct evidence that mycoplasmas could be transformed under laboratory conditions. These studies have led to numerous attempts to develop mutagenesis schemes ***(2–4)***, as well as cloning vectors for gene analysis and other functions in mycoplasmas ***(5–8)***. The three common methods to transform bacteria, chemical treatment with calcium chloride or polyethylene glycol (PEG), liposome-mediated delivery of encapsulated DNA, and electroporation-based procedures have all been used in mycoplasmas and will be described here.

Mycoplasmas are structurally simple prokaryotes lacking cell walls. With only a lipid membrane bilayer to penetrate, one would presume that they would be easy targets for transformation and other genetic manipulations. This has not been the case, however. For instance, mycoplasmas are incapable of taking up naked DNA without assistance either by chemical treatment or electrical fields. Conjugation between mycoplasmas has been reported ***(9,10)***, but the mechanism of genetic exchange is not understood and the genera capable of conjugation are limited. Transduction has not been reported in mycoplasmas, although several viruses have been identified in different genera. Thus, transformation is the only dependable mechanism of genetic manipulation in mycoplasmas.

Because of low transformation frequencies, typical transformation schemes used with mycoplasmas require microgram quantities of DNA. Therefore, routine DNA constructs must first be amplified in other bacterial cloning hosts

From: *Methods in Molecular Biology, Vol. 104: Mycoplasma Protocols*
Edited by: R. J. Miles and R. A. J. Nicholas © Humana Press Inc., Totowa, NJ

prior to introduction into mycoplasmas, since frequencies are too low for ligation mixtures to be used directly. Viral-based cloning vectors for *Spiroplasma citri* transform at higher frequencies ***(11)***, which might allow transformation directly from ligation mixtures, but no frequencies have been reported that begin to reach the frequencies observed with *Escherichia coli*. Thus, the systems that are available for mycoplasma transformation are laborious, time-consuming, and inefficient. Despite these limitations, mycoplasma transformation systems are becoming important tools for the mycoplasmologist. They offer the opportunity to study mycoplasma genetic phenomena in the homologous host. For instance, gene fusions with β-galactosidase ***(12,13)***, *tetM*, and *phoA* (Minion, unpublished) can now be constructed in some mycoplasma species.

The cationic lipid Lipofectin® (Gibco BRL Life Technologies, Grand Island, NY) can be used to enable transformation in mycoplasmas. LipofectACE® failed to support transformation of either *Acholeplasma oculi* or *Mycoplasma gallisepticum*, however (Minion, unpublished). Not all of the available lipid transfection reagents have been tested on mycoplasmas, but it might be worthwhile to explore this option if PEG and electroporation do not work in some species. The disadvantage with this system is that it is expensive; the advantage is that it is simple and reproducible. There are fewer variables that could go wrong.

Finally, mycoplasmas can be transformed with liposome-encapsulated plasmid DNA (Minion and Artiushin, unpublished). Most mycoplasmas have potent membrane-associated nucleases ***(14)***, which might interfere with transformation of some species. To counteract their effects, one could treat the mycoplasmas with a protease to remove the nucleolytic proteins, but this has not enhanced the transformation frequencies with *M. gallisepticum* and *A. oculi* (Minion, unpublished). The encapsulated DNA is resistant to nuclease degradation and could allow transformation of some nontransformable species, although this has not been demonstrated directly (Minion, unpublished).

2. Materials

1. PEG solutions are routinely buffered with Tris (tris[hydroxymethyl]aminomethane), in tissue-culture-grade water and autoclaved. The compositions of PEG solutions we have used successfully are as follows: 40% (w/v) PEG-8000 (Sigma) in 10 m*M* Tris-HCl plus 0.5*M* sucrose, pH 6.5; 70% PEG-8000 in 10 m*M* Tris-HCl, pH 6.5. Autoclave at 15 psi for 15 min to sterilize. (*See* **Note 1**.)
2. 0.1*M* $CaCl_2$ dissolved in water and autoclaved.
3. PBS: 10 m*M* Na_2HPO_4, 0.150 m*M* NaCl, pH 7.3 (autoclaved).
4. TE: 10 m*M* Tris-HCl, 0.1 m*M* EDTA, pH 8.0 (autoclaved).
5. Electroporation buffer: 8 m*M* HEPES (*N*-[2–hydroxyethyl]piperazine-N'-[2-ethanesulfonic acid]), 272 m*M* sucrose, pH 7.4 (autoclaved).
6. Mycoplasma growth media are unusually complex and contain serum (10–20% v/v), fresh yeast extract, and protein-rich base. We routinely use a PPLO-based

medium for the growth of *A. oculi* and *M. gallisepticum*. PPLO broth (Difco Laboratories, Detroit, MI) 25 g/L, 10% v/v heat-inactivated (56°C, 30 min) GG-free horse serum (Gibco BRL Life Technologies), 5% v/v fresh yeast extract ***(15)***, 0.5% w/v glucose, and 2.5 µg/mL of Cefobid (Pfizer, New York, NY), pH 7.8. Other species may require different media for growth. Several reviews are available for reference ***(15,16)***. (*See* also Chapters 4–7.)

7. RPMI 1640 (a defined animal cell-culture medium, Gibco BRL Life Technologies) plus 25 m*M* HEPES, pH 7.5.
8. Lipofectin® (Gibco BRL Life Technologies).
9. Liposomes are prepared from phosphatidylcholine (Sigma) with or without the addition of an equal molar ratio of cholesterol or from lipids extracted from the transforming mycoplasma species. To extract lipids from mycoplasmas, cells are grown to late-log phase, harvested by centrifugation, washed once in PBS, and the cell pellet extracted with a 5:1 chloroform/methanol mixture. The organic phase is evaporated to dryness, and the lipids are dissolved in chloroform and stored at –70°C under nitrogen. Liposomes (large unilamellar vesicles) containing plasmid DNA are prepared by reverse-phase evaporation using the method of Straubinger and Papahadjopoulos ***(17)***. Four micrograms of phosphatidylcholine or 10 µg of mycoplasma lipids are dissolved in 0.6 mL of ether. DNA solution (170 µL) is added, the solutions mixed by vortexing, and then the suspension is sonicated for 2 min. The ether component is evaporated under slowly reduced pressure, and the remaining material resuspended in 300 µL of PBS. To determine the amount of DNA internalized in liposomes, a portion of the liposome suspension is subjected to DNase I treatment and the amount of protected DNA estimated using agarose gel electrophoresis. The liposome suspension can be stored at 4°C under nitrogen for several months without adverse effects.

3. Methods

3.1. PEG Transformations

1. Mycoplasmas (*see* **Note 2**) are grown to mid- to late-log phase (*see* **Note 3**) in a suitable medium; PPLO broth is used for *M. gallisepticum* and *A. oculi*. For most species, this results in a slightly acidic medium of approx pH 6.8. For arginine utilizers, the medium turns basic and should be harvested at a pH of approx 7.8–8.0. Five milliliters of culture are required for each transformation.
2. The mycoplasma culture is centrifuged at 12,000*g* for 10–15 min at 4°C (*see* **Note 4**).
3. The pellet is resuspended in 0.1*M* $CaCl_2$ at one-twentieth of the original volume (*see* **Note 5**). Take care to break up any cell clumps by vigorous pipeting. Pipetmen® or other similar device is indispensable for this purpose.
4. The cell suspension is then incubated on ice for 30–60 min prior to transformation.
5. Two milliliters of PEG (40% at room temperature) are added to an Oak Ridge centrifuge tube (50 mL) for each transformation (*see* **Notes 6** and **7**). The volume of PEG can be adjusted (1.5–2.5 mL) to maximize transformation frequency for each mycoplasma strain or species. For some species, 70% PEG is used ***(18)***.

6. The transforming DNA (*see* **Note 8**) is placed in a sterile 1.5-mL microfuge tube and adjusted to a volume of 50 μL with 0.1*M* $CaCl_2$.
7. 250 μL of the mycoplasma suspension are then added to each microfuge tube containing 50 μL of DNA. Once mixed, the contents of the microfuge tube are then added to one Oak Ridge tube containing PEG. (Alternatively, the cells, $CaCl_2$, and PEG can be mixed directly in the Oak Ridge centrifuge tube.)
8. The tube is vortexed immediately for 5–10 s to mix the contents completely.
9. Immediately following vortexing, the PEG-mycoplasma suspension is diluted with 25 mL of PBS or TE buffer by shaking the tube up and down several times.
10. Repeat **steps 7–9** for other transformations while maintaining the transformation mixtures at room temperature.
11. The cell suspensions are then centrifuged at 12,000*g* for 30 min at 4°C.
12. The supernatant is decanted, and 1.5 mL of warm growth medium is added to each tube. Mycoplasma pellets may not be visible at this stage, because the mycoplasmas may be distributed along the entire side of the tube.
13. The tube is vortexed for 10–15 s, and the sides of the tube are carefully washed to remove any bound mycoplasmas (*see* **Note 9**).
14. The culture is then incubated at 37°C for 1.5 h.
15. A portion of the cell suspension is plated for total colony-forming units on nonselective media, and 100 μL of the culture are spread-plated directly on selective media.
16. The remaining culture is centrifuged in a sterile microfuge tube for 2 min. The pellet is resuspended in 100 μL of growth medium and spread-plated onto selective medium.

3.2. Electroporation

1. Mycoplasma cultures are grown to mid-log phase and harvested by centrifugation at 12,000*g* for 10 min at room temperature.
2. Resuspend the pellet in electroporation buffer to an original volume and immediately centrifuge again.
3. The cells are resuspended to approx 1/300 vol in the same buffer (room temperature). (*See* **Note 10**.)
4. The cells (60 μL) are placed on ice, and 10 μg of DNA are added. It is important that the DNA is dissolved in water or electroporation buffer to minimize salt concentrations and prevent arching during electroporation.
5. The cell–DNA mixture is transferred to a chilled cuvet with a 0.2-cm gap and immediately pulsed. Using a BTX model 600 electroporator, the settings are 2.5 kV charging voltage with resistance of 129 Ω, capacitance of 50 μF, and pulse length from 5–6 ms. Settings for a Bio-Rad Gene Pulser have been reported to be 2.5 kV, 100 Ω, 25 μF in a 0.2-cm gap cuvet *(4)*.
6. The cells are then gently resuspended in 1 mL of cold growth medium, incubated for 10 min at room temperature, and then incubated at 37°C for 2–3 h.
7. The cells are then mixed thoroughly, a portion of the cell suspension is plated for total colony-forming units on nonselective medium, and the remaining cells plated on selective media. Usually the cells from a single cuvet are spread plated onto three to four plates.

3.3. Transformation with Cationic Lipids (Lipid Transfection Reagents)

1. Mycoplasmas are grown to mid-log phase overnight and the cells are washed twice with PBS.
2. The resulting pellet is resuspended in RPMI 1640 plus 25 m*M* HEPES, pH 7.5.
3. Lipofectin (30 or 50 μL) is mixed with 10 μg plasmid DNA, sterile water is added to a final volume of 100 μL, and the mixture is incubated for 15 min at room temperature.
4. One milliliter of cells is added to the cationic lipid–DNA mixture, inverted to mix, and allowed to stand at room temperature. Various incubation times can be used here from 0–60 min, but no more than 10 min of incubation seemed to be necessary for *A. oculi* (Minion, unpublished).
5. Two milliliters of warm nonselective medium are added to the mycoplasma–cationic lipid–DNA mixture, which is then incubated at 37°C for 2 h and plated on selective media.

3.4. Transformation with Liposome-Encapsulated DNA

1. For transformation of mycoplasmas using liposome-encapsulated DNA, cultures are usually harvested in late-log phase, washed once with PBS, and then resuspended to $^1/_5$–$^1/_{10}$ vol in PBS.
2. 100 μL of the cell suspension are then mixed with 50 μL of a DNA:liposome suspension. Five percent PEG (final concentration from a 40% stock solution) can be added at this point.
3. The mycoplasma–liposome mixture is then incubated at 37°C for 10 min.
4. One milliliter of mycoplasma broth is added, and the cells are allowed to recover for 1–2 h at 37°C.
5. A portion of the cell suspension is plated on nonselective medium for total colony-forming units, and the remaining cells are plated on selective media.

4. Notes

1. PEG solutions are stable for at least a year, and a large batch should be made and titered on each strain or species under study to determine the exact volume giving maximum frequencies. Each lot will be slightly different, and the volume of PEG may change in the transformation mixtures.
2. Not all mycoplasmas can be transformed using PEG procedures. *Mycoplasma pneumoniae* loses viability when treated with PEG solutions ***(4)***, and so electroporation must be used for transformation with this species. This may be true of other species as well.
3. The growth phase of the mycoplasma culture is usually extremely important to the transformation efficiency. We have found that mid-log phase cultures have the highest transformation frequency with PEG procedures, but owing to the overall low transformation efficiencies, it is important to have large numbers of cells in the transformation mixture. Therefore, we usually let the cultures grow until

they reach a mid- to late-log phase. The optimal phase of growth for transformation may vary significantly between strains and species, and should be tested.

4. The presence of protein in the DNA preparations inhibits transformation. Therefore, it might be useful to wash mycoplasmas several times in PBS or similar buffer to remove as much serum protein contamination as possible.
5. For PEG procedures, we routinely resuspend our mycoplasmas in cold 0.1*M* $CaCl_2$. We have observed with *A. oculi* and *M. gallisepticum* that our transformation frequencies are increased three- to fivefold by doing this. This was also observed with transformation of *Mycoplasma mycoides* subsp. *mycoides* ***(18)***. This is not always effective, however, and a decrease in frequency might result with some species.
6. The presence of even small amounts of detergents usually results in failed experiments. For this reason, do not wash centrifuge tubes with detergents between experiments. Carefully wash them with deionized water, scrub with brushes, dry, and autoclave. Also, it is advantageous to set aside centrifuge tubes used only for mycoplasma transformations.
7. After encountering PEG, mycoplasmas become sticky and tend to adhere to the sides of the centrifuge tube during centrifugation. It is critical that the sides of the tube are carefully rinsed in order to recover transformants. Vortexing is not sufficient (although it does help). One should use a Pipetman® or similar pipeting device to wash the tube sides with growth medium. Care must be taken to maintain sterility during this step.
8. We routinely use semipurified DNA for transforming mycoplasmas. In some instances, the presence of RNA aids in transformation ***(19)***. Yeast tRNA (10 μg/transformation) can be added to the mixture if the transformation frequencies, are low. This is not necessary if RNA is present in the DNA preparation, and it is not always effective in increasing transformation frequencies since the addition of yeast tRNA had little effect on transformation frequencies of *M. gallisepticum* ***(13)***. The extraction of DNA from mycoplasmas is described in Chapter 17.
9. A P1000 Pipetman® (or similar device) is used to deliver the PEG solution. Because of the viscosity of the PEG solution, care must be taken to ensure that all of the solution has been dispensed. Use the same instrument for all transformations to ensure consistency between experiments. It is possible to have volume differences as large as 100 μL if care is not taken during this step. For some species, this will make a significant difference in transformation frequency ***(19)***.
10. If arching is a problem with electroporation, the cells can be washed three times in electroporation buffer, allowing the cells to rest for 15 min in 400 μL of buffer between washes ***(4)***. This allows ions to diffuse from the cells, thereby lowering the conductivity of the mixture.

References

1. Dybvig, K. and Cassell, G. H. (1987) Transposition of gram-positive transposon Tn*916* in *Acholeplasma laidlawii* and *Mycoplasma pulmonis*. *Science* **235,** 1392–1394.

2. Dybvig, K. and Woodard, A. (1992) Construction of *recA* mutants of *Acholeplasma laidlawii* by insertional inactivation with a homologous DNA fragment. *Plasmid* **28,** 262–266.
3. Mahairas, G. G. and Minion, F. C. (1989) Random insertion of the gentamicin resistance transposon Tn*4001* in *Mycoplasma pulmonis. Plasmid* **21,** 43–47 (Author's correction [1993] **30,** 177, 178).
4. Hedreyda, C. T., Lee, K. K., and Krause, D. C. (1993) Transformation of *Mycoplasma pneumoniae* with Tn*4001* by electroporation. *Plasmid* **30,** 170–175.
5. Knudtson, K. L. and Minion, F. C. (1993) Construction of Tn*4001 lac* derivatives to be used as promoter probe vectors in mycoplasmas. *Gene* **137,** 217–222.
6. Hahn, T.-W., Krebes, K. A., and Krause, D. C. (1996) Expression in *Mycoplasma pneumoniae* of the recombinant gene encoding the cytadherence-associated protein HMW1 and identification of HMW4 as a product. *Mol. Microbiol.* **18,** 1085–1094.
7. Renaudin, J., Marais, A., Verdin, E., Duret, S., Foissac, X., Laigret, F., and Bove, J. M. (1995) Integrative and free *Spiroplasma citri oriC* plasmids: Expression of the *Spiroplasma phoeniceum* spiralin in *Spiroplasma citri. J. Bacteriol.* **177,** 2870–2877.
8. Marais, A., Bove, J. M., and Renaudin, J. (1996) *Spiroplasma citri* virus SpV1-derived cloning vector: Deletion formation by illegitimate and homologous recombination in a spiroplasmal host strain which probably lacks a functional *recA* gene. *J. Bacteriol.* **178,** 862–870.
9. Mahairas, G. G., Jian, C., and Minion, F. C. (1990) Genetic exchange of transposon and integrative plasmid markers in *Mycoplasma pulmonis*. *J. Bacteriol.* **172,** 2267–2272 (Author's correction [1993] **175,** 3692).
10. Barroso, G. and Labarere, J. (1988) Chromosomal gene transfer in *Spiroplasma citri. Science* **241,** 959–961.
11. Gasparich, G. E., Hackett, K. J., Stamburski, C., Renaudin, J., and Bove, J. M. (1993) Optimization of methods for transfecting *Spiroplasma citri* strain R8A2 HP with the spiroplasma virus SpV1 replicative form. *Plasmid* **29,** 193–205.
12. Knudtson, K. L. and Minion, F. C. (1993) Use of *lac* gene fusions in the analysis of *Acholeplasma* upstream gene regulatory sequences. *J. Bacteriol.* **176,** 2763–2766.
13. Cao, J., Kapke, P. A., and Minion, F. C. (1994) Transformation of *Mycoplasma gallisepticum* with Tn*916*, Tn*4001*, and integrative plasmid vectors. *J. Bacteriol.* **176,** 4459–4462.
14. Minion, F. C., Jarvill-Taylor, K. J., Billings, D. E., and Tigges, E. (1993) Membrane-associated nuclease activities in mycoplasmas. *J. Bacteriol.* **175,** 7842–7847.
15. Freundt, E. A. (1983) Culture media for classic mycoplasmas, in *Methods in Mycoplasmology*, vol. 1, *Mycoplasma Characterization* (Razin, S. and Tully, J. G., eds.), Academic, New York, pp. 127–135.
16. Tully, J. G. (1983) General cultivation techniques for mycoplasmas and spiroplasmas, in *Methods in Mycoplasmology*, vol. 1, *Mycoplasma Characterization* (Razin, S. and Tully, J. G., eds.), Academic, New York, pp. 99–101.

17. Straubinger, R. M. and Papahadjopoulos, D. (1983) Liposomes as carriers for intracellular delivery of nucleic acids. *Methods Enzymol.* **101,** 512–527.
18. King, K. W. and Dybvig, K. (1991) Plasmid transformation of *Mycoplasma mycoides* subspecies *mycoides* is promoted by high concentrations of polyethylene glycol. *Plasmid* **26,** 108–115.
19. Mahairas, G. G. and Minion, F. C. (1989) Transformation of *Mycoplasma pulmonis*: demonstration of homologous recombination, introduction of cloned genes and the preliminary description of an integrating shuttle system. *J. Bacteriol.* **171,** 1775–1780 (Author's correction [1993] **175,** 3692).

26

Transposon Mutagenesis

LeRoy L. Voelker and Kevin Dybvig

1. Introduction

Transposon mutagenesis, although utilized extensively in cell-walled bacteria, has been used rarely within the class *Mollicutes* ***(1)***. The reasons for such limited use include a lack of the necessary techniques and transposons that contain suitable antibiotic markers. There has been limited success with methods of artificial transformation using the Gram-positive bacterial transposons Tn*916* and Tn*4001*. The most successful methods for transformation are a polyethylene glycol- (PEG) mediated procedure based on that used to transform Gram-positive bacterial protoplasts and electroporation. Species that have been transformed with these techniques include *Acholeplasma laidlawii*, *Acholeplasma orale, Mycoplasma arthritidis*, *Mycoplasma pulmonis*, *Mycoplasma capricolum*, *Mycoplasma mycoides*, *Mycoplasma gallisepticum,* and *Spiroplasma citri* ***(2,3)***. Transformation of mycoplasmas is a rather inefficient process usually requiring several micrograms of purified plasmid DNA to yield 10^{-6}–10^{-8} transformants/colony-forming unit (CFU) ***(3)***. Protocols for transforming mycoplasmas can be found in Chapter 25.

A simpler and more efficient method for transposon mutagenesis is to exploit the fertility properties of Tn*916* through the use of a conjugal donor, such as *Enterococcus faecalis*. Transfer of Tn*916* by this method has been shown for *M. arthritidis* ***(4)***, *Mycoplasma hominis* ***(5)***, and *M. pulmonis* ***(6)***. Mycoplasma cells are mixed with a donor strain of *Enterococcus faecalis* (e.g., strain CG110 ***[7]***), allowed to incubate for 2 h and assayed on selective medium. The simplicity of this procedure enables the rapid examination of numerous mycoplasma strains as recipients of Tn*916*.

The most likely explanation for failure of a mycoplasma strain to serve as a recipient to conjugal transfer of Tn*916* is the presence of restriction and

From: *Methods in Molecular Biology, Vol. 104: Mycoplasma Protocols*
Edited by: R. J. Miles and R. A. J. Nicholas © Humana Press Inc., Totowa, NJ

modification systems that are common barriers to gene transfer and prevalent within the mycoplasmas ***(8)***. Therefore, Tn*916* conjugation can be used to identify strains that lack restriction-modification systems and are amenable to artificial transformation methods.

2. Materials

1. Growth medium: choose a medium suitable for the particular mycoplasma being examined.
2. Selective medium plates (60 mm): growth medium supplemented with the appropriate concentration of tetracycline as determined (*see* **Subheading 3.1.**).
3. BHI medium: Dissolve 18.5 g of brain heart infusion broth (Difco, Detroit, MI) in 400 mL of distilled water. Add 2.5 g of yeast extract (Difco), bring the volume up to 500 mL and autoclave. For BHI plates, add 5 g of agar before autoclaving. Store the medium at 4°C.
4. *E. faecalis* CG110 ***(7)***.
5. Plasmid pAM120 ***(9)***.
6. Ampicillin: dissolve 500 mg in 10 mL of distilled water, filter through a 0.2-μm filter, dispense in 1-mL aliquots, and store at –20°C.
7. Tetracycline: dissolve 12.5 mg of tetracycline in 1 mL of 70% ethanol. Tetracycline is light-sensitive and needs to be stored in the dark at –20°C. Do not store the stock solution longer than 2 wk.

3. Methods

3.1. Determination of Tetracycline Concentration

The optimal concentration of tetracycline for use in selective medium needs to be determined for each mycoplasma species examined.

1. Prepare agar plates using the appropriate growth medium supplemented with tetracycline at the following concentrations: 2, 5, and 10 μg/mL (*see* **Note 1**).
2. Inoculate 3 mL of growth medium with 30 μL of the desired mycoplasma strain, and incubate at 37°C until late-log phase.
3. Pipet 100 μL of mycoplasma cells onto an agar plate at each of the tetracycline concentrations.
4. Incubate the plates at 37°C for 5–7 d in the dark.
5. Examine the plates for the appearance of colonies. Select the lowest concentration of tetracycline that prevents the growth of any mycoplasma colonies. If all of the plates show growth, repeat the procedure with higher concentrations of antibiotic. Conversely, if none of the plates has any growth, repeat with lower concentrations.

3.2. Mating of E. faecalis with Mycoplasmas

1. Inoculate 3 mL of growth medium with 30 μL of desired mycoplasma strain, and incubate at 37°C until mid- to late-log phase.

2. Inoculate 2 mL of BHI medium supplemented with 10 µg/mL tetracycline (*see* **Note 2**) with 1 µL of *E. faecalis* CG110, and incubate at 37°C overnight with 100 rpm agitation.
3. The following morning, dilute the CG110 overnight culture 1:10, 1:100, and 1:1000 in fresh BHI medium without tetracycline (*see* **Note 3**) and incubate at 37°C until the mycoplasma cells are ready. When the mycoplasma culture is ready, use the lowest dilution of CG110 that has visible growth as judged by turbidity. Remove a small sample to determine on BHI agar the number of *E. faecalis* CFU used.
4. Add 1 mL of mycoplasma cells (1×10^8 CFU) to a 1.5-mL microcentrifuge tube, and add 100 µL of CG110 cells (10^8 CFU/mL).
5. Centrifuge the cells at 16,000*g* for 3.5 min at room temperature.
6. Resuspend the pellet in 100 µL of growth medium supplemented with 50 µg/mL ampicillin. (*See* **Note 4**.)
7. Incubate at 37°C for 2 h (*see* **Note 5**) to allow for expression of the tetracycline gene.
8. Remove a small sample to determine on growth medium the number of mycoplasma CFU. Pipet the rest of the sample onto selective medium supplemented with 50 µg/mL ampicillin.
9. Incubate the plates at 37°C for 5–7 d in the dark.
10. Pick any colonies that develop into 2 mL of selective broth medium, and incubate at 37°C until late-log phase.
11. Analyze any possible transconjugants by Southern blot analysis using pAM120 as probe.

4. Notes

1. Selective medium plates can be stored at 4°C for a maximum of 2 wk, after which they must be discarded. Plates that are older than this will develop background colonies caused by the breakdown of the tetracycline.
2. The inclusion of tetracycline in the BHI medium for overnight growth prevents loss of Tn*916* from the cells.
3. The tetracycline is omitted from the medium at this point so that there is no carryover when mixed with the mycoplasma cells. The centrifugation step performed after the mycoplasmas are mixed with enterococcal cells increases cell density, which may help facilitate conjugal transfer.
4. Extra ampicillin is added to inhibit the growth of *E. faecalis* cells, which would otherwise quickly outgrow the mycoplasmas.
5. The length of incubation necessary for expression of the tetracycline gene is dependent on the particular mycoplasma species being examined. This time should be equal to or greater than one generation time.

References

1. Hedreyda, C. T. and Krause, D. C. (1995) Identification of a possible cytadherence regulatory locus in *Mycoplasma pneumoniae*. *Infect. Immun.* **63,** 3479–83.

2. Dybvig, K. (1993) The genetics and basic biology of *Mycoplasma pulmonis*: how much is actually *Acholeplasma*? *Plasmid* **30,** 176–178.
3. Dybvig, K. and Voelker, L. L. (1996) Molecular biology of mycoplasmas. *Ann. Rev. Microbiol.* **50,** 25–57.
4. Voelker, L. L. and Dybvig, K. (1996) Gene transfer in *Mycoplasma arthritidis*: transformation, conjugal transfer of Tn*916*, and evidence for a restriction system recognizing AGCT. *J. Bacteriol.* **178,** 6078–6081.
5. Roberts, M. C. and Kenny, G. E. (1987) Conjugal transfer of transposon Tn*916* from *Streptococcus faecalis* to *Mycoplasma hominis*. *J. Bacteriol.* **169,** 3836–3839.
6. Dybvig, K. (1990) Genetic manipulation of mycoplasmas, in *Recent Advances in Mycoplasmology* (Stanek, G., Cassell, G. H., Tully, J. G., and Whitcomb, R. F., eds.), Gustav Fischer Verlag, Stuttgart, pp. 43–46.
7. Gawron-Burke, C. and Clewell, D. B. (1982) A transposon in *Streptococcus faecalis* with fertility properties. *Nature* **300,** 281–284.
8. Maniloff, J., Dybvig, K., and Sladek, T. L. (1992) Mycoplasma DNA restriction and modification, in *Mycoplasmas: Molecular Biology and Pathogenesis* (Maniloff, J., McElhaney, R. N., Finch, L. R., and Baseman, J. B., eds.), American Society for Microbiology, Washington, DC, pp. 325–330.
9. Gawron-Burke, C. and Clewell, D. B. (1984) Regeneration of insertionally inactivated streptococcal DNA fragments after excision of transposon Tn*916* in *Escherichia coli*: strategy for targeting and cloning of genes from gram-positive bacteria. *J. Bacteriol.* **159,** 214–221.

27

Demonstration of Extrachromosomal Elements

LeRoy L. Voelker and Kevin Dybvig

1. Introduction

Mycoplasmas are parasites and pathogens of plants, insects, and animals, including humans. Although mycoplasmas are highly evolved parasites, they are not immune to being parasitized themselves. Within the class *Mollicutes,* extrachromosomal elements have been described for several genera, but although abundant in the spiroplasmas, are generally very rare ***(1,2)***. The plasmids identified so far are cryptic and have come from only two species, *Mycoplasma mycoides* subsp. *mycoides* and *Spiroplasma citri* ***(3)***. The best characterized are <2 kb in size, and contain genes only necessary for plasmid replication and maintenance. Sequence analysis indicates that they are related to a large family of Gram-positive bacterial plasmids that replicate by way of single-stranded DNA intermediates ***(4)***.

Numerous bacteriophages have been isolated from *Acholeplasma* and *Spiroplasma* species, but few from the genus *Mycoplasma* ***(1,5)***. Characterized phage genomes from mollicutes range from 2–40 kb, are single- or double-stranded DNA, and circular or linear. Both virulent and temperate bacteriophages have been isolated. The virulent life cycle, however, usually does not cause host lysis, since progeny phage particles are released by budding through the cell membrane as found with Ff phages and animal viruses. Since the majority of extrachromosomal elements within the mollicutes are bacteriophage genomes, one should not assume that any newly identified element is a plasmid and remain open to the possibility of a phage genome.

A plaque assay is necessary to differentiate a bacteriophage from a plasmid genome. The quickest and simplest assay for screening numerous host strains and suspect phage material is the megaplaque assay. In addition, the conditions

From: *Methods in Molecular Biology, Vol. 104: Mycoplasma Protocols*
Edited by: R. J. Miles and R. A. J. Nicholas © Humana Press Inc., Totowa, NJ

required to set up this type of assay are not as critical as those in the plaque-forming units (PFU) assay.

Once strains sensitive to phage infection have been identified, it must be proven that the suspected phage can actually replicate within the host strain. This is necessary, since bacterial cells that produce bacteriocins or bacteriocin-like substances can also produce zones of clearing similar to plaques on lawns of bacterial cells. This can be shown with a one-step growth curve where an increase in PFUs during the infection process indicates that the extrachromosomal element in question is in fact a bacteriophage.

Because most laboratories lack the necessary equipment or expertise for electron microscopic analysis of DNA, the use of various enzymatic treatments for characterization of extrachromosomal elements is described here. To prepare phage DNA for enzymatic characterization, it is essential first to purify phage particles from other cellular material. This can often be accomplished by precipitation of phage with polyethylene glycol (PEG). Plasmids can usually be purified from cell lysates by any of the myriad techniques currently available, such as column chromatography or CsCl/ethidium bromide ultracentrifugation.

2. Materials

1. Growth media: choose a medium suitable for the particular mycoplasma being examined.
2. Top agar: 0.91% NaCl, 0.606% Tris-HCl, pH 7.0. Adjust the pH to 7.0, add 0.7% select agar, autoclave, and store in a 65°C incubator until needed.
3. 5% Dienes stain: 0.05 g Bacto-Dienes stain (Difco, Detroit, MI) in 100 mL 95% ethanol.
4. PEG solution: 30% PEG (mol wt 8000), 1.6*M* NaCl. Store at 4°C.
5. TE buffer: 10 m*M* Tris-HCl, pH 8.0, 1 m*M* EDTA.
6. 1*M* $MgCl_2$.
7. DNase I: Make stock solutions of 10 mg/mL and 1 mg/mL in TE buffer.
8. Buffered phenol.
9. Phenol/chloroform (1:1).
10. Chloroform.
11. 1*M* Tris-HCl, pH 7.5.
12. RNase A: Make a 10 mg/mL stock in distilled water, boil for 30 min, aliquot, and store at –20°C.
13. S1 nuclease: Make 10 , 1 , and 0.1 U/µL stocks in 20 m*M* Tris-HCl, pH 7.5, 50 m*M* NaCl, 0.1 m*M* $ZnSO_4$, 50% (v/v) glycerol. Store at –20°C.
14. 10X S1 Nuclease buffer: 2*M* NaCl, 0.5*M* sodium acetate, pH 4.5, 10 m*M* $ZnSO_4$, 5% (v/v) glycerol.
15. 3*M* sodium acetate, pH 5.2: Dissolve 408.1 g of sodium acetate·$3H_2O$ in 800 mL of distilled water. Adjust the pH to 5.3 with glacial acetic acid, and bring the volume up to 1 L with distilled water.
16. Absolute ethanol.

3. Methods

3.1. Isolation of Bacteriophages

Stocks of suspect phage that are free of host cells must be prepared before further examination. If the suspect material is a liquid culture, this can be accomplished by centrifuging the culture at 10,000*g* for 10 min, collecting the supernatant, and filtering through a 0.2-µm filter. This filtrate can then be aliquoted and stored at 4°C indefinitely in most cases. If the suspect phage is on agar, the plate can be flooded with sterile growth medium, incubated at 4°C for several hours, and this wash treated as described above.

3.2. Plaque Assays

3.2.1. MegaPlaque Assay

1. Inoculate 3 mL of growth medium with 30 µL of candidate host strain (*see* **Note 1**), and incubate at 37°C until mid- to late-log phase (*see* **Note 2**).
2. Add 1.5 mL of top agar to a 17 × 100 mm glass culture tube and incubate at 42°C.
3. Add 50–200 µL of host cells (*see* **Note 3**) to the precooled top agar, vortex gently (avoid air bubbles), and immediately pour onto an agar plate prewarmed to 37°C.
4. Immediately spot 10 µL of suspect phage stock or uninoculated growth medium (control) into the center of the plate before the top agar hardens.
5. Let the plate incubate at room temperature for 15 min to allow the top agar to harden.
6. Incubate the plate at the appropriate temperature for the particular host examined. At 24 and 48 h examine the plate by incident light. A large zone of clearing (megaplaque) in the center of the test plate, but not the control plate is indicative of bacteriophage (*see* **Note 4**).

3.2.2. PFU Assay

1. Inoculate 3 mL of growth medium with 30 µL of candidate host strain (*see* **Note 1**) and incubate at 37°C until mid- to late-log phase (*see* **Note 2**).
2. Aliquot 1.5 mL of top agar into a 17 × 100 mm glass culture tube and incubate at 42°C.
3. Make serial 10-fold dilutions of the phage stock as far as 10^{-8} in 90 µL of sterile growth medium.
4. Add 10 µL diluted (or undiluted) phage stock to 50–200 µL of host cells (*see* **Note 3**), and incubate at 37°C for 30–60 min to allow adsorption of the phage to the host.
5. Add the phage-host cell suspension to the precooled top agar, vortex gently (avoid air bubbles), and immediately pour onto an agar plate prewarmed to 37°C.
6. Let the plates incubate at room temperature for 15 min to allow the top agar to harden.
7. Incubate the plates at the appropriate temperature for the particular host examined. At 24 and 48 h, examine the plates by incident light. Small zones of clearing (plaques) in the lawn of host cells are indicative of bacteriophage (*see* **Note 4**).
8. The dilution giving a reasonable number of plaques (30–300) should be chosen, and the number of PFU determined.

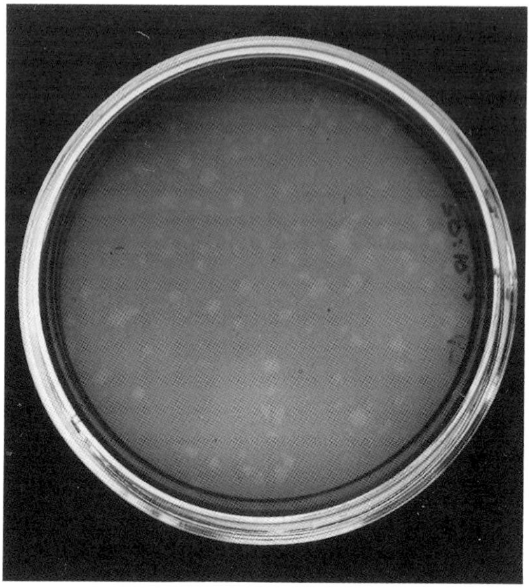

Fig. 1. Plaques of a mycoplasma bacteriophage. A lawn of *Mycoplasma arthritidis* strain PG61 infected with bacteriophage MAV1 *(8)* was stained with 5% Dienes stain to enhance the contrast for photography. Individual plaques can be seen in the lawn of mycoplasmal cells.

9. Individual plaques, if well separated, may be "picked" by stabbing the center of a plaque with a sterile large-orifice pipet tip.

3.2.3. Plaque Staining

If a photograph of individual plaques is desired, the cells can be stained to enhance the contrast between the cell lawn and the phage plaques (**Fig. 1**).

1. Flood the plate with 5% Dienes stain, and let stand at room temperature for 2–3 min.
2. Pour off the stain, and wash the plate with distilled water.
3. Flood the plate with distilled water, and let sit at room temperature for 2 min.
4. Continue washing the plate with distilled water until the wash water no longer has a blue color. Make sure the plate continually remains wet and is not allowed to dry before washing is complete.

3.3. Bacteriophage Replication

The ability of the suspected phage actually to replicate within the identified host strain can be examined with the following one-step growth curve. If there is an increase in PFU during the infection process, the extrachromosomal element in question is a bacteriophage rather than a bacteriocin-encoding plasmid.

1. Inoculate 3 mL of growth medium with 30 µL of host strain, and incubate at 37°C until mid- to late-log phase (*see* **Note 2**).
2. Infect 1.0 or 0.5 mL of log-phase cells with 0.5 mL of phage stock or a picked plaque, and incubate at 37°C for 30–60 min. The multiplicity of infection (MOI) should be ≤1.
3. Centrifuge the infected culture at 12,000*g* for 3.5 min, and discard the supernatant.
4. Resuspend the cell pellet in growth medium to give a final dilution of 1×10^{-5} in a volume of 25 mL to prevent reinfection by progeny phage.
5. Incubate the culture at 37°C with gentle shaking (100 rpm).
6. At 30-min intervals, remove 1 mL of culture, and centrifuge at 12,000*g* for 3.5 min or filter through a 0.2-µm syringe filter. Continue sampling for 6 h.
7. Assay the samples for PFU as described in **Subheading 3.2.2.**

3.4. Bacteriophage Purification and Characterization

3.4.1. Purification

1. Pour 1.0 mL of phage stock (*see* **Note 5**) into a 1.5-mL microcentrifuge tube.
2. Add 333 µL of PEG solution, mix, and incubate at room temperature for 1 h.
3. Centrifuge at 16,000*g* for 10 min, and discard the supernatant.
4. Centrifuge at 16,000*g* for 5 min, and carefully remove all remaining traces of PEG with a pipet tip. If intact phage is required for further manipulation, stop at this point.
5. Resuspend the phage pellet in 178 µL of TE buffer, add 2 µL of Triton X-100, 2 µL of 1*M* $MgCl_2$, 20 µL of DNase I (1 mg/mL stock), and incubate at 37°C for 1 h.
6. Add 200 µL buffered phenol, vortex, and centrifuge at 16,000*g* for 10 min.
7. Transfer the aqueous phase (top) to a new microcentrifuge tube, add 200 µL buffered phenol, vortex, and centrifuge as before.
8. Transfer the aqueous phase (top) to a new microcentrifuge tube, add 200 µL phenol/chloroform (1:1), vortex, and centrifuge as before.
9. Transfer the aqueous phase (top) to a new microcentrifuge tube, add 200 µL chloroform, vortex, and centrifuge as before.
10. Transfer the aqueous phase (top) to a new microcentrifuge tube, add 1/10 vol 3*M* sodium acetate, pH 3.5, add 2 vol of 100% ethanol, mix, and incubate at –70°C for 15 min.
11. Centrifuge the sample at 16,000*g* for 10 min, resuspend the pellet in 50 µL of TE buffer, and store at –20°C.

3.4.2. Genome Content

1. In separate 1.5 mL microcentrifuge tubes set up the following 100 µL reactions:

Reaction 1	Reaction 2
(89 – x) µL Tris-HCl, pH 7.5	(99 – x) µL Tris-HCl, pH 7.5
1 µL 1*M* $MgCl_2$	x µL Phage nucleic acid
x µL Phage nucleic acid	1 µL RNase A (10 mg/mL stock)
10 µL DNase I (10 mg/mL stock)	

2. Incubate the reactions at 37°C for 30 min.
3. Analyze the digestion products by conventional agarose-gel electrophoresis ***(6)***.

3.4.3. Genome Structure

Plasmids, although typically circular, have rarely been found as linear molecules, and bacteriophages commonly have linear genomes. In addition, bacteriophage genomes may be double-stranded (ds) or single-stranded (ss). Therefore, to characterize further any extrachromosomal element, its structure must be determined. The strandedness can be determined by treatment of the DNA with S1 nuclease, which specifically degrades ssDNA. Proper controls must also be included to ensure that dsDNA is not degraded under the specific condition used. By varying the concentration of S1 nuclease or the incubation time, the conformation of the DNA can also be obtained. This is because S1 nuclease will nick covalently closed circular (CCC) DNA to produce an open circular (OC) form, and more extensive digestion will further convert this OC form into linear (L) DNA ***(7)***. Because these forms all have different mobilities on agarose gels, circular molecules can be identified by the conversion from CCC to OC to L forms, and DNA resistant to S1 treatment is most likely linear.

1. In separate 1.5-mL microcentrifuge tubes, set up the following 100 μL reactions:
   (89 – x) μL distilled water
   10 μL 10X S1 nuclease buffer
   x μL Phage DNA
   1 μL S1 nuclease (10, 1, and 0.1 U/μL stocks)
2. Incubate the reactions at 37°C for 1–30 min.
3. Add 100 μL of phenol/chloroform (1:1), vortex, and microcentrifuge at 12,000*g* for 10 min.
4. Transfer the aqueous phase (top) to a new microcentrifuge tube, add 1/10 vol 3*M* sodium acetate, pH 3.5, add 2 vol of 100% ethanol, mix, and incubate at –70°C for 15 min.
5. Centrifuge the sample at 16,000*g* for 10 min, resuspend the pellet in 50 μL of TE buffer, and store at –20°C.
6. Analyze the digestion products by conventional agarose-gel electrophoresis.

4. Notes

1. Several factors should be considered when looking for a suitable host strain. The strain from which the phage was originally identified may be a lysogen and, because of superimmunity, be resistant to subsequent infection. Host range for phages is usually restricted to a particular species and in some cases to individual strains. As a result, numerous strains of the species from which the phage was isolated need to be examined. The growth rate of the mycoplasma strain and phage in question should be taken into account when deciding on volumes to use

for the PFU assay. The initial use of the megaplaque assay, in which cell density is not as critical, can overcome this problem. The megaplaque assay also allows for the rapid screening of many strains at once.

2. It is very important to grow the host cells to the mid- to late-log phase of the growth cycle. Cells that have grown past this point usually do not work and may give a false-negative result. The inclusion of phenol red in the growth medium can aid in determining the correct growth phase to harvest the cells.
3. The cell density used in producing lawns is critical. Lawns that become visible within 24 h of incubation will most likely be too dense and obscure any plaques that may be present. In addition, lawns not visible by 48 h are probably too sparse for plaque development. This can be controlled by starting cultures with smaller or larger inocula.
4. Although demonstration of plaque-forming ability is highly indicative of a bacteriophage, a complete characterization should include the use of electron microscopy to obtain photographs of actual phage particles. This is best done by negative staining of purified phage. This can be further complicated if the phage does not have a typical morphology.
5. Analyzing the genome of a previously unidentified bacteriophage can be a significant task. Obtaining a sufficient quantity of pure material to work with is often the limiting step. Phage purification requires determining infection conditions, which produce high titers of phage, and because the replicative form of the phage genome may be different from the form in phage particles, the phage must be purified from other host material. Many purification protocols utilize density gradients composed of media, such as cesium chloride, sucrose, glycerol, and so forth. Some mycoplasma phages have been found to be unstable in many if not all of these gradient media. The addition of 1% bovine serum albumin in all buffers has been used to maintain stability in some cases. The protocol given in **Subheading 3.4.1.** is a good substitute when gradient purification is not possible. A phage titer of 1×10^9 PFU/mL is required to obtain sufficient DNA to study in detail.

References

1. Maniloff, J. (1992) Mycoplasma viruses, in *Mycoplasmas: Molecular Biology and Pathogenesis* (Maniloff, J., McElhaney, R. N., Finch, L. R., and Baseman, J. B., eds.), American Society for Microbiology, Washington, DC, pp. 41–59.
2. Dybvig, K. and Voelker, L. L. (1996) Molecular biology of mycoplasmas. *Ann. Rev. Microbiol.* **50,** 25–57.
3. Dybvig, K. (1990) Mycoplasmal genetics. *Ann. Rev. Microbiol.* **44,** 81–104.
4. Gruss, A. and Ehrlich, S. D. (1989) The family of highly interrelated single-stranded deoxyribonucleic acid plasmids. *Microbiol. Rev.* **53,** 231–241.
5. Maniloff, J. (1988) Mycoplasma viruses. *Crit. Rev. Microbiol.* **15,** 339–387.
6. Ausubel, F. M., Brent, R., Kingston, R. E., Moore, D. D., Seidman, J. G., Smith, J. A., et al. (1994) *Current Protocols in Molecular Biology.* John Wiley, Inc., New York.

7. Beard, P., Morrow, J. F., and Berg, P. (1973) Cleavage of circular, superhelical simian virus 40 DNA to a linear duplex by S1 nuclease. *J. Virol.* **12,** 1303–1313.
8. Voelker, L. L., Weaver, K. E., Ehle, L. J., and Washburn, L. R. (1995) Association of lysogenic bacteriophage MAV1 with virulence of *Mycoplasma arthritidis. Infect. Immun.* **63,** 4016–4023.

28

Expression of Foreign Genes in *Acholeplasma laidlawii*

Tanja K Jarhede and Åke Wieslander

1. Introduction

The expression of foreign genes in mollicutes has been difficult owing to inefficient transformation systems, and, until recently, by the lack of suitable cloning vectors. Few genes, excluding those associated with vector functions, have successfully been cloned and expressed in mollicutes (*see* **Table 1**; *1–8*).

The mollicute species *Acholeplasma laidlawii* is often used for investigations of biological membranes. It has in biophysical terms the best-characterized membrane of all cells and organelles, especially with respect to the physical properties and phase equilibria of the lipids. Much less is known about the interactions between proteins and lipids in the membrane. To study these interactions more thoroughly, we needed to introduce and express genes in *A. laidlawii.* The plasmid pNZ18 ***(9)*** (*see* **Fig. 1A**) has been shown to replicate in *A. laidlawii* ***(10)***. The replicon of pNZ18 is from the cryptic *Lactococcus lactis* plasmid pSH71 and two antibiotic resistance genes from *Staphylococcus aureus* ***(11)***. The host range of plasmids with the pSH71 replicon is extremely broad. They replicate not only in many Gram-positive bacteria, but also in many strains of the Gram-negative *Escherichia coli* (***10,11*** and references therein). The origin of replication and the replication protein A of pNZ18 have sequence homologies to the corresponding sequences of the mycoplasma plasmid pADB201 ***(12)***, and this could be one reason for the ability of pNZ18 to replicate in *A. laidlawii.* However, pNZ18 does not appear to replicate in *Mycoplasma pulmonis* (Dybvig, personal communication). This may be peculiar to *M. pulmonis,* since other elements, such as Tn*4001,* which are not suitable for use with this bacterium, may be used with other mollicutes.

From: *Methods in Molecular Biology, Vol. 104: Mycoplasma Protocols*
Edited by: R. J. Miles and R. A. J. Nicholas © Humana Press Inc., Totowa, NJ

Table 1
Genes That Have Been Cloned and Expressed in Mollicutes

Species	Vector	Expressed gene	Reference
Spiroplasma citri	SpV1	CAT gene from *E. coli*	*(1)*
Acholeplasma oculi	Tn*4001*	*lacZ* from *E. coli*	*(2)*
S. citri	SpV1	fragment of the *M. pneumoniae* P1 gene	*(3)*
Mycoplasma mycoides subsp. *mycoides* LC	p2D4, pIKΔ	Erythromycin resistance gene from streptococcus	*(4)*
A. laidlawii	pNZ19	Spiralin gene from *S. citri*	*(5)*
A. laidlawii	pGIP312	α-Amylase gene from *Bacilius licheniformis*	*(5)*
S. citri	pOT1	Spiralin gene from *S. phoeniceum*	*(6)*
Mycoplasma pneumoniae	Tn4001	Truncated HMW1 gene from *M. pneumoniae*	*(7)*
A. laidlawii	pNZ18	PS synthase gene from *E. coli*	*(8)*

Some genes have been impossible to express in *Mollicutes* ***(5,13)***, and those that are expressed give only low amounts of proteins ***(3,5,6)***. There could be many reasons for expression problems, such as differences in promoter sequences and in codon usage, mRNA instability, inefficient translation, protein folding problems, and rapid protein degradation. However, problems with expression of foreign genes are thought most often to reside at the level of transcription and translation initiation. This is most often the reason why the majority of cloned *E. coli* genes could not be expressed in Gram-positive hosts. To achieve optimal expression of heterologous genes, it is often necessary to use species-specific promoters and ribosome binding sites (RBSs).

To isolate transcription and translation sequences that are functional in both *A. laidlawii* and *E. coli,* we used the promoter probe vector pGIP312 (*see* **Fig. 1B**) ***(5)***. This plasmid has the same replicon as pNZ18 *(see above),* and thus, the ability to replicate in both *E. coli* and *A. laidlawii.* The α-amylase gene in pGIP312 lacks transcription and translation-initiating sequences

Fig. 1. Restriction maps of plasmids. **A**: pNZ18 according to de Vos (personal communication). The chloramphenicol acetyltransferase gene *(Cam)* is derived from the plasmid pC194 and the kanamycin/neomycin nucleotidyl transferase gene *(kan/neo)* from pUB110 ***(11)***. The figure is reproduced from Sundström and Wieslander ***(10)*** with kind permission of Elsevier Science–NL, Sara Burgerhartstraat 25, 1055 KV Amsterdam,

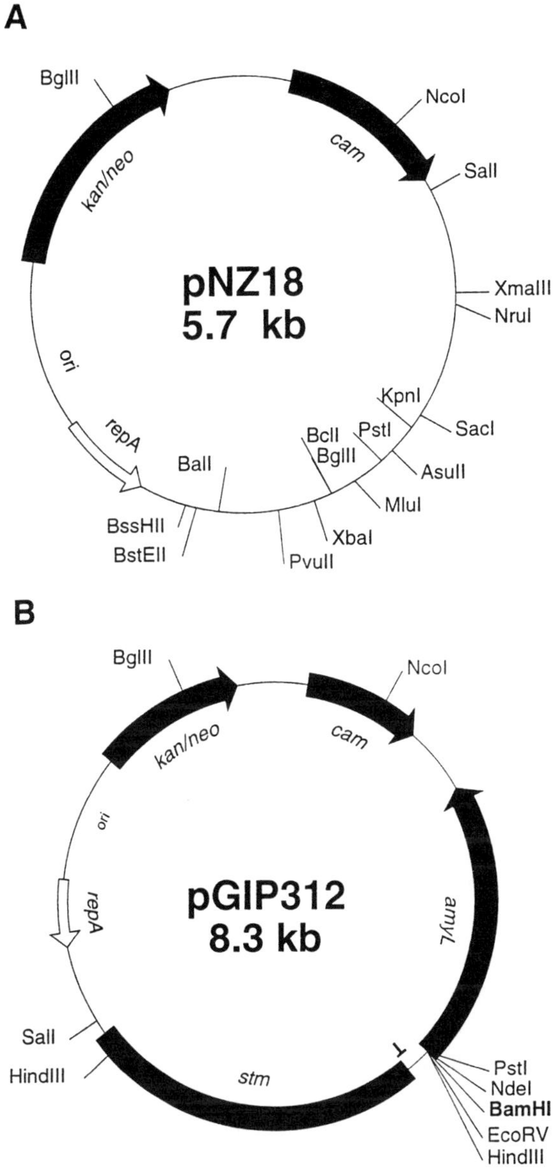

The Netherlands. **B**: pGIP312 according to Hols et al. ***(14)***. The streptomycin resistance gene *(Stm)* is derived from plasmid pMTL23P. Plasmid pGIP312 contains a truncated α-amylase gene **(amyL)** from *Bacillus licheniformis* ***(14a)***. The α-amylase gene lacks promoter, RBS, and the 5'-fragment encoding the signal peptide. The figure is reprinted from Jarhede et al. ***(5)*** with kind permission from the Society for General Microbiology.

(14). Random *A. laidlawii* DNA fragments were cloned in front of the truncated α-amylase gene in the vector. A DNA fragment containing promoter, RBS, and a start codon in the proper open reading frame in front of the α-amylase gene will induce the expression of this gene. Expression in *E. coli* and *A. laidlawii* is measured by the activity of the gene product, i.e., degradation of starch in agar plates, which results in a decrease in iodine staining.

The selected *A. laidlawii* promoter sequences were very useful for expression of the *E. coli* phosphatidylserine (PS) synthase in *A. laidlawii* ***(8)***. This enzyme could not be expressed in *A. laidlawii* from its endogenous *E. coli* promoter and RBS, but if instead corresponding *A. laidlawii* sequences were used, it was expressed. The sequence that gave the highest amount of PS in *A. laidlawii* was selected. Although the level of PS produced in vivo was maximally 1.6% of total lipids, it did influence the properties of the *A. laidlawii* membrane. The cells changed morphology and became more sensitive to antibiotics that are able to cross passively through the membrane.

The procedures described here for expression of foreign genes in *A. laidlawii* include the following steps. First, promoters and RBSs from *A. laidlawii* are selected in *E. coli* as discussed above. Those sequences that also are functional in *A. laidlawii* are then annealed in front of a foreign gene using the PCR method "splicing by overlap extension" ***(8,15)*** (*see* **Fig. 2**). In the first PCR, the foreign gene is amplified without its promoter, RBSs, and the part coding for the first amino acid. The promoters and RBSs selected from *A. laidlawii* are also amplified separately. The primers are designed to give PCR products with overlapping complementary ends (*see* **Fig. 2**). A second PCR with the two primers annealing at the nonoverlapping ends will amplify the hybrid gene consisting of promoter and RBS from *A. laidlawii* followed by the foreign gene. The gene construct is then cloned in plasmid pNZ18 and amplified in *E. coli* before transformation of *A. laidlawii*. Finally, the gene product is identified by immunoblotting or enzymatic assay.

Many of the procedures used in this chapter are standard procedures that can be found in, e.g., Ausubel et al. ***(16)*** or Sambrook et al. ***(17)***. These methods are only briefly described here, but methods that are more specific to mollicutes are described in detail. An extensive review of expression of genes in mollicutes can be found in the thesis by Jarhede ***(8)***.

2. Materials

2.1. Construction of an A. laidlawii *DNA library in pGIP312*

1. *A. laidlawii* strain 8195 ***(18)***, clone 2501 (*see* **Note 1**) ***(10)***.
2. Growth medium part A: 20 g tryptose (Difco) and 5 g NaCl in 800 mL H_2O. Sterilize by autoclaving. Store at 4°C.

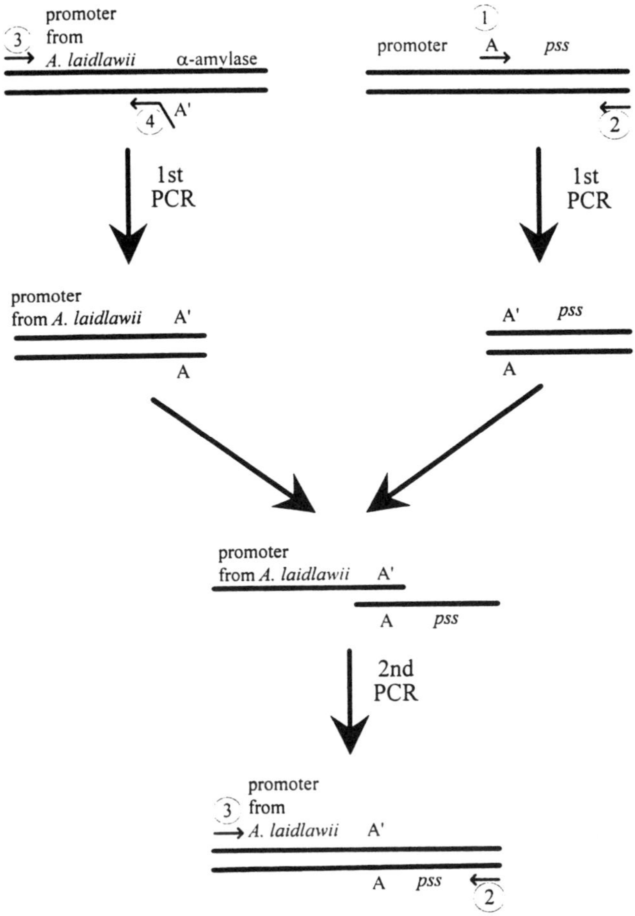

Fig. 2. PCR with overlapping primers. To the left is shown the PCR ampification of the promoters and RBSs selected from *A. laidlawii* with pGIP312. To the right in the figure is the amplification of the *E. coli pss* gene from plasmid pDD88 ***(18a)***. *See* **Notes 11** and **13** for a description of the primers. The sequence marked A' in primer 4 is complementary to the sequence A in primer 1. In the second PCR, the amplified promoter and gene are mixed and amplified with primers 2 and 3.

3. Growth medium part B: 4 g bovine serum albumin, 5 g Tris-HCl, 7 g glucose, 60 mg penicillin G in 200 mL H_2O. Sterilize by filtration. Add palmitic and oleic acids from sterile ethanolic stock solutions. The final concentration in complete growth medium (**item 4**) should be 75 μM of each fatty acid. Mix by magnetic stirring for 1–2 h at room temperature. Store at 4°C.
4. Complete growth medium for *A. laidlawii*: Mix 800 mL part A and 200 mL part B. Store at 4°C.

5. RapidPrep™ Genomic DNA Isolation Kits for cells and tissue (Pharmacia Biotech).
6. Restriction endonucleases, alkaline phosphatase, and T4 ligase.
7. Plasmid: promoter probe vector pGIP312 (*see* **Fig. 1B**) ***(5,14)***.

2.2. Selection of α-Amylase-Producing Colonies

1. *E. coli* strain MC1061 ***(19)***.
2. Reagents for transformation of *E. coli* ***(16,17,20)***.
3. Chloramphenicol: 34 mg/mL stock solution in ethanol. Sterilize by filtration and store at –20°C.
4. LA plates (10g tryptone, 5g yeast agar, 5g NaCl, 15g agar in 1 L distilled water) with appropriate antibiotics for selection of transformed *E. coli*.
5. Streptomycin: 12.5 mg/mL in H_2O. Sterilize by filtration and store at –20°C.
6. Soluble starch.
7. Filter for replica plating: Hybond-C, Amersham.
8. Iodine reagent: 0.3% (w/v) I_2 and 0.6% (w/v) KI.
9. Reagents for plasmid DNA isolation and agarose-gel electrophoresis.
10. PEG: 40% (w/v) polyethylene glycol 8000 in 0.5*M* sucrose and 0.01*M* Tris-HCl, pH 6.5. Autoclave the solution. Store at 4°C.
11. T buffer: 0.01*M* Tris-HCl, pH 8.0.
12. Neomycin: 50 mg/mL in H_2O. Sterilize by filtration, and store at –20°C.
13. 6% (w/v) agar. Sterilize by autoclaving. Store at 4°C. Melt and cool to 48°C before use.
14. Agar medium part A: tryptose and NaCl as in growth medium part A (**Subheading 2.1., item 2**), but dissolve in 600 mL, and sterilize by autoclaving.
15. Agar plates for *A. laidlawii*: Heat 600 mL agar medium part A and 200 mL growth medium part B (**Subheading 2.1., item 3**) to 48°C. Mix A, B, and 200 mL 6% agar solution. Add neomycin to 40 μg/mL (*see* **Note 2**) and pour plates. Store at 4°C for not more than 1 mo.

2.3. PCR, Cloning, and Detection

1. GeneAmpPCR reagent kit (Perkin Elmer Cetus).
2. Perkin-Elmer Cetus thermal cycler.
3. Paraffin oil.
4. Klenow enzyme.
5. Geneclean kit (Bio 101 Inc.).
6. Plasmid: pNZ18 (*see* **Fig. 1A**) ***(9)***.
7. Reagents for SDS-PAGE and Western blotting.
8. MAbs and/or polyclonal antibodies against the protein of interest.

3. Methods

3.1. Construction of an A. laidlawii DNA Library in pGIP312 (See Note 3)

1. Inoculate 400 mL complete growth medium with growing *A. laidlawii* cells (2% inoculation), and incubate without shaking at 30°C for 22 h.

2. Harvest the cells by centrifugation at 30,000*g* for 15 min at 5°C.
3. Wash the cells with isotonic NaCl (8.5 g/L) or phosphate buffer pH 7.5. Centrifuge as above.
4. Follow the instructions of RapidPrep™ Genomic DNA Isolation Kits for cells and tissue (Pharmacia Biotech) procedure D (*see* **Note 4**).
5. Measure the absorbance at 260 and 280 nm to determine the concentration and purity of the DNA ***(16,17)***.
6. Partially digest 20 µg genomic DNA with the restriction endonuclease *Sau*3A for 60 min at 37°C ***(16,17)***. The sizes of the fragments should be approx 100–400 bp. Ligate ***(16,17)*** the fragments into the unique *Bam*HI site (restricted and dephosphorylated ***16*** or ***17***) in front of the truncated α-amylase gene of pGIP312 (*see* **Fig. 1B** and **Note 5**).

3.2. Selection of α-Amylase Producing Colonies

1. Transform competent *E. coli* ***(20)*** with the ligation ***(16,17)*** (*see* **Note 6**). Select transformants on LA plates containing 10 µg/mL chloramphenicol plus 12.5 µg/mL streptomycin. Grow for 24 h at 37°C.
2. Transfer cells by replica filter plating to a new agar plate supplemented as in **step 1** plus 0.2% (w/v) soluble starch. Grow for 24 h at 37°C. Detect amylase-producing colonies by staining the starch with the iodine reagent ***(21)***. Clear halos appear around amylase-producing colonies (*see* **Note 7**).
3. Prepare plasmids (*see* **Note 8**) from positive colonies (pick them from the original agar plate).
4. Grow *A. laidlawii* strain 8195, clone 2501 (*see* **Note 1**), overnight to $OD_{610} = 0.1$.
5. Transform the cells by the polyethylene glycol (PEG) method ***(22)*** (*see* Chapter 25 in this book). PEG solution (2.5 mL) (*see* **Note 9**) is mixed with 10 µg of plasmid DNA and 250 µL washed cells (*see* **Subheading 3.1., step 3**) to give a final concentration of 36% (w/v). Incubate for 2 min. Dilute with 10 mL T buffer, centrifuge for 20 min at 18,000*g* and resuspend the cell pellet in 0.8 mL growth medium. Incubate for 2 h at 37°C. Spread 50 µL of the *A. laidlawii* cell suspension on each agar plate (14-cm diameter) containing 40 µg/mL neomycin for selection. Colonies can be seen after 4 d at 37°C.
6. Analyze the production of α-amylase in the transformants by starch-agar plates and iodine (*see* **step 2** and **Note 10**). The plasmids of positive clones are used in **Subheading 3.3.**

3.3. Gene Assembly by PCR, Splicing by Overlap Extension

1. Amplify the foreign gene with primers 1 and 2 (*see* **Fig. 2** and **Note 11**). Amplify separately the different promoters with primers 3 and 4 (*see* **Fig. 2** and **Notes 12** and **13**). The 100-µL PCR reaction mixture should contain 100 ng plasmid DNA, 1 µ*M* of each primer, 2.5 U AmpliTaq® DNA polymerase, 200 µ*M* of each dNTP, and 10 µL reaction buffer from GeneAmp®PCR reagent kit. Overlay the reaction mixture with paraffin oil. The PCR amplification is achieved with 15 cycles of: denaturation 1 min at 95°C, annealing for 1 min (*see* **Note 14**), and synthesis for

1.5 min at 72°C. Begin with a hot start for 5 min at 95°C before polymerase and primers are added, and end with a 7-min long synthesis step at 72°C.
2. To half of the amplified fragments, add 1/10 vol of 5 m*M* dNTP, and treat with 12.5 U of Klenow enzyme for 30 min at 37°C to get rid of extra 3'-bases ***(15)***.
3. Purify the products by agarose-gel electrophoresis and elute with Geneclean (*see* **Notes 15** and **16**).
4. Prepare a PCR mixture as above for the second PCR using 100 ng of the different transcription/translation initiation fragments combined with 100 ng of the gene fragment and primers 2 and 3 ***(15)***. The samples are subjected to 10–15 cycles of: 1 min at 95°C, 1 min at the annealing temperature (*see* **Notes 14** and **17**), and 2 min at 72°C.

3.4. Cloning in pNZ18

1. Cleave the PCR-amplified chimerical fragments with restriction enzymes corresponding to the sites incorporated in the primers (*see* **Notes 11** and **13**), e.g., *Sac*I (8 U/g DNA) and *Ecl*XI (*Xma*III) (20 U/µg DNA) for 30 min at 22°C followed by 3 h at 37°C for each enzyme. Ligate with the vector pNZ18 cleaved with the same enzymes (*see* **Notes 18** and **19**).
2. Transform competent *E. coli* ***(20)*** with the ligation mixture as described by Sambrook et al. ***(17)*** or Ausubel et al. ***(16)***. Select transformants on LA plates containing 10 µg/mL chloramphenicol.
3. Prepare plasmids from transformants (*see* **Note 8**).
4. Transform *A. laidlawii* strain 8195 clone 2501 using the polyethylene glycol (PEG) method (*see* **Subheading 3.2., steps 4** and **5** and **Note 20**).

3.5. Detection of Expression

The detection of the gene products can be done by Western blotting and/or activity measurements (*see* **Notes 21–23**). The measurement of activity depends on which enzyme has been cloned and will not be described here.

1. Pick colonies of transformed *A. laidlawii* with Pasteur pipets, and place in 1–2 mL prewarmed growth medium containing appropriate antibiotics. After about 2 d when turbidity is observed, inoculate to a larger volume.
2. Freeze a sample of the transformants in 15% glycerol at –70°C.
3. Preparation of samples, SDS-PAGE, and Western blotting are performed according to standard procedures (*see* Chapter 30; ***16,17,23***).

4. Notes

1. Clone 2501 is derived from a clone of strain 8195 that once had been transformed with pNZ18 and then been induced to lose the plasmid ***(10)***. Clone 2501 is transformed with a more than 1000-fold higher transformation frequency (repeatedly 4×10^{-4} transformants/CFU) in comparison to 8195.
2. It is important that the temperature of the combined medium is not more than 48°C when the neomycin is added. If too high a temperature is used, cells with-

out plasmid may survive on the agar plates. Their colonies will, however, not have the typical granular morphology ***(10)***.

3. For some genes, it may not be necessary to exchange the original promoter with a hostspecific promoter. To clone a gene with its original promoter, start at **Subheading 3.4.** We have expressed the spiralin gene of *Spiroplasma citri* in *A. laidlawii* without exchanging the promoter region ***(5)***. The expression in *A. laidlawii* and *E. coli* was, however, much lower than in *S. citri.* To achieve optimal expression of heterologous genes, it is often necessary to use species-specific promoters and RBSs.
4. The genomic DNA can also be prepared with other kits and with standard procedures ***(16,17)***.
5. Other promoter probe vectors can also been used to isolate and characterize promoter sequences from mollicutes. Simoneau and Labarère ***(24)*** selected *S. citri* sequences that promoted the expression of a tetracycline resistance gene. Knudtson and Minion ***(25)*** used a promoter-less *lacZ* as reporter gene for selection of *A. oculi* promoters. Both these selections were done in *E. coli*. Knudtson and Minion ***(2)*** have also carried out selections directly in *A. oculi* and *M. gallisepticum*. *LacZ* fusions were constructed with a derivative of the transposon Tn*4001*.
6. If the transformation frequency of *A. laidlawii* is high enough (more than about 10^{-4} transformants/CFU) it is preferable to do the selection of promoters directly in *A. laidlawii*. Transform *A. laidlawii* with the ligation mixture, and screen for amylase production in starch agar plates as in **Subheading 3.2., step 2** but grow the cells for several days.
7. 0.6% of the *E. coli* colonies presented a starch-debranching activity after transformation with pGIP312 containing *A. laidlawii* DNA.
8. Plasmids can be isolated with kits, e.g., Magic miniprep (Promega) or Qiagen tip 20 (Qiagen GmbH) or by standard procedures ***(16,17)***.
9. In our experience, it is best to weigh the PEG solution to give a final concentration of 36% (w/v). The PEG solution is too viscous to be taken up with, e.g., a Gilson pipet.
10. The production of α-amylase is much weaker in *A. laidlawii* than in *E. coli*. Grow the cells for several days on the starch agar plates.
11. Primer 1 should be complementary to the sequence just downstream of the start codon of the gene. Primer 2 is designed to be complementary to a sequence downstream of the gene and should be modified to contain a restriction site that is also found in the vector pNZ18, e.g., *Ecl*XI (*Xma*III). The sequences of the primers 1 and 2 used by Jarhede ***(8)*** for amplification of the PS synthase gene were 5'TTGTCAAAATTTAAGCGTAATAAACATCAACAACACC3'and 5'GAGCTTTATCCCGGCCG CTCCAGC3', respectively.
12. It can be advantageous to try several different promoters. Those that are best for expression of α-amylase are not necessarily best for other genes. Some gene products are deleterious if expressed in high amounts.
13. Primer 3 should be complementary to a sequence in the vector pGIP312 upstream of the insert and should be modified to contain a restriction site

that is also found in the vector pNZ18, e.g., *Sac*I. Primer 4 should be complementary to a sequence in pGIP312 downstream of the insert. This primer should also contain a sequence complementary to primer 1. We used 37-base overlaps between primers 1 and 4. The sequences of primers 3 and 4 used in Jarhede ***(8)*** were: 5'GCATAAAGC**GAGCT**CAATCAATCACC3' and 5'<u>GGTGTTGTTGATGTTTATTACGCTTAAATTTTGACAA</u>CGCGT-CGACGTCATATGGATC3', respectively. The underlined sequence in primer 4 is complementary to primer 1 (and thus to the 5'-end of the *pss* gene).

14. The annealing temperature depends on the length and base composition of the primers. It can be calculated by the formula 2 (A + T) + 4 (G + C) – 5°C. For the primers in **Notes 11** and **13**, the annealing temperatures were 55°C in the first PCR and 50°C in the second.
15. Use Qiaex (Qiagen GmbH) instead of Geneclean if the fragments are smaller than 500 bp.
16. It is important to purify the PCR products from primers before the second PCR step.
17. Use as high annealing temperature and as few cycles as possible to minimize errors. It is advisable to sequence the PCR products to ensure that they are right.
18. Plasmid pNZ19 can also be used it as only differs in the orientation of the cloning cassette (the SalI-BssHI fragment *see* **Fig. 1**).
19. The *Sal*I site of pNZ18 and pNZ19 has often been used for cloning of genes. We have successfully used the *Pst*I-*Hin*dIII sites of pNZ19 and the *Sac*I-*Xma*III sites of pNZ18 for cloning and expression of spiralin and PS synthase, respectively. It is important that the cloned gene is oriented in the same direction as the *rep*A gene of the vector. The *rep*A gene lacks efficient transcription termination ***(26)***, and transcription of this gene could otherwise interfere with the expression of the cloned gene ***(5)***.
20. It is advisable to prepare plasmids from transformed *A. laidlawii* and use restriction enzyme cleavages and agarose-gel electrophoresis to ensure that the plasmids are correct. We have, however, not had any problem with structural instability.
21. Sometimes it can also be necessary to analyze the transcription of the cloned gene with Northern blotting.
22. The expression of cloned genes has not been high enough in *A. laidlawii* to allow detection with Coomassie blue staining of proteins separated in polyacrylamide gels. To the best of our knowledge, this also applies to the other genes that have been expressed in mollicutes (**Table 1**).
23. We have detected the activity of *E. coli* PS-synthase in *A. laidlawii*. PS was identified by two-dimensional thin-layer chromatography followed by autoradiography, and by an in vitro assay ***(8)***.

Acknowledgment

We thank M. Le Hénaff for the selection of *A. laidlawii* promoters, W. M. de Vos (Ede, The Netherlands) for providing us with pNZ18, P. Hols (Louvain-La-Neuve, Belgium) for pGIP312, Bengt Harald Jonsson for discussion about PCR experiments, and V. Tegman for excellent technical assistance. This work was supported by the Swedish Natural Science Research Council.

References

1. Stamburski, C., Renaudin, J., and Bové, J. M. (1991) First step toward a virus-derived vector for gene cloning and expression in spiroplasmas, organisms which read UGA as a tryptophan codon: synthesis of chloramphenicol acetyltransferase in *Spiroplasma citri. J. Bacteriol.* **173,** 2225–2230.
2. Knudtson, K. L. and Minion, F. C. (1993) Construction of Tn*4001lac* derivatives to be used as promoter probe vectors in mycoplasmas. *Gene* **137,** 217–222.
3. Marais, A., Bové, J. M., Dallo, S. F., Baseman, J. B., and Renaudin, J. (1993) Expression in *Spiroplasma citri* of an epitope carried on the G fragment of the cytadhesin P1 gene from *Mycoplasma pneumoniae. J. Bacteriol.* **175,** 2783–2787.
4. King, K. W. and Dybvig, K. (1994) Mycoplasmal cloning vectors derived from plasmid pKMK1. *Plasmid* **31,** 49–59.
5. Jarhede, T. K., Le Hénaff, M., and Wieslander, Å. (1995) Expression of foreign genes and selection of promoter sequences in *Acholeplasma laidlawii. Microbiology* **141,** 2071–2079.
6. Renaudin, J., Marais, A., Verdin, E., Duret, S., Foissac, X., Laigret, F., et al. (1995) Integrative and free *Spiroplasma citri ori*C plasmids: expression of the *Spiroplasma phoeniceum* spiralin in *Spiroplasma citri. J. Bacteriol.* **177,** 2870–2877.
7. Hahn, T.-W., Krebes, K. A., and Krause, D. C. (1996) Expression in *Mycoplasma pneumoniae* of the recombinant gene encoding the cytadherence-associated HMW1 and identification of HMW4 as a product. *Mol. Microbiol.* **19,** 1085–1093.
8. Jarhede, T. K. (1996) Cloning and expression of genes in *Acholeplasma laidlawii.* Ph. D. thesis. Umeå University, Sweden.
9. de Vos, W. M. (1987) Gene cloning and expression in lactic streptococci. *FEMS Microbiol. Rev.* **46,** 281–295.
10. Sundström, T. K. and Wieslander, Å. (1990) Plasmid transformation and replica filter plating of *Acholeplasma laidlawii. FEMS Microbiol. Lett.* **72,** 147–152.
11. de Vos, W. M. and Simons, G. F.M. (1994) Gene cloning and expression systems in Lactococci, in *Genetics and Biotechnology of the Lactic Acid Bacteria* (Gasson, M. J. and de Vos, W. M., eds.), Blackie Academic and Professional, Glasgow, pp. 52–105.
12. de Vos, W. M., Kuiper, H., Lever, A., and Ventris, J. (1989) Heterogrammic replication of *Lactococcus lactis* plasmid pSH71 is regulated by a repressor-operator control circuit. ASM Annual Meeting, New Orleans, LA. Abstract H276.
13. Mahairas, G. G., Jian, C., and Minion, F. C. (1990) Development of a cloning system in *Mycoplasma pulmonis. Gene* **93,** 61–66.
14. Hols, P., Baulard, A., Garmyn, D., Dalplace, B., Hogan, S., and Delcour, J. (1992) Isolation and characterization of genetic expression and secretion signals from *Enterococcus faecalis* through the use of broad-host-range α-amylase probe vectors. *Gene* **118,** 21–30.

14a. Piggot, R. P., Rossiter, A., Ortlepp, S. A., Pembroke, J. T., and Ollington, J. F. (1984) Cloning in *Bacillus subtilis* of an extremely thermostable alpha amylase: comparison with other cloned heatstable alpha amylases. *Biochem. Biophys. Res. Commun.* **122,** 175–183.

15. Clackson, T., Güssow, D., and Jones, P. T. (1991) General application of PCR to gene cloning and manipulation, in *PCR. A Practical Approach* (McPherson, M. J., Quirke, P., and Taylor, G. R., eds.), IRL, Oxford, pp 187–214.
16. Ausubel, F. M., Brent, R., Kingston, R. E., Moore, D. D., Seidman, J. G., Smith, J. A., et al. (1992) *Short Protocols in Molecular Biology. A Compendium of Methods from Current Protocols in Molecular Biology,* 2nd ed. Greene Publishing and John, New York.
17. Sambrook, J., Fritsch, E. F., and Maniatis, T. (1989) *Molecular Cloning: A Laboratory Manual,* Cold Spring Harbor Laboratory Press, Cold Spring Harbor, New York.
18. Sladek, T. L., Nowak, J. A., and Maniloff, J. (1986) Mycoplasma restriction: Identification of a new type of restriction specificity for DNA containing 5-methylcytosine. *J. Bacteriol.* **165,** 219–225.
18a. DeChavigny, A., Heacock, P. N., and Dowhan, W. (1991) Sequence and inactivation of the *pss* gene of *Escherichia coli.* Phosphatidylethanolamine may not be essential for cell viability. *J. Biol. Chem.* **266,** 5323–5332.
19. Casadaban, M. J. and Cohen, S. N. (1980) Analysis of gene control signals by DNA fusion and cloning in *Escherichia coli. J. Mol. Biol.* **138,** 179–207.
20. Hanahan, D. (1985). Techniques for transformation of *E. coli,* in *DNA Cloning; A Practical Approach,* vol. 1 (Glover, D. M., ed), IRL, Oxford, pp. 109–135.
21. Smith, H., Bron, S., van EE, J., and Venema, G. (1987) Construction and use of signal sequence selection vectors in *Escherichia coli* and *Bacillus subtilis. J. Bacteriol.* **169,** 3321–3328.
22. Sladek, T. L. and Maniloff, J. (1983) Polyethylene glycol-dependent transfection of *Acholeplasma laidlawii* with mycoplasma virus L2 DNA. *J. Bacteriol.* **155,** 734–741.
23. Nyström, S., Johansson, K.-E., and Wieslander, Å. (1986) Selective acylation of membrane proteins in *Acholeplasma laidlawii. Eur. J. Biochem.* **156,** 85–94.
24. Simoneau, P. and Labarère, J. (1990) Construction of chimeric antibiotic resistance determinants and their use in the development of cloning vectors for spiroplasmas, in *Recent Advances in Mycoplasmology* (Stanek, G., Cassell, G. H., Tully, J. G., and Whitcomb R. F., eds.), Gustav Fischer Verlag, Stuttgart, pp. 66–74.
25. Knudtson, K. L. and Minion, F. C. (1994) Use of *lac* gene fusions in the analysis of Acholeplasma upstream gene regulatory sequences. *J. Bacteriol.* **176,** 2763–2766.
26. Simons, G., Buys, H., Holgers, R., Koenhen, E., and de Vos, W. M. (1990) Construction of a promoter-probe vector for lactic acid bacteria using the *lac*G gene of *Lactococcus lactis. Dev. Ind. Microbiol.* **31,** 31–39.

29

Mycoplasma Gene Expression in *Escherichia coli*

F. Chris Minion

1. Introduction

The expression of mycoplasma genes in *Escherichia coli* offers some interesting and unusual challenges compared to the expression of most other bacterial or eukaryotic cloned genes. For instance, the members of the genera *Mycoplasma*, *Ureaplasma*, and *Spiroplasma* all appear to utilize UGA (opal stop) codons as tryptophan coding codons ***(1)***. In addition, UGA seems to be the most common tryptophan coding codon in these genera as well, increasing the likelihood that many mycoplasma genes contain at least one or more UGA codons. This results in premature truncation of polypeptides during translation in normal *E. coli* hosts leading directly to the lack of expression of cloned mycoplasma genes, and the failure to identify cloned genes in genomic libraries using monospecific immunoreagents, such as monoclonal antibodies (MAbs). Using hyperimmune antisera, it has been possible to identify some cloned genes because of the probability of having antibodies directed against amino-terminal regions of the truncated polypeptides, and obviously, gene sequences lacking UGA codons are easily identified. This unusual codon usage, however, does prevent a systematic identification and analysis of the genes responsible for the complete antigenic repertoire of the mycoplasma genome.

The second unusual feature of mycoplasma DNA sequences is the unusually high AT content, which is above 70% for most species ***(2)***. *E. coli* is not very stringent in its recognition of promoter sequences during normal growth conditions, and consequently, AT-rich regions often serve as promoter sites ***(3)***. This results in aberrant transcription initiation independent of external promoter sequences that may be present in the cloning vector ***(4,5)***. Thus, it is extremely difficult to control the expression of most mycoplasma genes in *E. coli*. If the gene product (or partial gene product) is toxic or lethal to *E. coli,*

From: *Methods in Molecular Biology, Vol. 104: Mycoplasma Protocols*
Edited by: R. J. Miles and R. A. J. Nicholas © Humana Press Inc., Totowa, NJ

it may not be possible to clone the gene under normal circumstances. A high spontaneous mutant rate in the cloned sequence may arise owing to the toxicity of the product ***(5)***. Unfortunately, this would not be evident under normal cloning and screening conditions. The cumulative effect of the two processes, premature truncation and aberrant internal transcription/translational initiation, is the production of nonfunctional polypeptides or epitope-deficient partial gene products and an inability to identify some, if not the majority, of cloned mycoplasma gene products in *E. coli.*

Although it would be difficult to counteract the AT effect of mycoplasma gene sequences in *E. coli,* it is possible with the aid of UGA suppressors to produce full-length mycoplasma gene products. The most common approach has been to utilize an inducible modified tRNA, resulting in partial suppression of UGA codons during translation ***(6)***. This laboratory has been using the *trpT176* allele, which inserts tryptophan into UGA codons ***(7)***. Like all tRNA suppressor alleles, *trpT176* is only partially efficient in nonsense suppression ***(8)***. The efficiency of the *trpT176* allele has been shown to be context-dependent, with the efficiency at one site being reduced by two-thirds compared to another ***(7)***. How context (neighboring mRNA sequences) influences nonsense suppression is not known, since context could affect the efficiency of the tRNA suppression or the termination event or both. The structure of the mRNA might be affected locally, the interaction of the tRNAs on the ribosome might be affected, or codon/anticodon interactions might be stabilized through base stacking. Therefore, only a percentage (<40%) of the UGA codons are read through using this suppressor. Mycoplasma genes with only one or two UGA codons can be translated completely with the *trpT176* nonsense suppressor with high enough efficiency to allow immunodetection and possibly some functional analyses. However, those mycoplasma genes containing multiple UGA codons require a higher efficiency in UGA suppression. To circumvent this problem, an expression system was developed that utilized an inducible *trpT176* opal suppressor with a release factor 2 mutation, *prfB3*, which influenced termination events ***(9)***. The use of this system to screen genomic libraries and to enhance mycoplasma gene expression will be described here.

2. Materials

1. *E. coli*: ISM612 (*leu*[UGA] *lacZ659*[UGA] *trpA9605*[UGA] *his29*[UGA] *ilv thyA metB argH rpoB rpsL prfB3* Tn*10*)pISM3001; ISM614: LE392 (*supE44 supF58 hsdR514 galK2 galT22 metB1 trpR55 lacY1*) pISM3001 (either of these strains can be obtained from the author). Other strains that support λ growth can be substituted for ISM614.
2. Media: Luria broth (LB) (10 g tryptone, 5 g yeast extract, and 5 g NaCl in 1 L water): Superbroth (SB): (32 g tryptone, 20 g yeast extract, and 5 g NaCl in 1 L

water); agar for plates contains 15 g/L agar; soft agar overlays contain 7 g/L agar. For phage plates, 2.5 m*M* $CaCl_2$ is added to the autoclaved bottom agar, after cooling, and just prior to pouring.
3. Strain ISM612 is grown routinely in SB supplemented with 5 m*M* glucose, 10 μg/mL tetracycline, and 25 μg/mL chloramphenicol at 37°C. To prepare for λ growth, 0.2% maltose is added in lieu of glucose. ISM614 is grown in LB supplemented with 25 μg/mL chloramphenicol.
4. λ hosts are prepared from cells grown overnight, pelleted, and resuspended in 10 m*M* $MgSO_4$ in SB (ISM612) or 10 m*M* $MgSO_4$ in water (ISM614). The optical densities (OD_{600}) are adjusted to 0.8 for ISM612 and 0.3 for ISM614. The cells are maintained on ice until use. ISM614 can be maintained at 4°C for several days (4–5), but ISM612 should be prepared fresh daily.
5. SM buffer: 5.8 g NaCl, 2.0 g $MgSO_4 \cdot 7H_2O$, 6.05 g Tris-HCl, and 0.1 g gelatin in 1 L water. The pH is adjusted to 7.5 prior to autoclaving.
6. Chloramphenicol stock: 10 mg/mL in 100% ethanol:water (1:1). Store at –20°C.
7. Tetracycline stock: 10 mg/mL in sterile water, aliquoted, and stored at –20°C.
8. Maltose stock: 20% in water, autoclaved.
9. Calcium chloride stock: 2.5*M* $CaCl_2$ in water, autoclaved.
10. IPTG-impregnated nitrocellulose 85-mm circular filters are prepared by soaking sterile filters in 10 m*M* isopropyl-β-D-thiogalactopyranoside (IPTG) in water. The filters are allowed to dry completely on sterile filter paper.
11. Blocking solution: 10 m*M* Tris-HCl, 5% (w/v) dry milk, 140 m*M* NaCl, 0.05% Tween 20, pH 7.4. To preserve the solution, 0.01% sodium azide can be added (optional).
12. PBS: 10 m*M* Na_2HPO_4, 150 m*M* NaCl, pH. 7.4 (autoclaved).
13. TES: 50 m*M* Tris-HCl, 10 m*M* EDTA, 10% sucrose, pH 7.8 (autoclaved).
14. IPTG stock: 0.5*M* (1.2 g/10 mL water) filter-sterilized (0.22 μm). Store in aliquots at –20°C, and dilute to 10 m*M* with sterile water prior to use.
15. TS buffer: 10 m*M* Tris-HCl, 150 m*M* NaCl, pH 7.4 (autoclaved).

3. Methods

3.1. Screening of Genomic Libraries

1. The genomic library is prepared using any λ cloning vector, including replacement or insertion vectors. We routinely use Lambda ZAP® II (Stratagene, La Jolla, CA), Lambda FIX® II (Stratagene), or Lambda GEM-12 (Promega, Madison, WI). The presence of an inducible promoter within the cloning vector for gene expression does not seem to be necessary, but it could be important for some genes. This has not been rigorously tested.
2. The two *E. coli* λ hosts are grown overnight in antibiotic and maltose-supplemented media (0.2%): LB with 25 μg/mL chloramphenicol for ISM614, SB with 25 μg/mL chloramphenicol, and 10 μg/mL tetracycline for ISM612 (*see* **Note 1**).
3. The cells are resuspended in 10 m*M* $MgSO_4$ to the densities given in **Subheading 2, item 4**.

4. Phage diluted with SM buffer are mixed with 50 µL of the ISM614 cell suspension and incubated at 37°C for 15 min to allow attachment of the phage. The phage should be in a volume of <50 µL for maximum plating efficiency.
5. The ISM612 cell suspension (0.4 mL) is then added and mixed (*see* **Note 2**).
6. Three milliliters of LB soft agar equilibrated to 45°C are added. The cell suspension is mixed, and then poured onto LB agar phage plates. After the soft agar hardens, the plates are inverted and incubated overnight at 37°C.
7. One IPTG-saturated filter is placed on each plate (*see* **Notes 1** and **3**) and plates are allowed to incubate for an additional 5–6 h at 37°C.
8. The plates are cooled to 4°C, and the filters removed with sterile forceps (*see* **Note 4**) and placed into blocking solution. Prior to removal, the filter is marked relative to the plate for future reference. The plate is stored at 4°C until the filters are developed and the plaques are picked. Blocking of the filters is performed at room temperature for 2 h using blocking buffer.
9. Primary antisera diluted in blocking solution is then added to the filters. With hyperimmune antisera, dilutions vary from 1:500 to 1:2000. Hybridoma cell supernatants are usually diluted 1:1; ascites fluids are diluted 1:2000–1:4000. The exact dilution should be determined empirically for each primary and secondary antisera. The primary antisera is incubated for 2–4 h at room temperature or overnight at 4°C.
10. The filters are washed three times with TS buffer containing 0.05% Tween 20, incubating for 5 min between washes.
11. Secondary antibody is added and incubated for 2 h at room temperature. Secondary reagents, i.e., antibody conjugates or antibody-specific binding proteins (protein A), must be species- and antibody class-matched for every primary antibody. Do not use buffer containing sodium azide to dilute peroxidase enzyme conjugates.
12. The filters are then washed three times with TS buffer containing 0.05% Tween 20, and substrate is added to develop the filters. The substrate must be matched to the enzyme conjugate, and could produce a colored precipitate, chemiluminescence signal, or other signal. Radiolabeled secondary antibodies or antibody binding proteins can be used, but are not necessary for detection of immunoreactive plaques.
13. Immunoreactive spots are aligned to the original phage plate, and plaques are picked using a sterile Pasteur pipet by removing a plug from the center of the plaque (*see* **Note 5**). The plug is placed into 0.5 mL of SM buffer containing 20 µL of chloroform and stored overnight at 4°C to elute the phage.
14. Each phage should be plaque-purified at least twice by streaking or diluting phage for single plaques on soft agar overlays, picking, and eluting in SM buffer as described above (*see* **Note 5**). During the purification process, phage is tested for immunoreactivity after each step as described above.

3.2. Screening Individual Recombinant λ Phage for Mycoplasma Gene Products in E. coli

1. *E. coli* strain ISM612 prepared for λ growth is infected at a multiplicity of infection of 10:1 (phage:cells), by mixing phage and cells, incubated statically at 37°C for 15 min to allow attachment, and then shaken at 37°C .

2. After 30 min, IPTG is added (4 m*M* final concentration), and shaking at 37°C is continued.
3. Cultures are examined every 30 min following the addition of IPTG to assess lysis, and samples are collected to determine the time-course of antigen expression. Usually, the cells are incubated no longer than 5 h following IPTG addition. If the culture lyses, the supernatant can be assayed for the presence of mycoplasma gene products, but it is best if the infected cells are harvested just prior to lysis. The cells are then washed with PBS at least once by centrifugation (5 min, 12,000*g*), and resuspended in PBS or water.
4. For detection of cloned gene products, both immunological reagents and functional assays can be performed (*see* **Note 6**). For antigen detection, the samples are resuspended in water and are usually analyzed by immunoblotting (*see* Chapter 13).
5. Functional assays may require breakage of the cell wall. This can be achieved by resuspension of the cells in PBS (or other suitable buffer) followed by sonication to lyse the cells. Lysozyme can also be used to weaken the cell wall prior to sonication. Cells are washed and resuspended in TES buffer. Lysozyme is added to 2 mg/mL, and the cells are incubated on ice for 30 min prior to sonication. Lysis can also be achieved by passing the cells through a French pressure cell at 6000 psi.
6. Following centrifugation (10 min, 12,000*g*) to remove insoluble debris, functional assessment of enzymatic activity can be performed on the cell lysate.

3.3. Expression of Cloned Mycoplasma Genes Using Plasmid Vectors

1. *E. coli* containing a recombinant plasmid with cloned mycoplasma DNA is grown overnight in SB with antibiotics at 37°C.
2. The cells are diluted 5:1 in SB and incubated with shaking at 37°C for 2 h; 2.5 m*M* IPTG (final concentration) is then added, and the culture is shaken for an additional 5–6 h at 37°C.
3. If detection of gene expression requires a functional gene product, the cells are washed in TES buffer by centrifugation, treated with lysozyme, and sonicated (*see* **Subheading 3.2, step 5**). Alternatively, the cells can be resuspended in PBS and sonicated without lysozyme treatment if the presence of the enzyme will interfere with the assay or subsequent purification.
4. If the cells are to be assayed for the detection of mycoplasma antigen, they are washed by centrifugation and resuspended in water prior to analysis by immunoblot (*see* Chapter 13).

4. Notes

1. *E. coli* strain ISM612 grows well in SB at 37°C in the absence of induction, but once IPTG is added, the cells slow their growth. Consequently, it is advantageous to induce *trpT176* later in the growth cycle to maximize mycoplasma protein production.
2. ISM612 plaques λ with low efficiency. As much as a 4–5 log decrease in phage titer occurs if ISM612 was used as the sole plaquing host. By attaching the phage

initially to a "normal" λ host (e.g., ISM614, LE392, DH5α), maximum titers can be obtained in soft agar overlays. The extra "normal host" cells serve to enhance plaquing efficiencies and increase plaque size. They do not seem to affect antigen detection, since adequate numbers of opal suppressing hosts are available for that purpose.

3. Mark filters with a pencil in an asymmetrical pattern before placing them on the plates. After placing the filters on the plates, mark the outside of the Petri dish to correspond to the filters with a permanent marker. Alternatively, a pin could be used to stab through the filter and the agar below to mark the filters.
4. To remove filters from agar plates cleanly, chill at –20°C for 5–10 min (do not allow to freeze!). Flat blade forceps work best to remove the filters without tearing.
5. When isolating single, immunoreactive phage from plates, it is important to choose well-isolated plaques. When plaque-purifying, choose a dilution that gives well-isolated plaques as well.
6. Expression of mycoplasma genes in *E. coli* often results in various-sized products owing to premature truncation and/or aberrant transcription initiation. This may complicate immunoblot banding patterns, interfere with functional assays, or produce inhibitory products from alternative reading frames. The quantity of product produced may also be subject to codon usage patterns. Consequently, significant genetic engineering may be required to produce a specific mycoplasma gene product in large quantities in *E. coli*.

References

1. Muto, A., Andachi, Y., Yamao, F., Tanaka, R., and Osawa, S. (1992) Transcription and translation, in *Mycoplasmas: Molecular Biology and Pathogenesis* (Maniloff, J., McElhaney, R. N., Finch, L. R., and Baseman, J. B., eds.), American Society for Microbiology, Washington, DC, pp. 331–347.
2. Herrmann, R. (1992) Genomic structure and organization, in *Mycoplasmas. Molecular Biology and Pathogenesis* (Maniloff, J., McElhaney, R. N., Finch, L. R., and Baseman, J. B., eds.), American Society for Microbiology, Washington, DC, pp. 157–168.
3. Knudtson, K. L. and Minion, F. C. (1993) Use of *lac* gene fusions in the analysis of *Acholeplasma* upstream gene regulatory sequences. *J. Bacteriol.* **176,** 2763–2766.
4. Notarnicola, S. M., McIntosh, M. A., and Wise, K. S. (1990) Multiple translational products from a *Mycoplasma hyorhinis* gene expressed in *Escherichia coli*. *J. Bacteriol.* **172,** 2986–2995.
5. Jarvill-Taylor, K., VanDyk, C., and Minion, F. C. Characterization of a membrane nuclease encoding gene of *Mycoplasma pulmonis* and analysis of its expression in *Escherichia coli*. Submitted for publication.
6. Renbaum, P., Abrahamove, D., Fainsod, A., Wilson, G. G., Rottem, S., and Razin, A. (1990) Cloning, characterization, and expression in *Escherichia coli* of the gene coding for the CpG DNA methylase from *Spiroplasma* sp. strain MQ1(M.SssI). *Nucleic Acids Res.* **18,** 1145–1152.

7. Raftery, L., Egan, J., Cline, S., and Yarus, M. (1984) Defined set of cloned termination suppressors: in vivo activity of isogenetic UAG, UAA, and UGA suppressor tRNAs. *J. Bacteriol.* **158,** 849–859.
8. Brown, C. M., Stockwell, P. A., Trotman, C. N. A., and Tate, W. P. (1990) The signal for the termination of protein synthesis in prokaryotes. *Nucleic Acids Res.* **18,** 2079–2086.
9. Smiley, B. K. and Minion, F. C. (1993) Enhanced readthrough of opal (UGA) codons and production of *Mycoplasma pneumoniae* P1 epitopes in *Escherichia coli. Gene* **134,** 33–40.

30

Polyacrylamide Gel-Electrophoresis Separation of Whole-Cell Proteins

Michael F. Duffy, Amir H. Noormohammadi, Nina Baseggio, Glenn F. Browning, and Philip F. Markham

1. Introduction

Polyacrylamide gel electrophoresis (PAGE) of whole-cell proteins was first used for differentiation of mycoplasma species by Razin and Rottem *(1)* The method used was separation through 7.5% acrylamide gels containing 35% acetic acid and 5*M* urea. However, since the publication of Laemmli's method for sodium dodecyl sulfate PAGE (SDS-PAGE) *(2)*, most workers have adopted this approach.

The principal reasons for examining whole-cell proteins are to achieve sufficient separation on the basis of molecular mass to allow individual proteins to be identified. This technique is suitable for most applications and allows direct comparisons to be made between different samples run on the same gel. More efficient separations can be achieved using two-dimensional gel electrophoresis, utilizing separation on the basis of molecular mass and charge. This is more time-consuming and can be more demanding if multiple samples are to be compared. However, use of commercially available precast gel systems can allow highly reproducible results to be obtained.

In many cases, sufficient information can be obtained from direct staining of proteins in polyacrylamide gels, using either Coomassie blue or silver stains. In some cases, immunostaining is used to facilitate greater specificity. In these cases, unstained proteins are transferred to nitrocellulose or polyvinylidene difluoride (PVDF) membranes, and then probed with specific antisera or monoclonal antibodies (MAbs). Some information about the significance of disulfide bonds in proteins of interest may be gleaned by comparing reduced and unreduced samples separated on the same gel.

From: *Methods in Molecular Biology, Vol. 104: Mycoplasma Protocols*
Edited by: R. J. Miles and R. A. J. Nicholas © Humana Press Inc., Totowa, NJ

SDS-PAGE can also be used to purify proteins for subsequent experimental procedures (*see* Chapter 31). One example of this is excision of specific protein bands for production of monospecific antisera. Ground gel slices containing the protein of interest can be emulsified with adjuvant and used to vaccinate laboratory animals. Similar methods can also be used to produce small quantities of affinity-purified antibodies. The separated proteins are transferred to nitrocellulose or PVDF membrane, the proteins visualized by Ponceau S staining, and the region containing the protein of interest excised. This is then incubated with antisera, washed, and the antibody specifically binding to the protein on the nitrocellulose strip eluted. Proteins separated by SDS-PAGE and blotted onto PVDF membrane can also be used for amino-terminal peptide sequencing.

The following sections describe the preparation, use, and staining of SDS-polyacrylamide gels, following the method of Laemmli, blotting of these gels onto nitrocellulose membranes *(3)*, and methods for probing these immunoblotted proteins. These methods are adaptations of those described by Sambrook et al *(4)* and Harlow and Lane *(5)*. A method for separating proteins by two-dimensional gel electrophoresis is also described, based on the methods described in manufacturer's information sheets provided by Bio-Rad.

2. Materials

2.1. Mycoplasma Strains and Broth Media

We have successfully used the methods described here with *Mycoplasma gallisepticum, Mycoplasma synoviae, Mycoplasma pirum,* and *Mycoplasma pneumoniae*. Others have used these or similar techniques on a range of other *Mycoplasma* species. Broth and agar media used for growth will need to be chosen on the basis of the species used. Some guidance may be gained from Chapters 4–7.

2.2. SDS-Polyacrylamide Gels

1. Gel electrophoresis cell, including buffer chamber and lid, electrode assembly, combs, glass plates, spacers, and a gel casting stand (Mini-PROTEAN II cell or PROTEAN II xi cell, Bio-Rad) (*see* **Note 1**).
2. Power Pack capable of supplying voltages up to 200 V and current of up to 400 mA.
3. 30% Acrylamide/*bis*-acrylamide (37.5:1):146 g acrylamide, 4 g *bis*-acrylamide (*N,N'*-methylene-*bis*-acrylamide), distilled water to 500 mL. Mix until dissolved. Filter through Whatman No. 1 filter paper and store in the dark at 4°C. (*See* **Note 2**.)
4. 10% SDS: 50 g sodium dodecyl sulfate (SDS), distilled water to 500 mL. Store at room temperature. SDS may precipitate at lower temperatures, but the solution can be heated to redissolve the SDS before use.

5. Stacking Gel Buffer: 6.06 g tris(hydroxymethyl)-aminomethane (Tris base), distilled water to 100 mL. Adjust pH to 6.8 with hydrochloric acid. Store at 4°C.
6. Resolving gel buffer: 18.16 g Tris Base, distilled water to 100 mL. Adjust pH to 8.8 with hydrochloric acid. Store at 4°C.
7. Sucrose or butanol.
8. 10% Ammonium persulfate: 0.1 g ammonium persulfate, distilled water to 1 mL. Can be stored at 4°C for up to 1 wk.
9. *N,N,N',N'*-tetramethyl-ethylenediamine (TEMED).
10. Broth culture of mycoplasma species or strains of interest grown to early log phase. Volumes will vary, but typically 1.5 mL of culture is sufficient for one lane of a gel.
11. 10X Electrophoresis (electrode) buffer: 151.5 g Tris base, 721 g glycine, 50 g SDS, distilled water to 5 L. Store at room temperature.
12. Sample loading buffer: 0.2 g SDS, 1 mL glycerol, 1.25 mL 0.5 *M* Tris-HCl, pH 6.8 (stacking gel buffer), 500 μL β-mercaptoethanol, 0.1 mg bromophenol blue, distilled water to 10 mL. The β-mercaptoethanol should be deleted if examination of unreduced proteins is desired. Store at room temperature.
13. Protein molecular-mass standards: available from a number of suppliers and over a range of expected sizes. Prestained standards allow continuous monitoring of separation during electrophoresis (for example, Kaleidoscope Prestained Standards, Bio-Rad).
14. Coomassie blue staining solution: 0.1% (w/v) Coomassie brilliant blue, 10% (v/v) acetic acid, 40% (v/v) methanol in distilled water.
15. Destaining solution: 7% (v/v) acetic acid in distilled water.

2.3. Western Blotting Polyacrylamide Gels

1. Polyacrylamide gel containing the separated proteins from the species and/or strains of interest. The proteins may have been separated in one or two dimensions.
2. Electrophoretic blotting cell, including buffer chamber and lid, electrode assembly, cooling unit, fiber pads, and gel holding cassettes (Mini Trans-Blot Electrophoretic Transfer Cell or Trans-Blot Electrophoretic Transfer Cell, Bio-Rad).
3. Power Pack capable of supplying voltages up to 200 V and current of up to 400 mA.
4. Magnetic stirring rod and stirrer.
5. 10X Western transfer buffer: 151.5 g Tris base, 721 g glycine, distilled water to 5 L. Store at room temperature. (*See* **Note 6**.)
6. Methanol.
7. Polyvinylidene difluoride (PVDF) membrane (Immobilon-P; Millipore).
8. Blotting paper.

2.4. Probing Western Blots with Antibodies

1. Western blot of electrophoretically separated proteins.
2. PBS: 8 g NaCl, 0.2 g KCl, 1.84 g $Na_2HPO_4 \cdot 12H_2O$, 0.2 g KH_2PO_4, distilled water to 1 L.
3. 5% (w/v) Bovine serum albumin (BSA) in PBS.

4. PBST: PBS containing 0.5 mL Tween 20/L.
5. PBST containing 0.1% (w/v) BSA.
6. PBST containing 1% (w/v) BSA.
7. Primary antibody: Antiserum or MAbs against protein(s) of interest.
8. Secondary antibody: Horseradish peroxidase conjugated to antibody directed against the primary antibody.
9. Chromogen solution: Dissolve one 10-mg tablet of 3,3'-diaminobenzidine (Sigma) in 30 mL of 0.02*M* Tris, 137 m*M* NaCl. Adjust to pH 7.6 using HCl. Add 24 μL of 30% (w/v) H_2O_2 just before use.

2.5. Two-Dimensional Gel Electrophoresis

1. Capillary tube gel cell, including a tube gel adapter, sample reservoirs and stoppers, sample reservoir/capillary tube connectors, capillary tubes and a gel casting tube, and a tube gel ejector (Mini Tube Cell Module, Bio-Rad).
2. Power Pack capable of supplying voltages up to 1500 V.
3. Gel-electrophoresis cell, including buffer chamber and lid, electrode assembly, combs, glass plates, spacers, and a gel casting stand (Mini-PROTEAN II cell, Bio-Rad) (*see* **Note 1**).
4. Parafilm.
5. 30% Acrylamide/*bis*-acrylamide (37.5:1):146 g acrylamide, 4 g *bis*-acrylamide (*N,N'*-methylene-*bis*-acrylamide), distilled water to 500 mL. Mix until dissolved. Filter through Whatman No. 1 filter paper, and store in the dark at 4°C. (*See* **Note 2**.)
6. Urea.
7. 10% Triton X-100.
8. Bio-Lyte ampholytes 3-10 (Bio-Rad).
9. 10% Ammonium persulfate: 0.1 g ammonium persulfate, distilled water to 1 mL. Can be stored at 4°C for up to 1 wk.
10. TEMED.
11. Broth culture of mycoplasma species or strains of interest grown to early log phase. Volumes will vary, but typically 1.0 mL of culture is sufficient for one gel.
12. Urea lysis buffer: 2.85 g urea, 0.1 mL 10% Triton X-100, 300 μL Bio-Lyte ampholytes 3-10. Add distilled water to 5 mL, and store at –70°C in 200-μL aliquots.
13. 100X Upper chamber electrolyte: Dissolve 40 g NaOH in 100 mL distilled water. Dilute with distilled water and degas immediately before use.
14. Sample overlay buffer: 1.5 g urea, 1.0 mL 10% Triton X-100, 250 μL Bio-Lyte ampholytes 3-10. Add distilled water to 5 mL. Store in 200-μL aliquots at –70°C.
15. 100X Lower chamber electrolyte: Add 31.07 mL 93% phosphoric acid to 500 mL distilled water. Dilute with distilled water for use as a 1X solution.
16. 10% SDS: 50 g sodium dodecyl sulfate (SDS), distilled water to 500 mL. Store at room temperature. SDS may precipitate at lower temperatures, but the solution can be heated to redissolve the SDS before use.
17. Stacking Gel Buffer: 6.06 g tris(hydroxymethyl)-aminomethane (Tris base), distilled water to 100 mL. Adjust pH to 6.8 with hydrochloric acid. Store at 4°C.

18. Resolving gel buffer: 18.16 g Tris base, distilled water to 100 mL. Adjust pH to 8.8 with hydrochloric acid. Store at 4°C.
19. Sucrose or butanol.
20. 10X Electrophoresis buffer: 151.5 g Tris base, 721 g glycine, 50 g SDS, distilled water to 5 L. Store at room temperature.
21. Sample loading buffer: 0.2 g SDS, 1 mL glycerol, 1.25 mL 0.5*M* Tris-HCl, pH 6.8 (stacking gel buffer), 500 µL β-mercaptoethanol, 0.1 mg bromophenol blue, distilled water to 10 mL. Store at room temperature.
22. Coomassie blue staining solution: 0.1% (w/v) Coomassie brilliant blue, 10% (v/v) acetic acid, 40% (v/v) methanol in distilled water.
23. Destaining solution: 7% (v/v) acetic acid in distilled water.

3. Methods

3.1. SDS-Polyacrylamide Gels

Polyacrylamide gels are formed by crosslinking between acrylamide molecules using *bis*-acrylamide. The pore size of the gel depends on the acrylamide/*bis*-acrylamide ratio, and the amount of acrylamide and *bis*-acrylamide in solution during polymerization. For separation of proteins between 6.5 and 200 kDa (the usual range needed for examination of mycoplasma proteins), an acrylamide/*bis*-acrylamide ratio of 37.5:1 is used. Laemmli's method uses a large-pore-size stacking gel at lower pH than either the electrophoresis buffer or the resolving gel buffer to concentrate the proteins at a buffer interface. Although in the stacking gel the concentration at the interface does not allow any separation of the proteins, once this interface enters the resolving gel, the proteins begin to migrate according to molecular mass. Typically, the stacking gel is formed from 3% acrylamide, but the concentration in the resolving gel can range from 5–20% (w/v), with lower-concentration gels facilitating separation of larger-molecular-mass proteins. Polymerization is catalyzed by ammonium persulfate and TEMED.

1. Prepare glass plates by scrubbing with lint-free tissue paper wet with 70% ethanol. When all adherent material has been removed, assemble the gel sandwich by placing the spacers at either side of one glass plate and placing the second plate on top of spacers. Ensure that spacers and glass plates are aligned at either end, and clamp either side of the sandwich. Place the clamped sandwich into the pouring stand.
2. Prepare resolving and stacking gel mixtures, as detailed in **Table 1**, but do not add ammonium persulfate or TEMED until immediately before pouring the mixture into the gel sandwich (*see* **Note 3**). If both gel layers are to be poured at the same time (*see* **step 4**), add 0.1 g sucrose/mL to the resolving gel mixture.
3. Add the ammonium persulfate and TEMED to the resolving gel mixture, and gently swirl to mix, then pour the gel mixture into the gel sandwich, either with a pipet, taking care not to introduce the mixture too rapidly and trap air bubbles, or

Table 1
Composition of Gel Mixtures for SDS-PAGE[a]

	Resolving gel composition				
	Final acrylamide concentration of gel				
	5%	7.5%	10%	15%	20%
Resolving gel buffer (mL)	1.3	1.3	1.3	1.3	1.3
Distilled water (mL)	2.9	2.5	2.1	1.3	0.4
10% SDS (μL)	50	50	50	50	50
30% Acrylamide (mL)	0.8	1.3	1.7	2.5	3.3
10% Ammonium persulfate (μL)	16	16	16	13	9
TEMED (μL)	3	3	3	6	6
	Stacking gel composition				
Stacking gel buffer (mL)	1.25				
Distilled water (mL)	2.91				
10% SDS (μL)	25				
30% Acrylamide (mL)	0.8				
10% Ammonium persulfate (μL)	12.5				
TEMED (μL)	2.5				

[a]The volumes given are the amount of stock solutions added/5 mL of gel-mixture volume.

using a syringe, with its plunger removed, attached to a needle. Pour the resolving gel to a minimum height of 0.5 cm below the teeth of the comb when it is fully inserted.

4. Two alternative approaches may be adopted for layering the stacking gel on top of the resolving gel. The resolving gel may be poured, overlaid with butanol, and allowed to polymerize. The butanol is then poured off, the gel rinsed with distilled water, and the stacking gel poured on top and allowed to polymerize. Alternatively, both gels may be poured at the same time by dissolving 0.1 g sucrose/mL in the resolving gel mixture. The resolving gel is then poured, and immediately overlaid carefully with the stacking gel mixture (for instance, using a syringe attached to a 22-gage needle that has had its end bent) before the resolving gel polymerizes. Immediately after the stacking gel has been poured, the comb should be inserted.
5. At the completion of polymerization, a narrow zone of unpolymerized acrylamide can be observed around the teeth of the comb (*see* **Note 4**). If the comb is to be removed some time before loading samples onto the gel, the wells should be rinsed with distilled water to remove this acrylamide, since it will eventually polymerize and produce uneven well boundaries, which impair the resolution of individual bands on the gel.
6. Place the gel sandwich in an assembly apparatus. In most commercially available systems, the gel sandwich is part of the seal for the upper buffer chamber. Fill the upper chamber with 1 × electrode buffer, so that the upper

exposed surface of the gel is totally immersed. Examine the upper chamber for leaks before loading the protein samples.

7. Pellet the mycoplasma cells by centrifugation of 1.0 mL of broth at 13,000*g* for 5 min. Resuspend the cells in 100 µL sample loading buffer, and incubate the mixture in a boiling water bath for 5 min. For Western blots, the amount of culture used may be reduced up to fivefold.
8. After cooling, samples (typically 10 µL of the mixture prepared in **step 7**) and molecular-mass standards are loaded into individual wells in the stacking gel using a micropipet. Protein molecular-mass standards should be loaded into at least one well. Fill the lower gel chamber with 1X electrode buffer. If the apparatus allows it, the majority of the gel sandwich should be immersed in this lower chamber buffer, since it aids heat dispersal from the gel and ensures more even resolution across the gel, eliminating any smiling (band distortion) effects.
9. Apply current across the gel, ensuring that the positive electrode is in the lower chamber and the negative electrode in the upper chamber. Typically, a constant current of 20 mA is applied to large gels (20 cm × 20 cm × 1 mm) and a constant voltage of 200 V to small gels (8 cm × 8 cm × 0.5 mm). Continue electrophoresis until the bromophenol blue marker dye runs out of the bottom of the gel. The time taken for this will vary with the concentration of acrylamide, but as a general guide, large 7.5% acrylamide gels will take about 4 h and a small gel will take about 45 min.
10. After completion of electrophoresis, disassemble the gel sandwich, and gently ease the gel off the glass plates. If the proteins are to be blotted onto a membrane, the gel is not stained. If it is to be stained, it is immersed in Coomassie blue staining solution for 15 min with gentle rocking.
11. Destain the gel for 1–3 h in destaining solution. If optimal resolution is desired, destaining should be continued for 12 h with several changes of destaining solution. The protein bands are best visualized by examination over a light box. Gels can be dried onto blotting paper for long-term storage. An example of a Coomassie blue-stained gel is shown in **Fig. 1** (*see* **Note 5**).

3.2. Western Blotting Polyacrylamide Gels

1. Immerse the gel in 1X Western transfer buffer (*see* **Note 6**) containing 20% methanol for 5 min.
2. Cut a piece of PVDF membrane and two pieces of blotting paper to the same size as the gel, and immerse them in the same buffer as the gel.
3. Assemble a blotting sandwich by placing a piece of wet blotting paper on a sponge pad, and then placing the gel on the blotting paper, taking care to squeeze out any air bubbles between the gel and the blotting paper with gentle pressure using gloved fingers. Place the wet membrane on the gel, again taking care to squeeze out any bubbles. Then top this with the second piece of blotting paper and a second sponge pad.
4. Secure the sandwich and insert it into a blotting chamber, taking care to ensure that the membrane side of the gel is closest to the positive electrode.

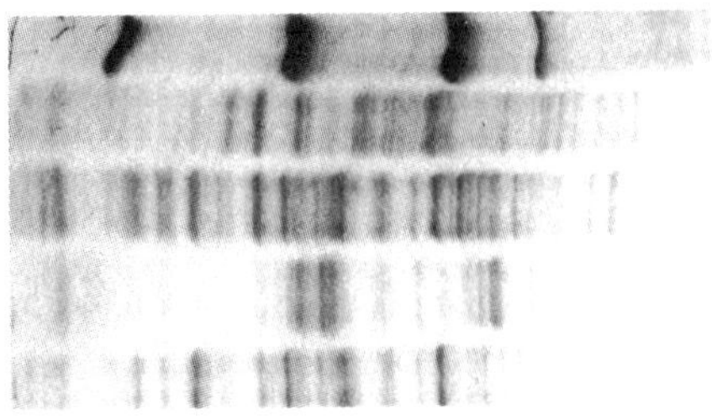

Fig. 1. Coomassie blue-stained whole-cell proteins of *M. gallisepticum* (MG whole cell) and *M. synoviae* (MS whole cell). These are compared to Triton X-114 fractionated proteins of *M. synoviae*, (MS hydrophobic phase and MS hydrophilic phase), a technique described in Chapter 31. Molecular-mass *(Mr)* markers are 97, 66, 44, and 31 kDa.

5. Fill the chamber with 1X Western transfer buffer containing 20% (v/v) methanol. The apparatus is generally cooled throughout the transfer, either by insertion of an ice block into the chamber (for miniblotting systems) or by placing a cooling coil supplied with running water in the chamber (for larger systems). Buffer is circulated in the chamber using a magnetic stirring rod.
6. Apply a constant potential difference of 100 V for 1 h in miniblotting systems or of 70 V for 16 h for larger systems. After transfer is complete, disassemble the sandwich, and wash the membrane briefly in distilled water. If the membrane is not to be used immediately, it can be dried, protein side up, on a piece of blotting paper.

3.3. Probing Western Blots with Antibodies

1. Incubate the membrane with 5% BSA in PBS for at least 2 h at room temperature on a rocking platform. This procedure reduces nonspecific binding of antibody to the membrane.
2. Wash the membrane three times for 5 minutes each in PBST/0.1% BSA.
3. Antisera or MAbs for probing Western blots may need to be titrated to determine optimal concentrations. Usually a dilution of 1/100 is used for initial assessment, using PBST/1% BSA as a diluent. Incubate the blot in the diluted antibody for 1 h at room temperature with continuous rocking.
4. Wash the blot three times for 5 minutes each in PBST/0.1% BSA. Add the secondary antibody (*see* **Note 7**) diluted in PBST/1% BSA. Again this needs to be titrated for individual experiments, but typically dilutions of 1/500–1/2000 are suitable. Incubate the blot in the diluted secondary antibody for 1 h at room temperature with continuous rocking.
5. Wash the membranes three times in PBST/0.1% BSA for 5 min each wash. The bound conjugated antibody is detected by the addition of chromogen solution. Color development is stopped by extensive washing with distilled water. (*See* **Note 8**.)

3.4. Two-Dimensional Gel Electrophoresis

In two-dimensional gel electrophoresis, the proteins are first separated on the basis of charge using isoelectric focusing, and then separated on the basis of size using SDS-PAGE. The isoelectric focussing is done in capillary tube gels. The system we have used is that supplied by Bio-Rad. The instructions provided by the manufacturer suggest two alternative methods for casting the gels, but we have found the batch casting method, with minor modifications as described below, to be the most reliable.

1. Seal the glass casting tube with Parafilm, and fill the casting tube completely with gel capillary tubes. The casting tube is then placed on a level surface.
2. Prepare the gel solution by adding and mixing the following components together:
   2.88 g urea.
   0.8 mL 30% acrylamide.
   1.2 mL 10% Triton X-100.
   2.3 mL distilled water.
   0.3 mL Bio-Lyte ampholytes 3–10.
   Warm the solution gently to dissolve the urea. Then add 12 µL 10% ammonium persulfate and 8 µL TEMED, and mix gently immediately before pouring the gel.
3. Slowly fill the casting tube with the gel solution using a syringe and needle, allowing the tubes to fill from the bottom. The casting tube is filled until all the capillary tubes are filled to a predetermined level. The gel is then allowed to polymerize.
4. Remove the Parafilm, and expel the tubes by pushing from the top. Gently rinse the acrylamide from the outside of the tubes and discard any containing air bubbles.
5. The top of the tube gel is attached to the sample reservoir with plastic tubing, and the sample reservoir then fitted into the gel apparatus. Other positions in the apparatus are plugged to ensure that the upper buffer chamber does not leak. The sample reservoir is filled with 1X upper chamber electrolyte, ensuring that all air bubbles are expelled.
6. Pellet the mycoplasma cells by centrifugation of 1.0 mL of broth at 13,000*g* for 5 min. Resuspend the cells in 100 µL urea lysis buffer. Use a Hamilton syringe to load between 10 and 50 µL of the protein sample onto the top of the gel in the capillary tube. (*See* **Note 9**.)
7. Overlay the sample with 20 µL of sample overlay buffer.
8. Assemble the gel apparatus, placing a stirring bar in the lower chamber to ensure circulation of the buffer during isoelectrofocusing.
9. Fill the lower tanks with 1X lower chamber electrolyte to cover the bottoms of the capillary tubes, and ensure that there are no air bubbles trapped in the bottom of the tubes. Expel any air bubbles with a syringe filled with 1X lower chamber electrolyte.
10. Apply power at 550 V for 10 min, and then increase to 1500 V and continue for a further 2 h.

11. While the first dimension is running, prepare the slab gel as described in **Subheading 3.1., steps 1–4**, but use a comb without teeth to produce a well that runs the full width of the gel. (*See* **Note 10**.)
12. Take the capillary tube from the isoelectrofocusing cell, and eject the gel using a 1-mL syringe filled with 1X electrophoresis buffer, carefully positioning it in place on top of the slab gel as it is ejected. This step must be performed carefully and can take a few attempts to master, since the pressure exerted to expel the gel must be relaxed as soon as it begins to slip from the tube, and the tube gel is easily broken (*see* **Note 9**).
13. Use a spatula to push the tube gel down carefully onto the top of the slab gel, ensuring that it makes complete contact. Carefully overlay the tube gel with 1X electrophoresis buffer and allow contact for 15 min. Assemble and run the second dimension as described in **Subheading 3.1., step 8**. The gel can be stained (**Subheading 3.1., steps 10** and **11**) or Western blotted (**Subheadings 3.2.** and **3.3.**) as described for conventional SDS-PAGE.

4. Notes

1. Equipment for SDS-PAGE is available from a number of commercial suppliers. In general, these allow for a range of spacer widths, from 0.5–1.5 mm, but the breadth and length of the gel are restricted by the apparatus. Typical dimensions range from 8.0 × 7.3 to 16 × 20 cm (breadth × length). Our laboratory has typically used equipment supplied by Bio-Rad, but other suppliers are available. This equipment allows easy construction of a leak-proof gel sandwich, which can then be filled with acrylamide. Polyacrylamide gels must be cast between glass plates, since oxygen inhibits polymerization. The two glass plates are separated by teflon spacers. The choice of plastic for spacers is significant, as some plastics can inhibit polymerization.
2. Unpolymerized acrylamide and *bis*-acrylamide, and TEMED are neurotoxic. Exposure to these chemicals should be minimized. Acrylamide and *bis*-acrylamide are particularly dangerous in their powdered form. In many cases, purchase of preprepared solutions from commercial suppliers may be preferable.
3. Many protocols suggest applying a vacuum to the mixtures to remove any dissolved oxygen, but we have not found this to be necessary.
4. The rate of polymerization is temperature-dependent, and can be enhanced by placing the poured gel in a 37°C incubator.
5. Silver stains may be used if more sensitive detection is required, but this is not usually necessary for staining whole-cell proteins of mycoplasmas.
6. This buffer is identical to electrophoresis buffer, except that it lacks SDS. If both procedures are to be performed, it may be more convenient to make up 10X Western transfer buffer as a stock and add SDS to electrophoresis buffer as needed.
7. In circumstances where an antispecies conjugate is not available, the secondary antibody may be substituted by horseradish peroxidase-conjugated protein A if the primary antibody is mammalian IgG.

8. Other chromogens are available that stain different colors (one example is fast red), which can be used to probe the same blot a second time with a different antibody, and light-emitting substrates are also available for use (for example, the ECL system supplied by Amersham). The deposition of the chromogen on the first conjugate blocks its activity with the second chromogen.
9. It is best to run several replicate tube gels with the same sample to provide spares, since the subsequent handling of the tube gel for the second-dimension separation often results in damage requiring the tube gel to be discarded.
10. A comb with two teeth may be preferred, one producing a well like that used for conventional SDS-PAGE, and the second running the remainder of the width of the gel to accommodate the tube gel. This allows molecular-mass standards to be run on the gel for comparison with the sample proteins.

References

1. Razin, S. and Rottem, S. (1967) Identification of Mycoplasma and other microorganisms by polyacrylamide-gel electrophoresis of cell proteins. *J. Bacteriol.* **94,** 1807–1810.
2. Laemmli, U. K. (1970) Cleavage of structural proteins during the assembly of the head of bacteriophage T4. *Nature* **227,** 680–685.
3. Burnette, W. N. (1981) "Western blotting": Electrophoretic transfer of proteins from sodium dodecyl sulphate-polyacrylamide gels to unmodified nitrocellulose and radiographic detection with antibody and radioiodinated protein A. *Anal. Biochem.* **112,** 195–203.
4. Sambrook, J., Fritsch, E. F., and Maniatis, T. (1989) *Molecular Cloning. A Laboratory Manual,* 2nd ed. Cold Spring Harbor Laboratory Press, Cold Spring Harbor, NY.
5. Harlow, E. and Lane, D. P. (1988) *Antibodies. A Laboratory Manual.* Cold Spring Harbor Laboratory Press, Cold Spring Harbor, NY.

31

Immunological and Biochemical Characterization of Membrane Proteins

Michael F. Duffy, Amir H. Noormohammadi, Nina Baseggio, Glenn F. Browning, and Philip F. Markham.

1. Introduction

Membrane proteins of mycoplasmas assume particular significance in investigations of the pathogenesis of disease. They are involved in adhesion to, and probably in other interactions with, the cells of the target tissues of the host *(1–5)*. They are frequently significant antigens in the immune response *(4–7)*, and in some cases have been observed to undergo high-frequency size and/or phase variation *(8–13)*. In addition, many membrane proteins are also important in metabolic pathways, including transmembrane transport *(14)*. The significance of future research on membrane proteins of mycoplasmas has been recently highlighted by the completion of the genomic sequences of *Mycoplasma genitalium* and *Mycoplasma pneumoniae* *(15,16)*. In both species, the products of a number of open reading frames could not be assigned a function by comparison to known prokaryotic genes. The vast majority of these unassigned products were putative membrane proteins.

The focus of this chapter is on identifying membrane proteins and characterizing some of their structural features. Establishing the functional role of membrane proteins requires a diverse range of methods, some of which are presented in the chapters in this volume on determination of enzyme activities, transposon mutagenesis, and detection and analysis of mycoplasma adhesins. However, as the foregoing indicates, detailed functional studies are clearly a priority for the future understanding of the complex relationship between mycoplasmas and their hosts. The methods that are described include those for establishing that a protein is exposed on the surface of the cell, purification of membrane proteins, identification of lipoproteins, identification of variable expression, construction

From: *Methods in Molecular Biology, Vol. 104: Mycoplasma Protocols*
Edited by: R. J. Miles and R. A. J. Nicholas © Humana Press Inc., Totowa, NJ

of an expression library for identification of the gene encoding the protein, and locating sequence features of membrane proteins.

2. Materials

2.1. Mycoplasma Strains and Broth Media

We have successfully used the methods listed here on *Mycoplasma gallisepticum, Mycoplasma synoviae, Mycoplasma pirum,* and *M. pneumoniae*. Others have used these or similar techniques on a range of other *Mycoplasma* species. Broth and agar media used for growth will need to be chosen on the basis of the species used. Some guidance may be gained from Chapters 4–7.

2.2. Trypsin Treatment of Intact Mycoplasma Cells

1. TS buffer: 50 m*M* Tris-HCl, 0.145*M* NaCl buffer, pH 7.4.
2. PBS: 8 g NaCl, 0.2 g KCl, 1.84 g $Na_2HPO_4 \cdot 12H_2O$, 0.2 g KH_2PO_4, distilled water to 1 L.
3. Trypsin: Diluted in TS buffer immediately prior to use.
4. Soyabean trypsin inhibitor: Diluted to 0.125% (w/v) in TS buffer.
5. Equipment and reagents for sodium dodecyl sulfate-polyacrylamide gel electrophoresis (SDS-PAGE): *See* Chapter 30.
6. Polyvinylidene difluoride (PVDF) membrane (Immobilon-P, Millipore).

2.3. Triton X-114 Phase Partitioning

1. Triton X-114.
2. PBS: *see* **Subheading 2.2., item 2**.
3. 6% (w/v) sucrose in PBS containing 0.06% (v/v) Triton X-114 (prepared as described in **Subheading 3.2., steps 1–5**).
4. Methanol/chloroform: Mix 4 mL of methanol with 1 mL of chloroform.
5. 8*M* urea in PBS.
6. Equipment and reagents for SDS-PAGE: *See* Chapter 30.
7. PVDF membrane (Immobilon-P, Millipore).

2.4. Radiolabeling Lipoproteins

1. 1 mCi of ^{3}H-palmitate or 20 μCi of ^{14}C-palmitate (NEN Dupont).
2. PBS: *see* **Subheading 2.2., item 2**.
3. Triton X-114.
4. 6% (w/v) Sucrose in PBS containing 0.06% (v/v) Triton X-114 (prepared as described in **Subheading 3.2., steps 1–5**).
5. Methanol/chloroform: Mix 4 mL of methanol with 1 mL of chloroform.
6. 8*M* urea in PBS.
7. Equipment and reagents for SDS-PAGE: *See* Chapter 30.
8. Autoradiographic film: Kodak XAR-5.
9. Amplify scintillant (Amersham).

2.5. Production of Monospecific Polyclonal Antisera

1. Triton X-114 phase-partitioned membrane proteins prepared as described in **Subheading 3.2.**
2. Equipment and reagents for SDS-PAGE: *See* Chapter 30.
3. Unsupported nitrocellulose membrane.
4. Ponceau S stain: 0.2% solution of Ponceau S (3–hydroxy-4-[2–sulfo-4-{sulfophenylazo}phenylazo]-2,7–naphthalene disulfonic acid) in 3% trichloroacetic acid and 3% sulfosalicylic acid.
5. Siliconized glass plates.
6. PBS: *see* **Subheading 2.2., item 2**.
7. Freund's complete adjuvant.
8. Laboratory rabbits.
9. Freund's incomplete adjuvant.
10. Western blot of Triton X-114 phase-partitioned membrane proteins prepared as described in **Subheading 3.2.**
11. Thiopentone.
12. Pentobarbitone.
13. PBST: PBS (*see* **Subheading 2.2., item 2**) containing 0.5 mL Tween 20/L.
14. Powdered skim milk: 5% (w/v) in PBST.
15. Serum diluent: 20% (v/v) fetal calf serum in PBST.
16. Elution buffer: 0.15*M* NaCl, 0.1*M* glycine at pH 2.6.
17. Neutralization buffer: 2*M* Tris-HCl, pH 7.5. Pretitrate the volume required to neutralize elution buffer.
18. Antibody diluent: 5% (v/v) fetal calf serum in PBST.

2.6. Immunoblotting of Colony Lifts

1. Mycoplasma agar media suitable for culture of the species of interest.
2. Circular pieces of supported nitrocellulose membrane (Hybond-C; Amersham).
3. 10% (w/v) Bovine serum albumin (BSA) in PBS (*see* **Subheading 2.2., item 2**).
4. PBST (*see* **Subheading 2.5., item 13**).
5. PBST containing 0.1% (w/v) BSA.
6. Rabbit monospecific antiserum against membrane protein of interest (*see* **Subheading 2.5.**).
7. PBST containing 1% (w/v) BSA.
8. Swine antirabbit immunoglobulin conjugated to horseradish peroxidase (Dako).
9. Substrate solution: Dissolve one 10 mg tablet of 3,3'-diaminobenzidine (Sigma) in 30 mL of 0.02*M* Tris, 137 m*M* NaCl. Adjust to pH 7.6 using HCl. Add 24 µL of 30% (w/v) H_2O_2 just before use.

2.7. Detection of Genes Encoding Membrane Proteins by Screening Expression Libraries

1. Genomic DNA from the species of interest (*see* Chapter 17).
2. Restriction endonuclease *Sau*3AI and the supplier's 10X digestion buffer.

3. Materials for agarose-gel electrophoresis: agarose, TPE electrophoresis buffer (20X stock contains 36 m*M* Tris-HCl, 1 m*M* Na_2EDTA, and 30 m*M* NaH_2PO_4 in distilled water), ethidium bromide (50 mg/mL in distilled water).
4. MicroSpin Sephacryl S-400 HR column (Pharmacia).
5. Phenol/chloroform/isoamyl alcohol (25:24:1).
6. 100% ethanol.
7. 3*M* sodium acetate.
8. Plasmid pGEX-4T-3, pGEX-3X (Amrad), and pGEX-1N DNA. This combination offers all three reading frames of the *Bam*HI cloning site. Unfortunately, pGEX-1N is not available commercially, and no other pGEX vector offers the third reading frame for the *Bam*HI cloning site.
9. Restriction endonuclease *Bam*HI and the supplier's 10X digestion buffer.
10. Bacterial alkaline phosphatase (BAP, Gibco BRL) and the supplier's 10X dephosphorylation buffer.
11. T4 DNA ligase and the supplier's 10X ligation buffer.
12. Electrocompetent *Escherichia coli* strain DH5α or JM109 cells. Inoculate 1 L of Luria broth (LB) broth with 10 mL of an overnight culture. Grow at 37°C with shaking to an A_{600} of 0.6. Chill on ice for 30 min and pellet the cells by centrifugation at 4000*g* for 15 min at 4°C. Remove the supernatant, and resuspend the cells in 1 L of sterile ice-cold distilled water. Pellet the cells by centrifugation at 4000*g* for 15 min at 4°C, discard the supernatant, and resuspend the cells in 0.5 L of sterile ice-cold water. Pellet the cells by centrifugation at 4000*g* for 15 min at 4°C, discard the supernatant, and resuspend the cells in 20 mL 10% ice-cold glycerol in water. Pellet the cells by centrifugation at 4000*g* for 15 min at 4°C, discard the supernatant, and resuspend the cells in 2 mL 10% ice-cold glycerol in water. Aliquot into 40-µL volumes and freeze on dry ice. These can be stored at –70°C for up to 6 mo.
13. 0.2-cm Electroporation cuvet (Bio-Rad).
14. Bio-Rad Gene Pulsor (Bio-Rad).
15. LB Broth: 10 g tryptone, 5 g yeast extract, 5 g NaCl, 1 g glucose, 1 L of distilled water. Autoclave and store at 4°C.
16. LB ampicillin agar plates. Add 15 g agar to 1 L of LB broth prior to autoclaving. Cool to 55°C in a water bath. Add 1 mL of 50 mg/mL ampicillin, then pour immediately into 9-cm diameter sterile Petri dishes (approx 30 mL/plate). The plates can be stored at 4°C, but should be used within 1 wk. The 50 mg/mL ampicillin stock is made in sterile distilled water, filtered through a 0.4-µm pore diameter filter and stored in 1-mL aliquots at –20°C.
17. Circular pieces of supported nitrocellulose (Hybond-C, Amersham).
18. LB ampicillin/IPTG agar: Spread 60 µL of 1*M* isopropyl-β-thiogalactopyranoside (IPTG) on an LB ampicillin agar plate about 2 h prior to use. The 1*M* IPTG stock is made in sterile distilled water, filtered through a 0.4-µm pore diameter filter, and stored at –20°C.
19. Chloroform.
20. Lysis buffer: 100 m*M* Tris-HCl, pH 7.8, 150 m*M* NaCl, 2 m*M* lysozyme.

21. PBST (*see* **Subheading 2.5., item 13**).
22. Blotting paper.
23. PBST containing 0.1% (w/v) BSA.
24. Rabbit monospecific antiserum against membrane protein of interest (*see* **Subheading 3.4.**)
25. PBST containing 1% (w/v) BSA.
26. Swine antirabbit immunoglobulin conjugated to horseradish peroxidase (Dako).
27. Substrate solution: 10 mg/mL of 3,3'-diaminobenzidine (Sigma) in 0.02*M* Tris, 137 m*M* NaCl. Adjust the pH to 7.6 using HCl. Add 0.03% (v/v) H_2O_2 just before use.

3. Methods

3.1. Trypsin Treatment of Intact Mycoplasma Cells

Trypsin treatment of intact cells is used to distinguish proteins that are exposed on the surface of intact cells from those that are protected from degradation by trypsin by the intact cell membrane. In addition to establishing that the moiety is exposed on the surface, this method also confirms that it is a protein that is being investigated. The presence of repetitive sequences within the protein may also be deduced by the appearance of a regular ladder of peptide fragments ***(11)***.

1. Grow the species of interest to late-log phase in 20 mL of broth media.
2. Harvest the mycoplasma cells by centrifugation at 12,000*g* for 30 min at 4°C.
3. Resuspend the pelleted cells in 10 mL TS buffer, then harvest the cells by centrifugation at 12,000*g* for 30 min at 4°C (*see* **Note 1**).
4. Wash the cells twice more with TS buffer as described in **step 3**.
5. Resuspend the washed cells in 600 μL of TS buffer, and divide this suspension into six 100-μL samples.
6. Prepare a fresh dilution series of trypsin (15.125, 31.25, 62.5, 125, and 250 μg/mL) in TS buffer.
7. Add 100 μL from each of the trypsin dilutions to five separate cell suspension aliquots, and add 100 μL of TS buffer only to the sixth sample.
8. Incubate all six samples at 37°C for 30 min.
9. Add 200 μL of 0.125% (w/v) soybean trypsin inhibitor in TS buffer to each sample to stop trypsin digestion.
10. Pellet the cells in each sample by centrifugation at 12,000*g* for 30 min at 4°C and resuspend them in 150 μL of TS buffer.
11. Take 20 μL of each cell suspension, and separate the proteins by SDS-PAGE (*see* Chapter 30). Visualize the proteins either by Coomassie blue staining or immunostaining after transfer to PVDF membrane (*see* **Note 2**).
12. Compare the intensity of each protein band in the undigested control with those in the different trypsin treated samples. Surface-exposed proteins will either be greatly reduced in intensity, shifted to a different molecular size (**Fig. 1**) (especially if only part of the protein is exposed), or completely absent (*see* **Notes 3** and **4**).

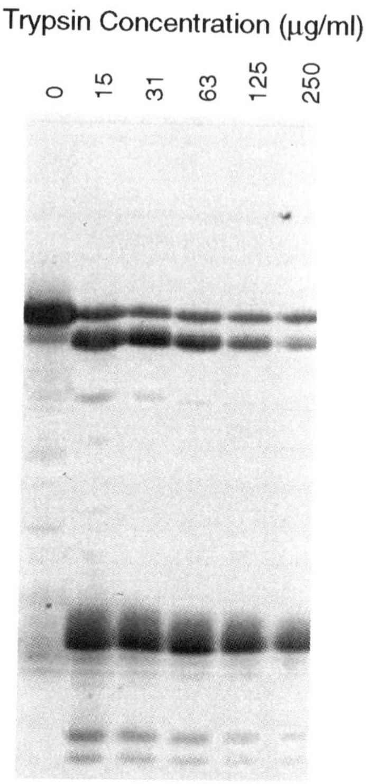

Fig. 1. Trypsin treatment of intact *M. gallisepticum* cells. *M. gallisepticum* cells were treated with a series of concentrations of trypsin as labeled. Whole-cell proteins were separated by SDS-PAGE, blotted onto PVDF membrane, and probed with an MAb specific for the major membrane protein, pMGA. The degradation of the 67-kDa protein into a series of smaller fragments is clearly apparent.

3.2. Triton X-114 Phase Partitioning

Proteins embedded in the cell membrane, or anchored to it by an acyl moiety are amphiphilic, and thus can be selectively solubilized using surfactants. Triton X-114 has been most commonly used to partition amphiphilic mycoplasma proteins, but other surfactants may also be suitable *(17)* and may be more selective, hence enabling better purification of specific membrane proteins. The partial purification of membrane proteins enhances investigation of their antigenic characteristics.

1. Dissolve 10 mL Triton X-114 in 500 mL of PBS, and mix until the solution is homogeneous.
2. Incubate the solution at 37°C for 18 h.
3. Remove and discard the upper aqueous phase. Add PBS to give a volume of 500 mL. Incubate the solution at 37°C for 18 h on a rocking platform.

4. Repeat **step 3** twice.
5. Decant the lower phase, and use this as a stock solution of 11.4% (v/v) Triton X-114 in PBS ***(18)***.
6. Grow a 20-mL culture of the species of interest to late-log phase, and harvest the cells by centrifugation at 12,000*g* for 30 min at 4°C (*see* **Note 5**).
7. Dissolve the cells in 500 µL of PBS containing 0.5% Triton X-114, and incubate the solution on ice for 60 min.
8. Centrifuge the solution at 12,000*g* for 30 min at 4°C.
9. Load the supernatant carefully onto a 1-mL 6% (w/v) sucrose cushion containing 0.06% (v/v) Triton X-114 in PBS.
10. Incubate the solution in a water bath at 37°C for 9 min.
11. Centrifuge the solution at 300*g* for 7 min at 37°C.
12. Aspirate the supernatant, which contains the hydrophilic components, and set it aside.
13. Resuspend the oily pellet, which contains the hydrophobic components, in 500 µL of ice-cold PBS.
14. Precipitate the proteins from the hydrophobic fraction by adding 500 µL methanol/chloroform and centrifuging the mixture at 9000*g* for 1 min at 4°C (*see* **Note 6**).
15. Aspirate the supernatant.
16. Resuspend the proteins in 100 µL 8*M* urea in PBS, and repeat **steps 14** and **15**.
17. Resuspend the proteins in 100 µL 8*M* urea in PBS.
18. Separate the proteins contained in the hydrophobic fraction (from **step 17**) and the hydrophilic faction (from **step 12**) using SDS-PAGE. Membrane proteins will be seen in the hydrophobic fraction. The proteins may be transferred to PVDF membrane and immunostained to enhance detection of the proteins of interest (**Fig 2**).

3.3. Radiolabeling Lipoproteins

Many membrane proteins of mycoplasmas are lipoproteins. These can be most easily identified by in vivo incorporation of radiolabeled fatty acids. Jan et al. ***(7)*** compared the relative efficiencies of incorporation of different radiolabeled fatty acids, and found palmitic acid to be most efficiently incorporated into lipoproteins. Either ^{3}H-palmitate or ^{14}C-palmitate can be used, and although ^{3}H-palmitate appears less expensive, the reduced amount of ^{14}C-palmitate required for adequate labeling negates the apparent economic advantage. Some protocols published previously have suggested that the medium used for labeling must be depleted of lipids. However we, and other workers ***(19)***, have not found this to be essential.

1. Grow a 20-mL culture of the species of interest to mid-log phase, and harvest the cells by centrifugation.
2. Resuspend the cells in 2 mL of fresh culture medium.
3. Add 1 mCi of ^{3}H-palmitate or 20 µCi of ^{14}C-palmitate dissolved in 20 µL of ethanol to the resuspended cells (*see* **Note 7**).

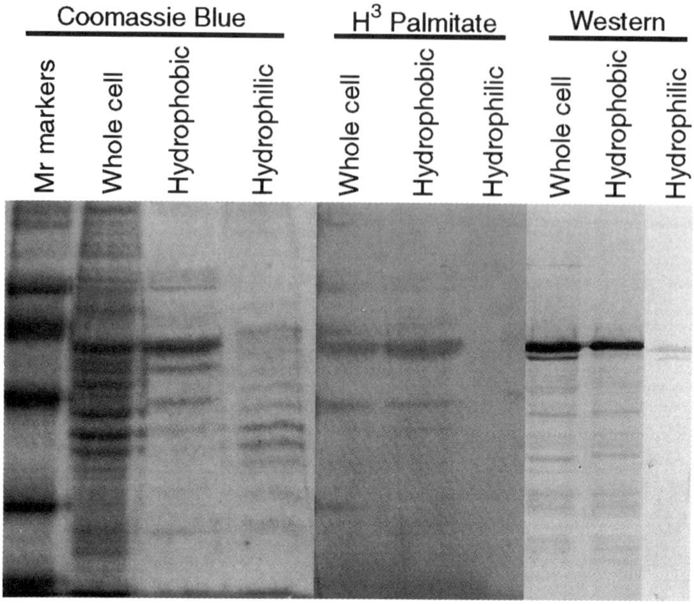

Fig. 2. Coomassie blue-stained, ^{3}H palmitate labeled, and immunostained whole-cell and Triton X-114 partitioned proteins of *M. gallisepticum*. Lipoproteins were labeled by growth in ^{3}H palmitate. Then an aliquot was partitioned into hydrophobic and hydrophilic proteins. Whole-cell proteins, and proteins in the hydrophobic and hydrophilic fractions were separated by SDS-PAGE in two identical gels. One gel was Western blotted and probed with rabbit anti-pMGA antibody (raised against immuno-affinity-purified protein). The selective purification of pMGA and several other less prominent proteins into the hydrophobic fraction is apparent in both the Coomassie blue-stained and the immunostained gels. The ^{3}H palmitate labeling indicates that pMGA is acylated. The molecular mass *(M_r)* markers are 110, 70, 45, 30, and 22 kDa.

4. Incubate the culture at 37°C for 18 h.
5. Harvest the mycoplasma cells by centrifugation at 8000*g* for 20 min.
6. Wash the cells by resuspension in 2 mL of PBS, followed by centrifugation at 8000*g* for 20 min. Repeat this procedure three times.
7. Resuspend the cells in 2 mL of PBS, and reserve 1 mL for analysis of whole-cell proteins by SDS-PAGE (*see* Chapter 30).
8. Partition the proteins using Triton X-114 as described in **Subheading 3.2., steps 7–18**.
9. Separate the proteins in both the hydrophilic and hydrophobic fractions of the partitioning step using SDS-PAGE (*see* Chapter 30).
10. Incubate the gel in Amplify scintillant according to the manufacturer's instructions, and dry it under vacuum.
11. Expose autoradiographic film to the dried gel for 2 wk at –70°C (**Fig. 2**) (*see* **Note 8**).

3.4. Production of Monospecific Polyclonal Antisera

Identification of antigenically related membrane proteins (for example, size variants), characterization of genes encoding membrane proteins using expression libraries, and characterization of variable expression all depend on specific serological reagents. Although monoclonal antibodies (MAbs) are usually the ideal reagents, monospecific polyclonal sera are often suitable and usually less laborious to produce. The protein of interest is partially purified by Triton X-114 phase partitioning, and then further purified by SDS-PAGE. The region of the gel containing the protein is excised, or alternatively, the proteins are transferred to a membrane and the region of the membrane bearing the protein excised. This purified protein is then used to immunize rabbits to produce the monospecific antiserum.

1. Use Triton X-114 phase partitioning to purify partially and concentrate membrane proteins.
2. Separate the hydrophobic fraction using SDS-PAGE (*see* Chapter 30).
3. Identify the protein of interest by either Coomassie blue staining or Western blotting onto nitrocellulose (*see* Chapter 30 for these methods). If using the gel for injection, cut a strip from the side of the gel, and stain it to enable the region containing the protein band to be identified. If using a Western blot, cut a narrow strip from the blot for immunostaining, and stain the remainder of the blot by immersion in Ponceau S (destain the blot with several changes of PBS) to facilitate localization of the protein band identified by immunostaining.
4. Cut the region containing the protein band from the gel or membrane using a scalpel blade, taking care to cut as close to the band as possible.
5. Wash the gel slice several times with distilled water, and then grind it between two siliconized glass plates. Add a drop of sterile water to the polyacrylamide to enable it to be drawn into a 1-mL syringe through an 18-gage needle. If the protein has been transferred onto a membrane, this can be treated by placing it in 1.0 mL of PBS and sonicating the mixture until the membrane has been dispersed (*see* **Note 9**).
6. Dispense the solution into four aliquots, and store three of these at –70°C until required for booster inoculations.
7. Mix the first aliquot with an equal volume of Freund's complete adjuvant, and emulsify the mixture by sonication on ice until a drop placed on iced water does not disperse.
8. Collect a 1-mL sample of blood from a rabbit (for later comparison) by incision of a peripheral ear vein, and then inject 0.5 mL of the inoculum prepared in **step 7** into the semimembranosus and/or semitendinosus muscles of one hindleg.
9. Four weeks after the first inoculation, prepare the booster inoculum by mixing an aliquot of the gel-purified protein with an equal volume of Freund's incomplete adjuvant, and emulsify by sonication as in **step 7**.

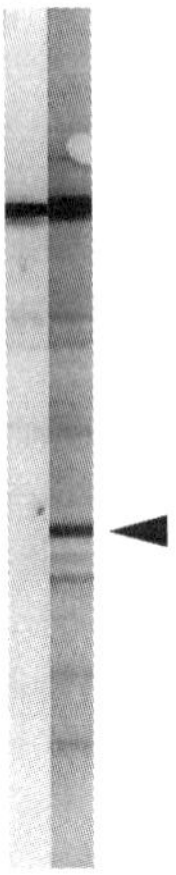

Fig. 3. Comparison of rabbit monospecific antisera raised against the 116-kDa membrane protein of *M. pneumoniae* before and after affinity purification using the method of Beall and Mitchell ***(20)***. Lane A shows a Western blot of whole-cell proteins of *M. pneumoniae* probed with the affinity-purified antibody, and lane B shows the same preparation probed with the sera prior to purification. The loss of reactivity with a protein of about 50 kDa (arrowed) is clearly apparent.

10. Inject 0.5 mL of the inoculum into the semimembranosus and/or semitendinosus muscles of one hindleg.
11. Two weeks after the booster inoculation, collect a 1-mL sample of blood from the rabbit by incision of a peripheral ear vein, and assess the antibody response by probing a Western blot of Triton X-114 partitioned membrane proteins with the serum.
12. If the antibody response is not sufficient to detect the protein of interest, **steps 9–11** can be repeated to boost the immune response of the rabbit further (*see* **Note 10**).
13. When the level of antibody in the rabbit's serum is sufficient, as assessed by Western blotting, the rabbit is anesthetized with approx 20 mg/kg thiopentone, and blood collected by cardiac puncture using a 16-gage needle attached to a 50-mL syringe. It is generally possible to collect 100 mL of blood using this method. When no more blood can be collected, the rabbit is euthanized with an intracardiac injection of pentobarbitone.

    This serum is generally sufficient for most purposes, including staining mycoplasma colony blots, but in some situations, further purification may be necessary (**Fig. 3**). This is particularly the case when the serum is required for screening expression libraries for identification of the gene encoding the protein. The method of Beall and Mitchell ***(20)*** is a convenient approach for the affinity purification of small quantities of polyclonal antisera.
14. Separate Triton X-114 partitioned membrane proteins by SDS-PAGE using a gel prepared with a comb with a single tooth that extends across most of the width of the gel.

15. Transfer the proteins onto PVDF membrane, and stain the membrane by immersion in Ponceau S stain. Destain the membrane by washing several times in PBS.
16. Locate the protein band of interest, and cut it from the membrane.
17. Block the membrane by incubation in 5% powdered skim milk in PBST for 3 h at room temperature, and then wash in PBST for 30 min at room temperature.
18. Incubate the strip of membrane bearing the protein of interest in 200 μL of the serum diluted 1/10 in serum diluent for 2 h at room temperature.
19. Wash the membrane three times for 10 min each with PBST.
20. Elute the antibody by incubating the membrane with 2 mL of elution buffer for 3 min.
21. Neutralize the pH of the eluted antibody immediately by adding a pretitrated volume of neutralization buffer.
22. Dilute the purified antibody in 4.4 mL of antibody diluent, and use for immunostaining without further dilution (*see* **Note 11**).

3.5. Immunoblotting of Colony Lifts

A number of membrane proteins of mycoplasmas have been observed to undergo high-frequency size and/or phase variation. Size variation can be detected by using monospecific antisera or MAbs to stain Western blots of whole-cell or Triton X-114 partitioned proteins of different clones of the same strain, but phase variation is best observed by immunostaining blots of mycoplasma colonies growing on agar.

1. Inoculate appropriate mycoplasma agar media with a dilution series of broth-cultured mycoplasmas that have been passed through a 0.4-μm filter.
2. Incubate the plate until colonies are clearly visible.
3. Carefully place a circular piece of nitrocellulose membrane on the plate, and mark it for orientation by piercing with a 22-gage needle. Leave the membrane in place for 10 s, and then gently lift it off and allow the blotted cells to dry on the membrane at room temperature.
4. Add sufficient 10% BSA in PBS to cover the membrane, and incubate at room temperature for 30 min on a rocking platform.
5. Add a 1/500 dilution of rabbit antisera against the membrane protein in PBST/1% BSA, and incubate at room temperature for 1 h on a rocking platform.
6. Wash the membranes three times (5 min each wash) in PBST/0.1% BSA.
7. Incubate the membrane at room temperature for 1 h on a rocking platform in a 1/1000 dilution of swine antirabbit immunoglobulin horseradish peroxidase conjugate in PBST/1% (w/v) BSA.
8. Wash the membranes three times (5 min each wash) in PBST/0.1% BSA.
9. Incubate the membrane on a rocking platform in a substrate solution.
10. Stop the color development by extensive washing in distilled water.
11. Examine the stained colonies using a dissecting microscope. Phase-variable proteins will be detectable as variation in staining between individual colonies or sectorial variation in staining within single colonies (**Fig. 4**).

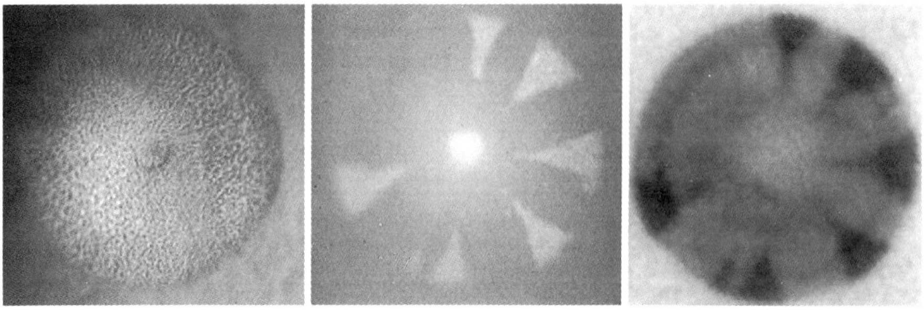

Unstained Colony | Fluorescent Antibody Stained Colony | Antibody Stained Colony Blot

Fig. 4. Immunostaining of a colony blot of *M. gallisepticum*. The three panels show a photograph of the unstained colony, the colony stained with a fluorescein-conjugated MAb against the major surface protein pMGA, and a blot of the same colony stained with horseradish peroxidase-conjugated MAb. The panels demonstrate the concordance between the original colony and the blot taken from it in detecting phase-variable expression of the membrane protein. The high frequency of variation in expression of some membrane proteins is also evident.

3.6. Detection of Genes Encoding Membrane Proteins by Screening Expression Libraries

Identification of the genes encoding membrane proteins can rely on partial peptide sequencing, and then the use of these sequences to design oligonucleotide probes to screen genomic libraries. Although we have had some success with this method, we have found screening of expression libraries to be more consistently rewarding, since many membrane proteins are amino-terminally blocked, and the quantity of purified protein needed to sequence internal peptide fragments is often difficult to obtain. Various methods have been used to identify the genes encoding mycoplasma membrane proteins. In view of the use of TGA as a tryptophan codon by mycoplasmas, it is unlikely that a complete protein will be expressed in *E. coli*. However, if an expression library is constructed from a partial digest of genomic DNA using a frequently cutting endonuclease, such as *Sau*3AI, and the library screened with a monospecific polyclonal antiserum, the probability of detecting at least part of the gene is high. This segment can then be used to detect larger fragments containing the gene, generally by screening libraries constructed from complete digests of genomic DNA with less frequently cutting enzymes. We have found the pGEX vector series useful, both for its high level of inducible expression, and the capacity to purify fusion proteins using glutathione beads or affinity columns. Although the lack of a commercial source of pGEX-1N, which offers the third reading frame for cloning, may be a hindrance to some workers, it is frequently not necessary to clone into more than one reading frame to identify the gene of interest.

1. Prepare genomic DNA of the species of interest (*see* Chapter 17).
2. Mix 10 µg of genomic DNA in 400 µL of *Sau*3AI digestion buffer, and keep on ice.
3. Add 60 µL of the DNA/restriction buffer mixture to one microfuge tube and 30 µL to each of nine other tubes, and keep on ice.
4. Add 4 U of *Sau*3AI to the first tube, mix, and transfer 30 µL from this mixture to the second tube. Mix the second tube and transfer 30 µL from it to the third tube and so on until the ninth tube is reached. Discard 30 µL from this tube. The 10th tube is kept as a negative control.
5. Incubate the reactions for 2 h at 37°C.
6. Stop the reactions by incubating all tubes at 70°C for 15 min.
7. Load 20 µL from each digest onto a 1.2% agarose gel, and separate the fragments by electrophoresis at 7 V/cm for 15 min in TPE buffer.
8. Identify the digest that has the majority of incomplete digest products concentrated at about 1.5 kbp, and use this to calculate the optimal concentration of restriction endonuclease/µg of genomic DNA.
9. Digest 20 µg of genomic DNA using this optimal concentration of enzyme in 100 µL of reaction buffer for 2 h at 37°C, and then stop the reaction by incubation at 70°C for 15 min.
10. Place 50 µL of sterile distilled water on top of a Pharmacia MicroSpin Sephacryl S-400 HR column, and spin the column in a centrifuge at 400*g* for 5 min.
11. Place 50 µL of the digested genomic DNA on the column, and spin the column in a centrifuge at 400*g* for 5 min. The eluate contains fragments >100 bp.
12. Add sufficient sterile distilled water to the eluate to make it up to 300 µL, and add 300 µL of phenol:chloroform:isoamyl alcohol, mix, and centrifuge at 18,000*g* for 10 min. Remove the upper aqueous phase, and add 30 µL of 3*M* sodium acetate and 700 µL of 100% ethanol. Centrifuge at 18,000*g* for 15 min, and discard the supernatant. Dry the pellet, and then resuspend it in 20 µL sterile distilled water.
13. Add 10 µg pGEX DNA to 40 U *Bam*HI in 30 µL of the appropriate reaction buffer, and incubate at 37°C for 2 h.
14. Take 20 µL of the digestion mixture, and mix with 15 µL of 10X bacterial alkaline phosphatase buffer, 115 µL of sterile distilled water, and 100 U of bacterial alkaline phosphatase. Then incubate for 1 h at 65°C. Add 150 µL sterile distilled water to the mixture, and add 300 µL of phenol:chloroform:isoamyl alcohol, mix, and centrifuge at 18,000*g* for 10 min. Collect the upper aqueous phase, and place it in a clean tube. Add 300 µL of phenol:chloroform:isoamyl alcohol, mix, and centrifuge at 18,000*g* for 10 min. Collect the upper aqueous phase, place it in a clean tube, and add 20 µL of 3*M* sodium acetate and 500 µL of 100% ethanol. Centrifuge at 18,000*g* for 15 min, and discard the supernatant. Dry the pellet, and resuspend it in 20 µL sterile distilled water.
15. Add 0.1 µg of digested and phosphatase-treated pGEX DNA and 0.5 µg of the size-selected *Sau*3AI-digested genomic DNA to a 20-µL reaction containing the

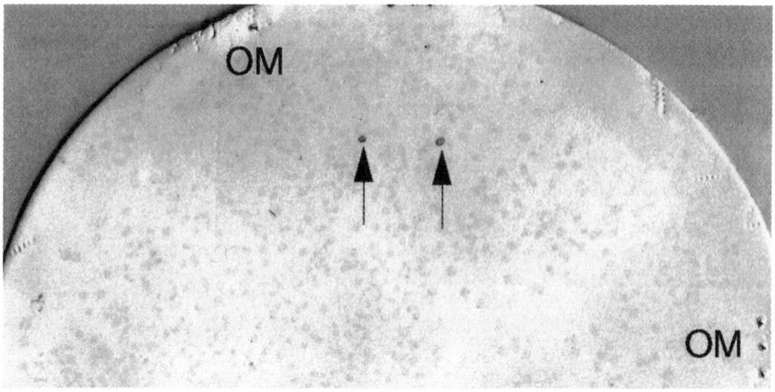

Fig. 5. Use of monospecific antisera to probe an expression library of *M. synoviae* genomic DNA. Two immunoreactive colonies are arrowed. OM indicates the orientation marks made by piercing the membrane with a needle at the time of blotting (1 at the top, and 3 at the right).

ligation buffer supplied by the manufacturer and 1 U of T4 DNA ligase. Incubate at room temperature for 4 h (see **Notes 12** and **13**).

16. Thaw 40 µL electrocompetent *E. coli* cells on ice, and add 1–2 µL of the DNA ligation mix.
17. Pipet the cells carefully into the bottom of an ice-chilled electroporation cuvet. Tap to distribute the solution, and remove bubbles.
18. Using a Gene Pulsor, subject the cell mixture to a potential difference of 2.5 kV using a capacitance of 25 µF in conjunction with an impedance of 200 Ω.
19. Add 0.8 mL of LB broth without antibiotic to the cuvet. Then transfer the cell suspension to a closed tube, and incubate for 60 min at 37°C.
20. Spread 200 µL of the broth on each of 4 LB ampicillin agar plates, and incubate the plates overnight at 37°C.
21. Place a circular piece of supported nitrocellulose carefully on the agar plate, and leave in place for 10 s, ensuring that the full surface of the plate makes contact with the membrane and that the membrane becomes saturated. Orient the plate by placing a series of recognizable holes through the membrane and the agar with an 18-gage needle (**Fig. 5**).
22. Place the filters colony side up on fresh LB ampicillin/IPTG agar plates, and incubate at 37°C for 3 h to induce expression of fusion proteins. Place the original agar plate back in the incubator for 6 h to allow the colonies to regrow.
23. Remove the membrane from the plate, and place it in a plastic box colony side up on paper towels dampened with water. Place an open glass Petri dish containing chloroform in the box, and place a lid on the box. Incubate at room temperature for 15 min.
24. Place the filter colony side up in a Petri dish containing two blotting paper disks saturated with lysis buffer, and incubate at 37°C for 30 min.

25. Place the filter in PBST, and incubate for 30 min at room temperature.
26. Repeat **step 25** with fresh PBST.
27. Place filters colony side down on two layers of blotting paper, and press down with a gloved hand. Peel the filter from the blotting paper. Bacterial debris, which can interfere with immunostaining, remains stuck to the blotting paper. Filters can be stored at 4°C.
28. Immunostain the colonies as described in **Subheading 3.5., steps 4–11**.
29. Identify immunoreactive clones (**Fig. 5**), and use these as template in DNA sequencing to identify the sequence encoding the membrane protein. The forward primer will generally provide sequence through the expressed gene fragment (*see* **Note 14**).

3.7. Sequence Characteristics of Mycoplasma Membrane Proteins

In general, a membrane protein will need a signal sequence to direct its translocation to the cell membrane. The two methods by which mycoplasma membrane proteins attach to the cell membrane are either using one or more membrane-spanning segments, or by a lipid moiety attached to an aminoterminal cysteine residue.

Identification of putative signal sequences, membrane-spanning segments, and lipoprotein signal sequences can be achieved by using the PSORT WWW server available at http://psort.nibb.ac.jp/, but unfortunately, many of the established mycoplasma lipoproteins are not recognized as such by the algorithms used for prediction, and many membrane proteins are not predicted with certainty. The Prosite ***(21,22)*** motif for bacterial lipoproteins states that there be at least one lysine or arginine residue in the first seven positions, that a cysteine residue must lie between positions 15 and 35 of the sequence, that the residue preceding it should be A, G, or S, preceded by I, V, M, S, T, A, G, or Q, preceded by L, I, V, M, F, S, T, A, or G, preceded by L, I, V, M, F, S, T, A, or G, and that the six residues preceding these four should not be D, E, R, or K. This motif is able to detect most, but not all, experimentally established mycoplasma lipoproteins. Thus, failure of either of these methods to predict a membrane protein should be interpreted carefully.

The consensus for the acylation signal sequence in **Fig. 6** has been derived from the published sequences for experimentally established lipoproteins from several taxonomically distant mycoplasmas ***(23–32)***, rather than from the putative lipoproteins present in the *M. genitalium* and *M. pneumoniae* genomic sequence. The most commonly occurring residues in each position are listed on the first line, with other residues occurring in that position listed vertically beneath the most common residue in order of decreasing frequency of occurrence (allowing for situations where multiple lipoproteins from the same species have been characterized).

```
MKKSKFKKFL...9-16(X)..IAASCQK
 NRKLLLLKF            VVIA SQ
 RNLNKIFLI            ALSK TS
  IMIMS AV            SSYG NT
    FR  IT            TT   GD
     S  GS                  E
     I   K
         G
```

Fig. 6. Mycoplasma acylation consensus sequence. The most commonly occurring residue at each site in the different experimentally confirmed lipoproteins of all mycoplasmas is listed in the first line. The second most common is shown in the second line, and less common residues are shown in lines 3–8. The cysteine residue at the amino-terminus of the mature protein is between 24 and 31 residues from the amino-terminus of the prolipoprotein.

4. Notes

1. Adherent mycoplasmas, such as *M. pneumoniae,* can be washed by rinsing the flask with PBS three times before scraping the cells from the flask and recovering them by centrifugation.
2. If antisera or an MAb directed against the protein of interest is available, the specificity and sensitivity of the assay will be greatly assisted by immunostaining. The antisera may be monospecific antisera or, alternatively, serum from an infected animal. Use of antisera will also assist in identification of smaller, unexposed portions of the protein (usually membrane-spanning and/or intracytoplasmic regions).
3. An alternative method is to use different incubation times, and the same trypsin concentration (for example, 100 µg/mL of trypsin for 5, 10, 20 and 40 min).
4. Other methods for identifying surface-exposed proteins use labeling of intact mycoplasma cells. One example of such an approach is the use of surface biotinylation using *N*-succinyl biotin ***(33)***.
5. This step may need to be varied to achieve similar yields for different species. For instance, 500- to 700-mL cultures would be needed for *M. pneumoniae,* and subsequent steps are also performed in larger volumes.
6. The original method of Wessel and Flugge ***(34)*** used a much more extended procedure for precipitation of proteins to remove detergent, but we have found this simplified procedure as effective as the original version.
7. It may be necessary to concentrate the labeled palmitate by evaporation. We have found that addition of ethanol to 1% is tolerated by *M. pneumoniae*, but that addition to 2.5% inhibits growth.
8. Alternatively, proteins can be transferred to a PVDF membrane, autoradiographic film exposed to the membrane, and immunostaining subsequently used to identify proteins of interest. The substrate used for staining blocks transmission of the radiation emitted by ^{3}H and thus immunostaining must be performed after exposure.

9. An alternative method, which we have found more difficult, is to cut the gel slice into smaller pieces with a scalpel blade, then mix the pieces with 0.5 mL of PBS, and pass it through a syringe fitted with an 18-gage needle several times. Once the solution passes through easily, replace the 18-gage needle with a 22-gage needle, and again pass the solution through the needle until it can be easily ejected.
10. There is an increasing risk with each boost that an antibody response will develop against minor contaminating proteins, thus decreasing the specificity of the antiserum, so it is best to minimize the number of booster injections.
11. The antibody solution can be collected after incubation with blots and reused at least four times, facilitating screening of multiple colony blots. The membrane strips used for purification can also be reused at least six times. After each use, they should be washed for 5 min in PBST and then blocked with 5% powdered skim milk in PBST for 30 min.
12. The vector-to-insert ratio may need to be adjusted and can be assessed by performing trial ligation reactions and choosing that ratio that gives the largest number of transformed *E. coli*.
13. There is a risk that adventitious ligation of more than one *Sau*3AI-digested fragment may be ligated into a single vector molecule, which may be confusing during sequence analysis. All analysis of data should assume that this may have occurred. This may be circumvented by partially filling the 5'-overhanging ends of *Sau*3AI-digested DNA by incubating the DNA in a reaction mixture containing Klenow fragment of DNA polymerase and dGTP and dATP. This prevents religation of the *Sau*3AI ends, but allows ligation of these ends to the ends of an *Xho*I-digested vector.
14. Promiscuous translation of mycoplasma DNA has been observed, and thus, the expressed fragment may lie elsewhere in the fragment. This can be recognized by analyzing the expressed immunoreactive protein for reactivity with antisera against both the protein of interest and glutathione S-transferase. In cases where promiscuous translation has occurred, the immunoreactive protein will not be fused to glutathione S-transferase, and thus, antibody against the expressed protein of interest and glutathione S-transferase will not bind the same band on immunoblots.

References

1. Bredt, W., Feldner, J., and Kahane, I. (1981) Adherence of mycoplasmas to cells and inert surfaces: phenomena, experimental models and possible mechanisms. *Isr. J. Med. Sci.* **17,** 586–588.
2. Kahane, I., Granek, J., and Reisch-Saada, A. (1984) The adhesins of *Mycoplasma gallisepticum* and *M. pneumoniae*. *Ann. Microbiol. (Inst. Pasteur)* **135A,** 25–32.
3. Dallo, S. F., Su, C. J., Horton, J. R., and Baseman, J. B. (1988) Identification of P1 gene domain containing epitope(s) mediating *Mycoplasma pneumoniae* cytoadherence. *J. Exp. Med.* **167,** 718–723.
4. Forsyth, M. H., Tourtellotte, M. E., and Geary, S. J. (1992) Localization of an immunodominant 64 kDa lipoprotein (LP 64) in the membrane of *Mycoplasma gallisepticum* and its role in cytadherence. *Mol. Microbiol.* **6,** 2099–2106.

5. Markham, P. F., Glew, M. D., Brandon, M. R., Walker, I. D., and Whithear, K. G. (1992) Characterization of a major hemagglutinin protein from *Mycoplasma gallisepticum*. *Infect. Immun.* **60,** 3885–3891.
6. Futo, S., Seto, Y., Okada, M., Sato, S., Suzuki, T., Kawai, K., et al. (1995) Recombinant 46-kilodalton surface antigen (P46) of *Mycoplasma hyopneumoniae* expressed in *Escherichia coli* can be used for early specific diagnosis of mycoplasmal pneumonia of swine by enzyme-linked immunosorbent assay. *J. Clin. Microbiol.* **33,** 680–683.
7. Jan, G., Fontenelle, C., Le, H. M., and Wroblewski, H. (1995) Acylation and immunological properties of *Mycoplasma gallisepticum* membrane proteins. *Res. Microbiol.* **146,** 739–750.
8. Boyer, M. J. and Wise, K. S. (1989) Lipid-modified surface protein antigens expressing size variation within the species *Mycoplasma hyorhinis*. *Infect. Immun.* **57,** 245–254.
9. Talkington, D. F., Fallon, M. T., Watson, H. L., Thorp, R. K., and Cassell, G. H. (1989) *Mycoplasma pulmonis* V-1 surface protein variation: occurrence in vivo and association with lung lesions. *Microb. Patho.* **7,** 429–436.
10. Rosengarten, R. and Wise, K. S. (1990) Phenotypic switching in mycoplasmas: phase variation of diverse surface lipoproteins. *Science* **247,** 315–318.
11. Rosengarten, R. and Wise, K. S. (1991) The Vlp system of *Mycoplasma hyorhinis:* combinatorial expression of distinct size variant lipoproteins generating high-frequency surface antigenic variation. *J. Bacteriol.* **173,** 4782–4793.
12. Behrens, A., Heller, M., Kirchhoff, H., Yogev, D., and Rosengarten, R. (1994) A family of phase- and size-variant membrane surface lipoprotein antigens (Vsps) of *Mycoplasma bovis*. *Infect. Immun.* **62,** 5075–5084.
13. Droesse, M., Tangen, G., Gummelt, I., Kirchhoff, H., Washburn, L. R., and Rosengarten, R. (1995) Major membrane proteins and lipoproteins as highly variable immunogenic surface components and strain-specific antigenic markers of *Mycoplasma arthritidis*. *Microbiology* **141,** 3207–3219.
14. Gilson, E., Alloing, G., Schmidt, T., Claverys, J. P., Dudler, R., and Hofnung, M. (1988) Evidence for high affinity binding-protein dependent transport systems in gram-positive bacteria and in Mycoplasma. *EMBO J.* **7,** 3971–3974.
15. Fraser, C. M., Gocayne, J. D., White, O., Adams, M. D., Clayton, R. A., Fleischmann, R. D., et al. (1995) The minimal gene complement of *Mycoplasma genitalium*. *Science* **270,** 397–403.
16. Himmelreich, R., Hilbert, H., Plagens, H., Pirkl, E., Li, B.-C., and Herrmann, R. (1996) Complete sequence analysis of the genome of the bacterium *Mycoplasma pneumoniae*. *Nucleic Acids Res.* **24,** 4420–4449.
17. Brenner, C., Jan, G., Chevalier, Y., and Wroblewski, H. (1995) Evaluation of the efficacy of zwitterionic dodecyl carboxybetaine surfactants for the extraction and the separation of mycoplasma membrane protein antigens. *Anal. Biochem.* **224,** 515–523.
18. Bordier, C. (1981) Phase separation of integral membrane proteins in Triton X-114 solution. *J. Biol. Chem.* **256,** 1604–1607.

19. Jan, G., Fontenelle, C., Verrier, F., Le, H. M., and Wroblewski, H. (1996) Selective acylation of plasma membrane proteins of *Mycoplasma mycoides* subsp. *mycoides* SC, the contagious bovine pleuropneumonia agent. *Curr. Microbiol.* **32,** 38–42.
20. Beall, J. A. and Mitchell, G. F. (1986) Identification of a particular antigen from a parasite cDNA library using antibodies affinity purified from selected portions of Western blots. *J. Immunol. Methods* **86,** 217–223.
21. Bairoch, A. (1993) The PROSITE dictionary of sites and patterns in proteins, its current status. *Nucleic Acids Res.* **21,** 3097–3103.
22. Bairoch, A. and Bucher, P. (1994) PROSITE: recent developments. *Nucleic Acids Res.* **22,** 3583–9.
23. Yogev, D., Rosengarten, R., Watson, M. R., and Wise, K. S. (1991) Molecular basis of Mycoplasma surface antigenic variation: a novel set of divergent genes undergo spontaneous mutation of periodic coding regions and 5' regulatory sequences. *EMBO J.* **10,** 4069–4079.
24. Wise, K. S., Kim, M. F., Theiss, P. M., and Lo, S. C. (1993) A family of strain-variant surface lipoproteins of *Mycoplasma fermentans. Infect. Immun.* **61,** 3327–3333.
25. Cleavinger, C. M., Kim, M. F., and Wise, K. S. (1994) Processing and surface presentation of the *Mycoplasma hyorhinis* variant lipoprotein VlpC. *J. Bacteriol.* **176,** 2463–2467.
26. Markham, P. F., Glew, M. D., Sykes, J. E., Bowden, T. R., Pollocks, T. D., Browning, G. F., et al. (1994) The organisation of the multigene family which encodes the major cell surface protein, pMGA, of *Mycoplasma gallisepticum. FEBS Lett.* **352,** 347–352.
27. Ferris, S., Watson, H. L., Neyrolles, O., Montagnier, L., and Blanchard, A. (1995) Characterization of a major *Mycoplasma penetrans* lipoprotein and of its gene. *FEMS Microbiol. Lett.* **130,** 313–319.
28. Futo, S., Seto, Y., Mitsuse, S., Mori, Y., Suzuki, T., and Kawai, K. (1995) Molecular cloning of a 46–kilodalton surface antigen (P46) gene from *Mycoplasma hyopneumoniae*: direct evidence of CGG codon usage for arginine. *J. Bacteriol.* **177,** 1915–1917.
29. Foissac, X., Saillard, C., Gandar, J., Zreik, L., and Bove, J. M. (1996) Spiralin polymorphism in strains of *Spiroplasma citri* is not due to differences in post-translational palmitoylation. *J. Bacteriol.* **178,** 2934–2940.
30. Lysnyansky, I., Rosengarten, R., and Yogev, D. (1996) Phenotypic switching of variable surface lipoproteins in *Mycoplasma bovis* involves high-frequency chromosomal rearrangements. *J. Bacteriol.* **178,** 5395–5401.
31. Simmons, W. L., Zuhua, C., Glass, J. I., Simecka, J. W., Cassell, G. H., and Watson, H. L. (1996) Sequence analysis of the chromosomal region around and within the V-1-encoding gene of *Mycoplasma pulmonis*: evidence for DNA inversion as a mechanism for V-1 variation. *Infect. Immun.* **64,** 472–479.
32. Zhang, Q. and Wise, K. S. (1996) Molecular basis of size and antigenic variation of a *Mycoplasma hominis* adhesin encoded by divergent vaa genes. *Infect. Immun.* **64,** 2737–3744.

33. Djordjevic, S. P., Eamens, G. J., Scarman, A. L., and Chin, J. C. (1996) Characterisation of *Mycoplasma hyopneumoniae* membrane proteins and identification of discrete antigens which significantly reduce lung pathology in vaccinated pigs challenged with a virulent strain of *M. hyopneumoniae. IOM Lett.* **4,** 282–283.
34. Wessel, D. and Flugge, U. I. (1984) A method for quantitative recovery of protein in dilute solutions in the presence of detergents and lipids. *Anal. Biochem.* **138,** 141–143.

32

Detection and Analysis of Mycoplasma Adhesins

Konrad Sachse

1. Introduction

To study the first stages of interactions between bacteria and mammalian host cells, it is necessary to locate the surface antigens involved in cytodhesion processes. The identification of a bacterial adhesin is normally based on experimental evidence of adherence inhibition in vitro by the purified protien or, more often, a monoclonal antibody (MAb) directed against this protein, as well as visual data from immune electron microscopy. Alternatively, the adhesive function can be demonstrated via the construction or selection of mutants lacking the protein in question. The considerable experimental effort required to identify a single adhesion factor is one of the reasons why only a few such proteins have been found for most bacterial species so far, although in view of the complex nature of the attachment process, a larger number of adhesins can be expected to be involved.

In the first part of the present chapter, a simple screening procedure is described that is designed to recognize putative adhesion factors from Western blots of SDS-PAGE-separated whole-cell bacterial proteins. Western blot membranes are incubated with a suspension of ^{35}S-labeled host cells in conditions of defined stringency *(1)*. The resulting binding pattern, which reveals individual bacterial proteins capable of interacting with eukaryotic cell receptors, is visualized by autoradiography.

The present procedure was found to be simple and useful for detecting mycoplasmal proteins involved in cytadhesion or to verify findings obtained by other methods. For example, results obtained with *Mycoplasma pneumoniae* are presented in **Fig. 1**. A Western blot of whole-cell proteins of strain M 129 was stained with Coomassie blue (lane 1), incubated with A 427, a human lung carcinoma cell line, and autoradiographed (lane 2). In the autoradiographic

From: *Methods in Molecular Biology, Vol. 104: Mycoplasma Protocols*
Edited by: R. J. Miles and R. A. J. Nicholas © Humana Press Inc., Totowa, NJ

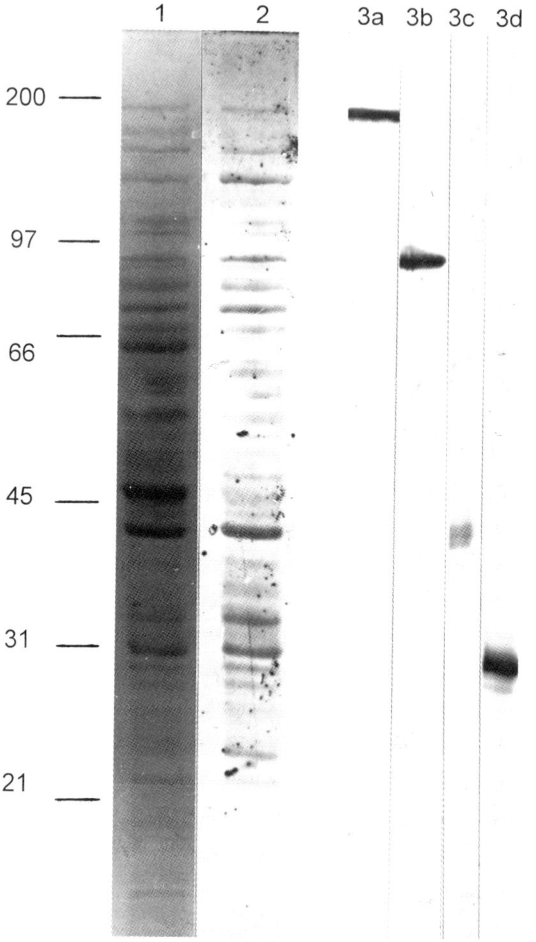

Fig. 1. Screening for *M. pneumoniae* adhesins using the Western blot adherence assay. Whole-cell proteins of strain M 129 were separated by SDS-PAGE in a 12% gel, blotted onto Immobilon P, Coomassie blue-stained (lane 1), incubated with ^{35}S-labeled A427 cells, and autoradiographed (lane 2), and immunostained with specific antisera against the cytadhesion-related proteins P1 (lane 3a), P90 (lane 3b), P40 (lane 3c), and P30 (lane 3d). The positions of mole-wt markers (in kDa) are indicated at the left-hand margin.

pattern, the surface-located proteins P1, P90, P40, and P30, which are known to be cytoadherence-related antigens ***(2,3)***, are clearly represented as intense bands, thus confirming their capability of establishing specific links with mammalian host cells. The method is distinguished both by simplicity and versatility. It is particularly well suited as a screening technique, since practically all protein antigens of the microorganism are present on the membrane after Western transfer. For closer scrutiny, membrane proteins or other protein fractions can be tested, as well as blots of two-dimensional gels. The stringency of conditions for mammalian cell attachment to immobilized epitopes can be varied by altering the ionic strength of the binding buffer and also via the shaking frequency. An appropriate host cell line should be chosen for each microorganism, if possible, related to their pathogenic pathway in vivo (*see* **Note 1**).

Limitations of the assay are owing to the fact that as a consequence of denaturation, some adherence epitopes may not be functional on the blot membrane, so that the particular adhesin would not be recognized. In this situation, an additional protocol for renaturation of blotted proteins can be followed in an attempt to re-establish native structures of adhesion epitopes. On the other hand, unspecific binding of host cells to cytosolic proteins can produce false-positive reactive bands if whole-cell lysates are examined. This can be avoided by using preparations of surface or membrane proteins of the microorganism instead.

Once a putative adhesin has been identified, more evidence is required to confirm its adhesive function. For this purpose, a tissue-culture plate assay can be used, where mycoplasmas interact with a lawn of mammalian cells. Participation of a particular bacterial surface protein in cytoadhesion can be established and quantitatively evaluated by demonstrating the capability of a corresponding MAb to reduce adherence rates, i.e., by means of an adherence inhibition assay. In this case, the mycoplasma cells in suspension should be allowed to adsorb the antibody to their surface during a preincubation period before transferring the mix to host cell-coated wells. The use of affinity-purified MAbs or monospecific antisera (immunoglobulin fraction) is strongly recommended in order to exclude nonspecific interference of albumins and other proteins present in crude sera. Some antibodies tend to facilitate aggregation of mycoplasma cells and may not be suitable for this assay. To illustrate the potential of the tissue-culture plate assay, experimental data from investigations on *M. pneumoniae* are given in **Fig. 2**. Specific antisera to proteins P1, P90, P40, and P30 were found to reduce adherence rates, thus confirming their suggested involvement in cytoadherence.

The test can also be conducted in a competitive mode using the putative adhesin in a purified form as competitor for attachment to host cell receptors. If participation of structural subunits or functional groups is to be examined, oligopeptides, or oligo- or monosaccharides can be used, as well as any soluble low-mole-wt compound carrying the respective functional group(s).

Finding the relevant concentration range where the inhibiting effect of the antibody or competitor can be measured is often a matter of trial and

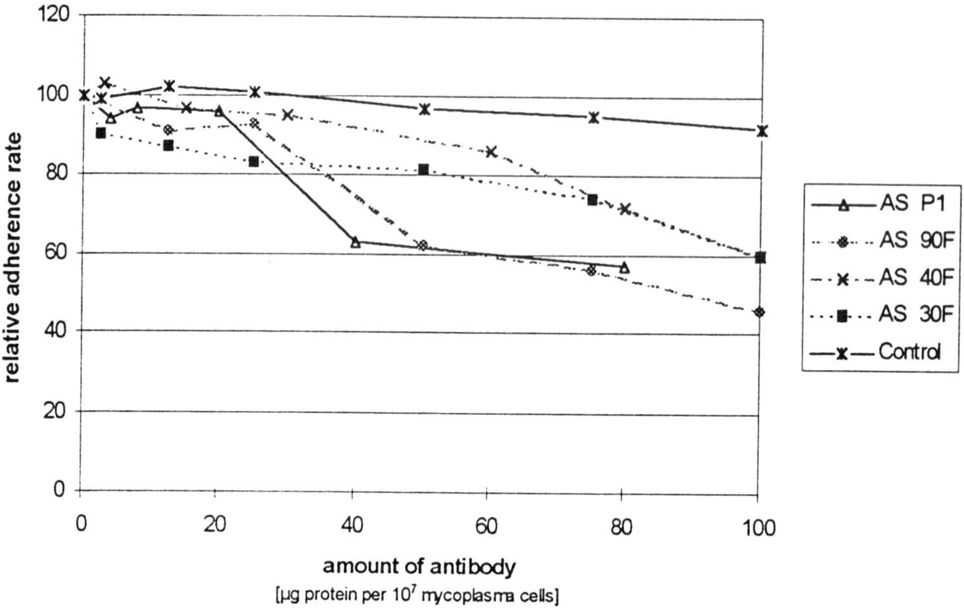

Fig. 2. Reduction of *M. pneumoniae* cytadhesion to A427 cells by specific antisera raised against the cytadhesion-related antigens P1, P90, P40, and P30. A rabbit preimmune serum was used as control.

error. Wherever possible, calculations or estimates based on molar ratios of interacting molecular partners should be made prior to the selection of initial experimental parameters.

2. Materials

2.1. Western Blot Adherence Assay

1. Reagents for standard SDS-PAGE and Western blotting.
2. Western blotting membrane: PVDF (Immobilon P, Millipore, Eschborn, Germany).
3. Horizontal shaker, e.g., IKA-VIBRAX-VXR (Janke & Kunkel, Staufen, Germany).
4. Coomassie blue staining solution: 0.1% (w/v) Coomassie brillant blue R-250 in water/methanol/acetic acid (50/40/10, v/v/v).
5. Destaining solution: Water/methanol/acetic acid (37.5/37.5/25, v/v/v).
6. Phosphate-buffered saline (PBS): 12 m*M* Na_2HPO_4, 12 m*M* NaH_2PO_4, and 145 m*M* NaCl, pH 7.0.
7. PBST: PBS containing 0.3% (v/v) Tween-20.
8. Blocking solution for membranes: PBST containing 2% (w/v) skimmed milk powder. Add milk powder immediately before use.
9. 25-cm^2 cell-culture flasks (Costar, Bodenheim, Germany).

10. Tissue-culture media: depends on the cell line, e.g., MEM + 10% fetal calf serum (FCS) for EBL, DMEM/RPMI (1:1, v/v) + 10% FCS for HeLa, DMEM containing 4.5 g/L glucose + 10% FCS for A 427.
11. Gentamicin (Sigma, Diesenhofen, Germany).
12. Stock solution for metabolic labeling of tissue-culture cells: L-^{35}S-methionine, in vivo cell labeling grade, >370 TBq/mmol (Amersham, Braunschweig, Germany, #SJ.1015).
13. 10X Trypsin-EDTA: Dissolve 0.8 g NaCl, 0.08 g KCl, 0.1 g glucose, 0.058 g $NaHCO_3$, 0.05 g trypsin (Boehringer Mannheim, Germany), and 0.02 g EDTA in 10 mL, and adjust pH to 7.2–7.3. Store portions of 0.5 mL in Eppendorf tubes at –20°C until use.
14. Protease inhibitor: 100 m*M* Pefabloc SC (Boehringer Mannheim).
15. Buffer A: 0.05*M* Tris-HCl, pH 7.2, 0.1*M* NaCl, and 1 m*M* $CaCl_2$. Store at 4°C.
16. Film for autoradiography: Hyperfilm-βmax (Amersham).
17. X-ray developer and photographic fixer.
18. Stripping buffer: 100 m*M* sodium-2-mercaptoethanesulfonate, 2% (w/v) SDS, and 62.5 m*M* Tris-HCl, pH 6.7.

2.2. Adherence Assay in Tissue-Culture Plates

1. Standard liquid broth for the respective mycoplasma species.
2. Stock solution for metabolic labeling of mycoplasmas: [9,10(*n*)-^{3}H]Palmitic acid, 1.5–2.2 TBq/mmol (Amersham # TRK 909).
3. Buffer A: as above.
4. Tissue-culture plates: 24-wells, flat-bottom cavities (e.g., # 662160, Greiner, Frickenhausen, Germany).
5. Blocking solution: 0.1% (w/v) bovine serum albumin in buffer A.
6. MAb: Must be purified by affinity chromatography. Prepare solution with defined protein content in buffer A.
7. SDS solution: 1% (w/v) in water.
8. Cocktail for liquid scintillation counting (LSC), e.g., Rotiszint eco plus (Roth, Karlsruhe).
9. LSC vials, e.g., Zinsser Polyvials, 20 mL (Zinsser, Frankfurt/M. Germany).

3. Methods

3.1. Western Blot Adherence Assay (Identification of Putative Adhesins) (See Note 2)

3.1.1. Preparation of Tissue-Culture Cells

1. Incubate a suspension of trypsinized tissue-culture cells in 25-cm^2 cell-culture flasks containing 5 mL of the respective medium and 50 μg/mL gentamicin at 37°C and 5% CO_2 for 18 h.
2. Add 2 μL (0.74 MBq) of ^{35}S-methionine labeling solution and continue incubation for 18 h. The final cell density should be in the order of 10^5/mL.
3. Remove medium and wash cells once with 5 mL PBS. Collect liquids in a special container for radioactive waste.

4. Trypsinize cells by adding 5 mL of 10X Trypsin-EDTA and gently shaking the flask for 1 min (*see* **Note 5**).
5. Discard about 90% of the liquid and knock the flask several times vertically onto your palm, so that cells begin to flow downward and settle on the bottom.
6. Add 50 µL of protease inhibitor and thoroughly resuspend cells in 4 mL of buffer A using a pipet.

3.1.2. Processing of Western Blots (See **Notes 3** and **5**)

1. Carry out SDS-PAGE and Western blotting according to your standard protocols.
2. Visualization of protein bands: Place the membrane into 20 mL of staining solution for 10 min, and remove background stain by transferring it into 20 mL of destaining solution. Repeat destaining step, if necessary, until optimal contrast between bands and background is achieved.
3. Photograph protein pattern or record it otherwise, e.g., using an image-processing device.
4. Incubate membrane in 20 mL of blocking solution for 1 h. Coomassie blue-stained bands will slowly disappear. (After this step, the membrane can be wrapped in plastic film and stored at –20°C for several weeks.)
5. Incubate membrane with tissue culture cells prepared as described in **Subheading 3.1.1.** The cell suspension may be diluted with a few milliliters of buffer A, if necessary. Shake horizontally (e.g., 90 strokes/min at an amplitude of 3.5 cm) at 37°C for 4 h.
6. Remove cell suspension (radioactive waste), and wash membrane three times in 20 mL PB-ST for 5 min.
7. Place membrane into a cassette for autoradiography, and cover it with plastic film so that it will not dry during exposure. Place autoradiography film on top (in the darkroom), close cassette, and expose for about 4 d.
8. Develop autoradiography.
9. If identification of certain protein bands on the blot by immunostaining is intended: Remove adherent tissue-culture cells by incubating the membrane in 20 mL of stripping buffer at 50°C for 30 min. Wash two times in PBST, carry out blocking as in **step 4**, wash three times in PBST, and proceed with your standard protocol of immunostaining (*see* **Note 6**).

3.2. Adherence Assay in Tissue-Culture Plates

3.2.1. Preparation of Host Cell Cultures

1. Inoculate each cavity of the tissue-culture plate with 0.5 mL of freshly trypsinized cells (density 2.5×10^5 cells/mL) in culture medium containing 50 µg/mL of gentamicin.
2. Incubate at 37°C and 5% CO_2 for 20 h.
3. Check the uniformity of the cell lawn microscopically. In each cavity, there should be a final number of 2×10^5 cells forming a confluent monolayer.
4. Remove culture medium from each well using a pipet.

5. Block potential sites of nonspecific adhesion by adding 200 µL of blocking solution to each well and shaking at 37°C for 15 min.
6. Remove buffer from the wells prior to the adherence assay.

3.2.2. Preparation of Mycoplasma Cultures

1. Inoculate 200 mL of broth with the respective mycoplasma strain. Add 60 µL of ^{3}H-palmitic acid labeling solution. Incubate at 37°C until early log phase.
2. Harvest mycoplasmas by centrifugation at 6000*g* for 20 min.
3. Wash pellets twice with buffer A in the centrifugation tube.
4. Thoroughly resuspend mycoplasmas in 1.6 mL of buffer A using a pipet, and transfer suspension into an Eppendorf tube.

3.2.3. Adhesion (See **Note 7**)

1. If an antibody is examined for its inhibitory capacity: Pipet 100-µL aliquots of the mycoplasma suspension (from the previous step) into 16 Eppendorf tubes. Add different volumes of antibody solution in buffer A to each tube. Add buffer A to a final volume of 200 µL in all cases. Put at least two parallel trials of each antibody concentration and four blanks without antibody. Incubate at 4°C for 2 h. Subsequently, transfer the content of the tubes into wells of tissue-culture plates prepared as in **Subheading 3.2.1.**
2. If a competitive assay is conducted: Pipet 100-µL aliquots of the mycoplasma suspension into the wells of tissue-culture plates prepared as in **Subheading 3.2.1.** Add different amounts of protein or low-mole-wt substance (solutions in buffer A), and add buffer A to make up to a final volume of 200 µL in each well.
3. Seal plates with parafilm, and shake horizontally (90 strokes/min at an amplitude of 3.5 cm) at 37°C for 30 min.
4. Remove liquid from the wells using a pipet.
5. Wash wells twice with 500 µL of cold buffer A.
6. To solubilize attached mycoplasmas, add 100 µL of SDS solution and 500 µL of buffer A to each well. Seal plates and shake slightly at room temperature for 2 h.
7. Transfer content of each well into an LSC vial. Add 10 mL of LSC cocktail and measure β-irradiation.

4. Notes

1. The tissue-culture cell line should be selected individually for each mycoplasma species on the basis of pathogenesis. Although many receptors are common to different mammalian cells, it will facilitate interpretation of results and their eventual transfer to in vivo conditions if a cell line from a natural target organ of the host animal is used in the assays. We used, for instance, A 427, a human lung carcinoma cell line, as host cells for *Mycoplasma pneumoniae,* continuous embryonic bovine lung cells (EBL) for *Mycoplasma bovis,* and HeLa cells for *Mycoplasma hominis.* Each line should be checked for the absence of mycoplasma contamination before being used in an experimental series.

2. An important feature of the Western blot adherence assay is the possibility that a single Western blot, one- or two-dimensional, can be Coomassie blue stained, photographed, destained, incubated with labeled host cells, autoradiographed, stripped and reprobed with the respective antibody, and immunostained in succession. As a consequence, the whole-cell protein pattern, the host cell binding pattern, and immunostains are produced from the same master blot, thus highly facilitating assignment of bands and identification of putative adhesion proteins. The choice of a suitable PVDF membrane is crucial, since protein binding rates vary considerably and, most important, not all products allow reversible Coomassie blue staining.
3. Mammalian cells to be used in the Western blot adherence assay are trypsinized after incubation in order to remove them from the wall of the plastic culture flask. To attain a maximum cell yield, it is important to let the digestion proceed to a point when the majority of cells are neither under- nor overdigested. This point is reached when the cell lawn on the wall appears as a gray layer. Although a 1-min exposure to the enzyme is often sufficient, there is some variation depending on the cell line used and the specific activity of the trypsin.
4. All washing operations of membranes, as well as staining, destaining, blocking, and stripping, should be carried out at room temperature on a horizontal shaker with moderate shaking.
5. Western blots to be used in the adherence assay should be made from larger gels of approx 14-cm length. The use of the minigel format (6–8 cm length) is not recommended as the amount of immobilized protein, and the resolution of bands is usually not sufficient.
6. Blots incubated with tissue-culture cells may be reprobed several times with different MAbs, thus ensuring reliable identification and assignment of individual bands. In our hands, this works most effectively with chemiluminescence-based immunostaining techniques, such as Amersham's ECL.
7. For inhibition assays and competitive studies, 24-well tissue-culture plates have proven suitable in terms of sensitivity and practicability, since, owing to slow growth, mycoplasmas generally do not attain high cell densities in liquid culture and overall adherence rates tend to be comparatively low. Thus, the proportion of actually adhering cells in an average mycoplasma population was 6.9% for *M. bovis* ***(1)***, 5.5–7% for *Ureaplasma urealyticum* ***(4)***, and 2.71% for *Mycoplasma arthritidis* ***(5)***. For strongly adhering species of other bacteria, the assay can be miniaturized and carried out in 96-well microtiter plates. With the present protocol, it is convenient to run three series in parallel (different strains, antibodies, or substances), so that each series requires 16 wells. Under standard conditions of the culture plate assay, 10^8 CFU of mycoplasmas interact with 2×10^5 host cells in each cavity.

References

1. Sachse, K., Grajetzki, C., Rosengarten, R., Hänel, I., Heller, M., and Pfützner, H. (1996) Mechanisms and factors involved in *Mycoplasma bovis* adhesion to host cells. *Zbl. Bakt.* **284,** 80–92.

2. Razin, S. and Jacobs, E. (1992) Mycoplasma adhesion. *J. Gen. Microbiol.* **138,** 407–422.
3. Layh-Schmitt, G. and Herrmann, R. (1992) Localization and biochemical characterization of the ORF6 gene product of the *Mycoplasma pneumoniae* P1 operon. *Infect. Immun.* **60,** 2906–2913.
4. Saada, A. B., Terespolski, Y., Adoni, A., and Kahane, I. (1991) Adherence of *Ureaplasma urealyticum* to human erythrocytes. *Infect. Immun.* **59,** 467–469.
5. Washburn, L. R., Hirsch, S., and Voelker, L. L. (1993) Mechanisms of attachment of *Mycoplasma arthritidis* to host cells in vitro. *Infect. Immun.* **61,** 2670–2680.

33

Transmission Electron Microscopy and Immunogold Staining of Mollicute Surface Antigens

Gunna Christiansen and Svend Birkelund

1. Introduction

Mollicutes include the smallest, free-living microorganisms. They are cell-wall-less bacteria, many of which produce disease in humans, animals, plants, or insects. Morphologically, mollicutes are extremely variable and pleomorphic, even in pure culture. Some organisms assume a predominantly spherical (0.3–0.8 μm in diameter) or flask-shaped appearance, whereas others may form filaments or branched structures. Genetically, mollicutes are related to Gram-positive bacteria, but the lack of a cell wall prevents a typical staining reaction. Thin sections reveal a simple ultrastructure consisting of cell membrane and cytoplasm, including ribosomes and a characteristic prokaryotic nucleoid.

The mollicute cell membrane contains essentially all the cellular lipids, and a substantial fraction of the cellular proteins. In a number of genera (*Anaeroplasma*, *Entomoplasma*, *Mycoplasma*, *Spiroplasma*, and *Ureaplasma*), the membrane contains cholesterol; since mollicutes are incapable of synthesizing sterols, exogenous cholesterol is required for growth. Except for the presence of cholesterol, the chemical composition of mollicute membranes is similar to that of other prokaryotic membranes, consisting primarily of lipids and proteins. The molar ratio of lipid-to-protein is approx 60:1. By two-dimensional electrophoresis, it has been estimated that the *Acholeplasma laidlawii* membrane contains approx 140 different polypeptides, 40 of which are surface-exposed, as shown by iodination *(1)*. Some of the membrane proteins are membrane-bound "housekeeping" enzymes. A substantial number of the membrane proteins are lipid-modified, approx 25 compared to 10 in other bacteria *(2)*.

From: *Methods in Molecular Biology, Vol. 104: Mycoplasma Protocols*
Edited by: R. J. Miles and R. A. J. Nicholas © Humana Press Inc., Totowa, NJ

Electron microscopy is of importance in studying the size, structure, and complex morphology of mollicute cells *(3,4)*. It has also been applied to studies of the cell surface. Analysis of the mollicute cell surface is of particular interest, since it interacts directly with host cells and the host's immune system. Mollicute surface proteins may have specific functions in pathogenicity (e.g., as adhesins) and are frequently variable, thus enabling mollicute cells to evade immune recognition by the host. They may also provide targets for serologic identification. More extensive knowledge of the mollicute cell surface will improve understanding of their pathogenicity and may aid development of new vaccines.

To investigate the structure of the mollicute surface, monoclonal antibodies (MAbs) or monospecific polyclonal antibodies (PAbs) are important tools. The antibodies can be used in immunogold staining to localize mollicute surface antigens by immunoelectron microscopy techniques. MAbs against surface-exposed epitopes can be obtained following immunization of mice with whole undisrupted killed mollicute cells *(5)*. A monospecific, polyclonal antibody (PAb) can be obtained either by absorption procedures, from a nonspecific antiserum, or by immunization with a purified cell membrane component or purified recombinant fusion protein. Detection of antibodies bound to the cell surface is done either with G–protein or A–gold protein complexes, or with colloidal gold-labeled secondary antibodies *(6)*. Thus, detailed information concerning the localization of surface epitopes may be obtained *(7–10)*.

Analysis for surface exposure of eukaryotic cell membrane components is usually performed on thin sections. However, the small size of mollicute cells allows immunological analysis of the microorganisms directly adsorbed to the surface of an electron microscopy grid. This approach is more convenient, much faster, and can be performed without prior fixation.

Immunogold labeling followed by electron microscopic analysis provides a means of investigating: the surface exposure of antigens, the role of antigens in adhesion, analysis of the structure of specific cell membrane components and their interdependent relationships, antigen variation and crossreactivity, and growth-dependent induction of specific antigens. The technique is thus widely used in parallel with other molecular biological methods and with negative staining. Both negative staining and immunogold labeling of mollicutes are described in this chapter. *Mycoplasma hominis* is used as a specific example.

2. Materials

1. Growth medium: The growth medium should be appropriate for the mollicute strain being studied (*see* Chapters 4–7). For the growth of *M. hominis* strains, which are arginine-metabolizing and glucose-nonfermenting, BEa medium may be used. This is composed of: heart infusion broth (Difco, Detroit, MI), 2.2% (w/v);

horse serum, 15.5% (v/v); fresh yeast extract, 1.9% (w/v); thallium acetate, 0.008%; benzylpenicillin, 40 U/mL; L-arginine, 0.3% (w/v); and phenol red, 0.0023% (w/v). The pH is adjusted to 7.2, and the medium sterilized by filtration.

2. MAbs: The following procedure is suitable for *M. hominis* ***(5)***. Cells are cultivated in BEa medium, harvested by centrifugation at 20,000*g* for 30 min at 4°C, and washed three times in phosphate buffered saline (PBS: NaCl 8.5 g/L, K_2HPO_4 1.21 L g/L, KH_2PO_4 0.34 g/L, adjusted to pH 7.4 with HCl). Immunization of Balb/c mice is then carried out as follows: At days 1, 4, 8, 11, 15, and 18, mice are inoculated subcutaneously with cells (20 μg protein/immunization) suspended in Freund's incomplete adjuvant (Difco). At days 90 and 91, cells suspended in PBS (50 μg of protein) are inoculated intraperitoneally, and at day 93, the spleen cells obtained from the mouse are harvested and fused with 8-azoguanine-resistant myeloma cell line NS-1 cells, in the presence of 50% (w/v) polyethylene glycol (PEG-1000, pH 8.0). Cells are subcloned ***(5)***, and secreted antibodies from the hybridoma cells are detected by enzyme-linked immunosorbent assay (ELISA). MAbs are characterized by immunoblotting ***(11)***.
3. Monospecific polyclonal antibodies: Rabbits are immunized intramuscularly with 10 μg purified fusion protein emulsified in Freund's incomplete adjuvant (Difco) on days 1, 5, 7, 11, 13, and 18. Intravenous booster injections without Freund's incomplete adjuvant are given on days 39, 49, and 55. On day 68, the rabbits are killed and bled ***(12)***.
4. Secondary antibodies for immunoelectron microscopy: Goat antimouse conjugated with 5 or 10 nm gold; goat antirabbit conjugated with 10 nm gold (BioCell, Nota Bene Scientific, Copenhagan, Denmark).
5. Grids: 400 mesh copper or nickel grids (Gibco BRL, Grand Island, NY) coated with a thin carbon film and glow discharged ***(13)***, or grids covered with a parlodian film and similarly carbon-coated and glow-discharged ***(14)***. To coat grids with parlodian (Struers, Copenhagan, Denmark), prepare a 1% (w/v) solution in 1-pentanol. Allow one drop to spread on a 20-cm^2 aqueous surface. Then place the grids on the parlodian film and lift off using a microscope slide.
6. Wash buffers: PBS (*see* **item 3**) is used for all washes unless otherwise stated. Blocking and incubations of *Mycoplasma* cells with antibodies are done in 0.5% ovalbumin in PBS. Removal of nonspecifically bound immunogold is done in 1% gelatin (Sigma, from cold-water fish) in PBS (*see* **Notes 1** and **2**).
7. Fixing buffer (for fixation of cells adsorbed to the surface of grids): PBS with formaldehyde 0.5% (v/v) and glutaraldehyde 0.25% (w/v).
8. Stains: For negative staining, grids are stained with 1% (w/v) phosphotungstic acid (PTA), adjusted to pH 7.5 with NH_4OH. For immunoelectron microscopy, grids are stained either with 0.3% (w/v) PTA or with 1% (w/v) ammonium molybdate (*see* **Note 3**).

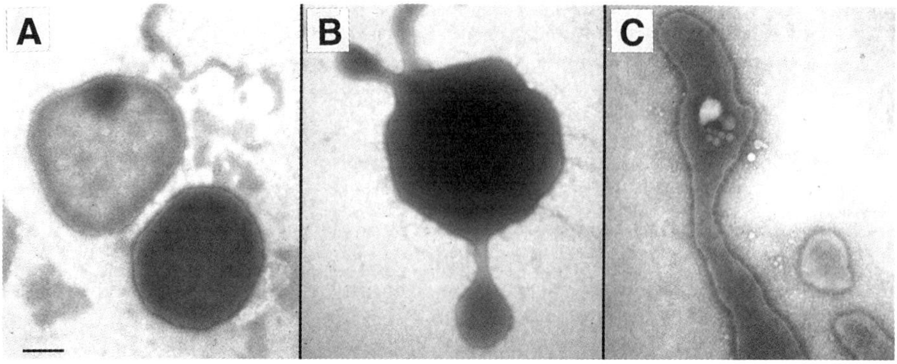

Fig. 1. *M. hominis* PG21 negative-stained with **(A)** and **(B)** 2% ammonium molybdate, and **(C)** 1% PTA. Bar, 0.2 μm.

3. Methods

3.1. Cultivation of Microorganisms

For electron and immunoelectron microscopy, exponentially growing, fresh mollicute cultures should be used.

1. Inoculate 200 μL of a frozen culture (stored at –70°C) into 1.8 mL medium.
2. Incubate at 37°C (or other appropriate temperature) until growth is evident. In media containing phenol red, this may be indicated by a color change. For example, growth of *M. hominis* in BEa medium raises the pH as a result of arginine hydrolysis, and the medium changes color from orange (pH 7.2) to purple (pH 7.6).
3. Transfer 200 μL of this culture to fresh medium (1.8 mL), and incubate until growth or a medium color change is observed.
4. As soon as growth is evident, harvest cells by centrifugation (20,000*g*), and resuspend in PBS.

3.2. Negative Staining

1. A 10-μL drop of cell suspension in PBS is placed for 1 min on top of a freshly glow-discharged 400-mesh copper grid.
2. The grids are stained with three drops of 1% PTA (15 s with each drop) and blotted dry by lightly applying to filter paper.
3. The fixed-cell preparations are examined by electron microscopy. We use either a JEOL JEM 100B at 60 kV or a JEOL 1010 at 40 kV. **Figure 1** shows negative stained *M. hominis* PG21 cells. Individual cells may be round, oval, or filamentous.

3.3. Immunogold Staining

1. A 10-μL drop of cell suspension is placed for 1 min on top of a freshly glow-discharged 400-mesh nickel grid (*see* **Notes 4** and **5**).

2. After washing in three drops of PBS, the grid is incubated for 5 min at room temperature in PBS containing 0.5% ovalbumin.
3. Fixation (optional) may be carried out by placing the grid in a drop of fixing buffer for 10 min prior to blocking with 0.5% ovalbumin in PBS.
4. The grids are transferred to a 50-μL drop of MAb or PAb. MAbs (tissue-culture supernatants) are diluted 1:5 in PBS with 0.5% ovalbumin. PAbs are diluted 1:100 in PBS with 0.5% ovalbumin. Grids are incubated for 30 min at 37°C.
5. Following incubation with the primary antibodies, the grids are washed three times in drops of PBS.
6. The grids are then incubated with the secondary antibody for 30 min at 37°C. The secondary antibodies (10 nm colloidal-gold-labeled goat antimouse or goat antirabbit antibodies) are diluted 1:20 in PBS containing 1% gelatin *(5)* (*see* **Note 6**).
7. To remove excess unbound colloidal gold, grids are washed twice (15 min each wash) at 37°C in PBS containing 1% gelatin, and three times in PBS (5 min each wash) (*see* **Note 7**).
8. The grids are stained with two drops of 0.3% PTA at pH 7.0 (*see* **Note 3**), blotted dry, and examined by electron microscopy (*see* **Subheading 3.2., step 3**).

Figure 2 shows electron micrographs of *M. hominis* PG21 cells reacted with MAb (MAb 552, *see* **Note 8**) and labeled with colloidal gold. The micrographs were taken using a JEOL JEM 100B electron microscope at 40 kV with 6.5 × 9 cm Agfa Scientia EM film (23D56 P3 AH; Agfa-Gevaert, Belgium). Without fixation (**Fig. 2A**), the label was predominately located at the periphery of the cells. After fixation, immunogold particles were positioned equally over the cell (**Fig. 2B**) indicating a high degree of lateral mobility of the target protein. Immunoelectron microscopy of MAb 552-induced aggregates shows a similar localization of labels without (**Fig. 3A**) and with fixation (**Fig. 3B**), indicating that formation of aggregates does not require lateral movement of the proteins. The application of immunogold labelling to studies of phase variation in the expression of surface proteins is shown in **Fig. 4** for *M. hominis* 7488 cells. The cells in this figure were reacted with PAb 135, which recognizes an epitope within the N-terminal region of a 120 kDa membrane antigen *(12,15)*. Differences in expression of the epitope among cells is evident.

3.4. Immunogold Double Labeling

1. Proceed as in **Subheading 3.3., steps 1–3**, then transfer grids to a 50-μL drop containing both MAb and PAb. MAbs (tissue-culture supernatants) are diluted 1:5, and PAbs diluted 1:100, in PBS with 0.5% ovalbumin. Incubate for 30 min at 37°C.
2. Following incubation with MAb, wash the grid three times in drops of PBS.
3. Incubate grids with a mixture of secondary antibodies (10 nm colloidal-gold-labeled goat antimouse antibodies and 5 nm goat antirabbit antibodies; BioCell), diluted 1:20 in PBS with 1% (w/v) gelatin for 30 min at 37°C (*see* **Note 6**).

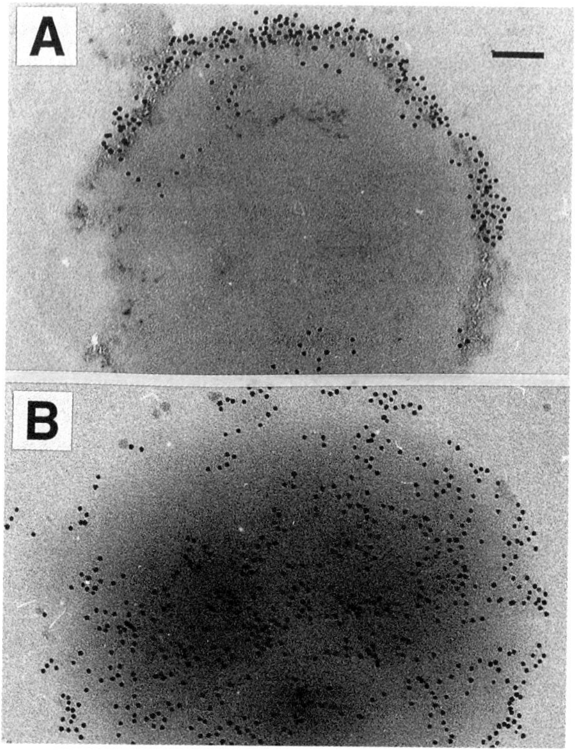

Fig. 2. Immunogold staining of *M. hominis* PG21. Cells from an exponentially growing culture were adsorbed onto the surface of a parlodion-carbon-coated, glow-discharged supporting film, reacted with MAb 552 and visualised with colloidal-gold-conjugated secondary antibodies. **(A)** No fixation, **(B)** Fixation with 0.25% glutaraldehyde prior to incubation with MAb 552. Bar, 0.1 μm.

4. To remove excess unbound colloidal gold, the grid is washed twice for 15 min at 37°C in PBS containing 1% (w/v) gelatin followed by three washes for 5 min each in PBS.
5. The grids are stained with two drops of 1% ammonium molybdate, blotted dry, and examined by electron microscopy (*see* **Subheading 3.2., step 3**) (*see* **Note 9**).

Electron micrographs showing double labelling of *M. hominis* 7488 cells with MAb 552 and PAb 135 are given in **Fig. 5**. The micrographs were taken using a JEOL 1010 electron microscope at 40 kV. The reaction with PAb 135 is much more pronounced on the unfixed cells (**Fig. 5A**) than on the fixed cells (**Fig. 5B**), indicating that some of the epitopes

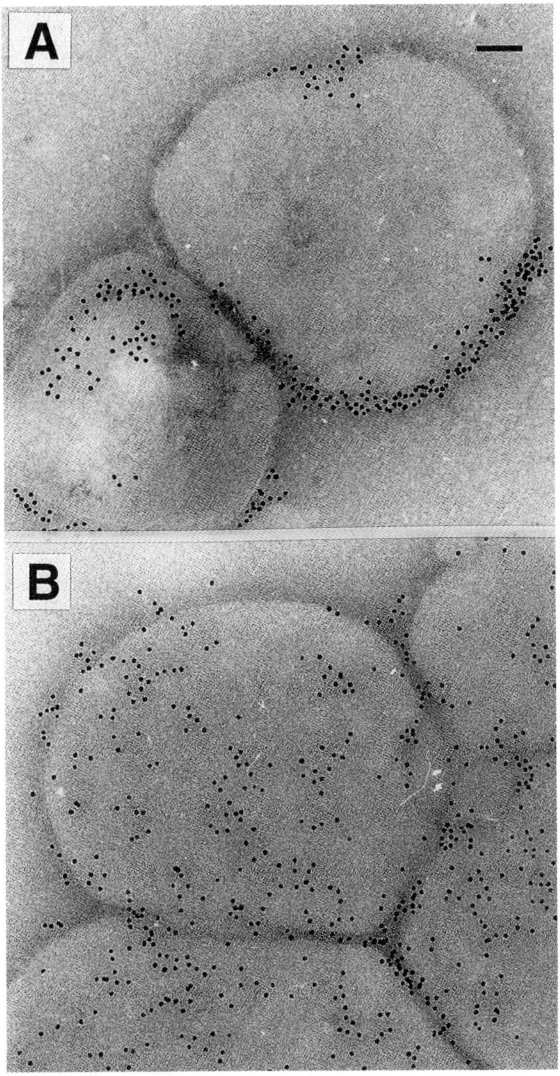

Fig. 3. Immunoelectron microscopy of MAb-induced aggregates of *M. hominis* PG21. Cells were cultivated in the presence of MAb 552 (1:10) overnight and adsorbed to the surface of a parlodion-carbon-coated, glow-discharged supporting film. Attached antibodies were visualized with colloidal-gold-conjugated secondary antibodies. **(A)** No fixation, **(B)** fixation with 0.25% glutaraldehyde prior to incubation with colloidal-gold-conjugated antibodies. Bar, 0.1 μm.

are sensitive to fixation (*see* **ref.** ***15***). Lateral migration as seen in **Figs. 2** and **3** was not as pronounced in the double-labeled samples.

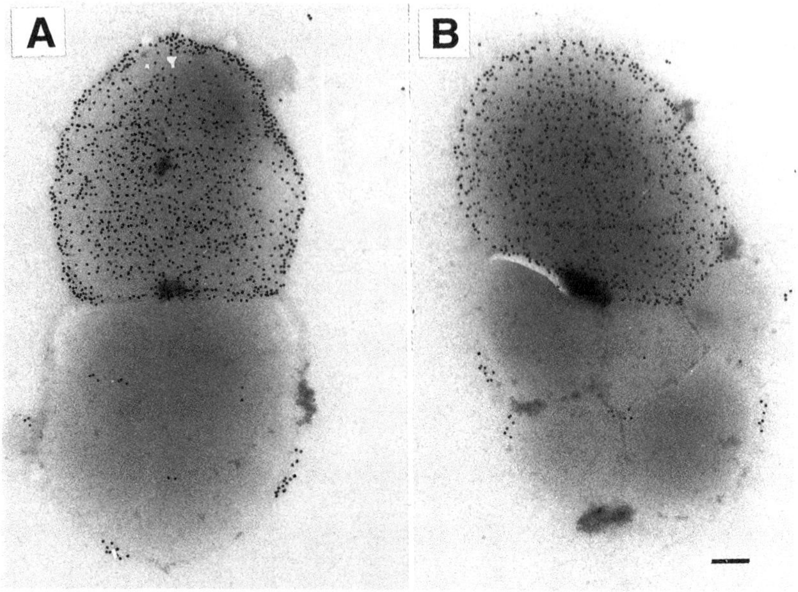

Fig. 4. Immunogold staining of *M. hominis* 7488. Cells were adsorbed to a carbon-coated, glow-discharged supporting film, reacted with PAb 135 and visualised with 10 nm colloidal-gold-conjugated goat anti-rabbit secondary antibodies. (*See* text for explanation). Bar, 0.1 μm.

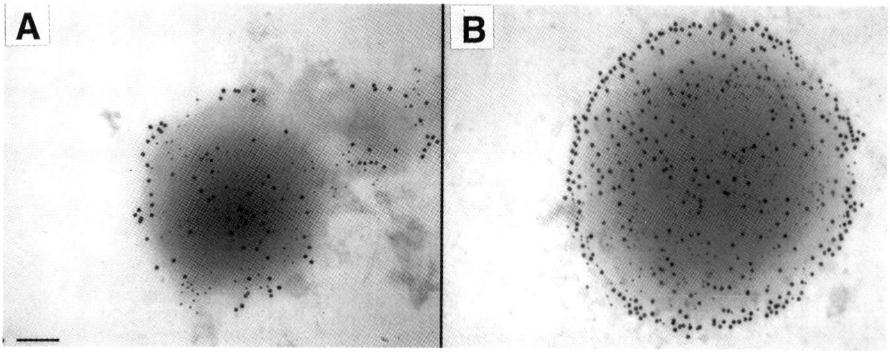

Fig. 5. Double-immunogold-labeling of *M. hominis* 7488. Cells were absorbed to a carbon-coated, glow-discharged supporting film and reacted with PAb 135 and MAb 552. Bound primary antibody was visualised with 10 nm colloidal-gold-conjugated goat antirabbit and 5 nm colloidal-gold-conjugated goat antimouse secondary antibodies, respectively. **(A)** No fixation, **(B)** fixation with 0.25% glutaraldehyde prior to incubation with gold-conjugated antibodies. Bar, 0.1 μm.

4. Notes

1. For successful immunoelectron microscopy, it is important that all buffers are sterilized by membrane filtration, since any particles will interfere with the results.
2. The buffers described for blocking and for removal of unbound gold conjugates give reproducible results and a very clean background. They have been selected from a variety of buffers frequently used for immunoelectron microscopy.
3. The concentration of the staining solutions is lower for immunoelectron microscopy than for negative staining. It is important not to obscure the bound gold conjugates.
4. All immunoreactions are done with the mollicute cells adsorbed to the surface of the grids. This reduces aggregation of cells and prevents loss of material during washing procedures.
5. We use nickel grids for immunoelectron microscopy. Nickel is more inert than copper and, thus, will not interfere with the immunoreactions. The holes in the 400-mesh grids are of sufficient size for analysis of mollicute cells and support the thin carbon film well during handling procedures.
6. As secondary antibodies, we use goat antirabbit and goat antimouse conjugated with 5 or 10 nm gold. Different brands of antimouse conjugates work well, and in general, antimouse conjugates give no background problems. This is in contrast to the antirabbit conjugates, of which only the BioCell products worked well in our hands.
7. In cases where there is too much unbound gold conjugate, additional washes with PBS containing gelatin can be used.
8. Ladefoged et al. ***(16)*** produced MAbs against a variety of *M. hominis* membrane proteins. Only one (MAb 552) reacted with all the 26 *M. hominis strains* tested. The epitope recognized by MAb 552 was shown by immunogold electron microscopy to react with a surface-localized *M. hominis* PG21 membrane protein.
9. To process double-immunogold-labeled micrographs from the Jeol 1010 electron microscope, we take images using a Kodak 1.4 slow scan camera, and transfer to a SUN Sparc 10 Station with an SDV digital video camera interface. Image analysis is then done using Hips image processing software (Sharp Image Software, NY) and the program Adobe Photoshop (Adobe Systems, San Jose, CA). Magnification is determined by the transfer of images of replica gratings in a similar manner.

Acknowledgment

This work was supported by grants from The Danish Health Science Research Foundation (grant numbers 12-0850-1 and 12-1620-1), The Danish Natural Science Foundation, The NOVO Foundation, Fonden til Lægevidenskabens Fremme, Aarhus University Research Foundation, and the John and BirtheMeyer Foundation.

References

1. Archer, D. B., Rodwell, A. W., and Rodwell, E. S. (1978) The nature and location of *Acholeplasma laidlawii* membrane proteins investigated by two-dimensional gel electrophoresis. *Biochim. Biophys. Acta* **513,** 268–283.
2. Nyström, S., Johansson, K. -E., and Wieslander, Å. (1986) Selective acylation of membrane proteins in *Acholeplasma laidlawii. Eur. J. Biochem.* **156,** 85–94.
3. Freundt, E. A. (1960) Morphology and classification of PPLO. *Ann. NY Acad. Sci.* **79,** 312–325.
4. Hu, P. C., Cole, R. M., Huang, Y. S., Graham, J. A., Gardner, D. E., Collier, A. M., et al. (1982) *Mycoplasma pneumoniae* infection: role of a surface protein in the attachment organelle. *Science* **216,** 313–315.
5. Birkelund, S., Lundemose, A. G., and Christiansen, G. (1988) Chemical cross-linking of *Chlamydia trachomatis. Infect. Immun.* **56,** 654-659.
6. Hayat, M. A. (1990) *Colloidal Gold. Principles, Methods, and Applications,* vol. III. Academic, San Diego.
7. Fuerst, J. A. and Perry, J. W. (1988) Demonstration of lipopolysaccharide on sheathed flagella of *Vibrio cholerae* O:1 by protein A-gold immunoelectron microscopy. *J. Bacteriol.* **170,** 1488–1494.
8. Robinson, E. N., Jr., McGee, Z. A., Kaplan, J., Hammond, M. E., Larson, J. K., Buchanan, T. M., et al. (1984) Ultrastructural localization of specific gonococcal macromolecules with antibody-gold sphere immunological probes. *Infect. Immun.* **46,** 361–366.
9. Robinson, E. N., Jr., McGee, Z. A., Buchanan, T. M., Blake, M. S., and Hitchcock, J. P. (1987) Probing the surface of *Neisseria gonorrhoeae*: simultaneous localization of protein I and H8 antigens. *Infect. Immun.* **55,** 1190–1197.
10. Robinson, E. N., Jr., Clemens, C. M., McGee, Z. A., and Cannon, J. G. (1988) Immunoelectron microscopic localization of outer membrane proteins II on the surface of *Neisseria gonorrhoeae. Infect. Immun.* **56,** 1003–1006.
11. Birkelund, S., and Andersen, H. (1988) Comparative studies of mycoplasma antigens and corresponding antibodies, in *Handbook of Immunoblotting of Proteins*, vol. II (Bjerrum, O. J. and Heegaard, N. H. H., eds.), CRC, Boca Raton, FL, pp. 25–33.
12. Nyvold, C., Birkelund, S., and Christiansen, G. (1997) The *Mycoplasma hominis* P120 membrane protein contains a 216 amino acid hypervariable domain that is recognized by the human humoral immune response. *Microbiology* **143,** 675–688.
13. Griffith. J. D. and Christiansen, G. (1978) Electron microscope visualization of chromatin and other DNA-protein complexes. *Ann. Rev. Biophys. Bioeng.* **7,** 19–35.
14. Klemm, P. and Christiansen, G. (1987) Three *fim* genes required for the regulation of length and mediation of adhesion of *Escherichia coli* type 1 fimbriae. *Mol. Gen. Genet.* **208,** 439–445.
15. Christiansen, G., Mathiesen, S. L., Nyvold, C., and Birkelund, S. (1994) Analysis of a *Mycoplasma hominis* membrane protein, P120. *FEMS Microbiol. Lett.* **121,** 121–128.
16. Ladefoged, S., Hauge, S., Andersen, H., Birkelund, S., and Christiansen, G. (1990) Use of monoclonal antibodies for detection of antigen variation in *Mycoplasma hominis. Zbl. Bacteriol. Suppl.* **20,** 634–639.

Index